People, Cultures and Societies: Exploring and Documenting Diversities

The series proposes to capture the diversities in people and their communities in India. It provides a unique and innovative resource for anthropological knowledge, philosophy, methods, and tools to understand, analyse and formulate sustainable, innovative solutions to address socio-cultural issues in India.

India is a repository of varied cultures and diversities. With the globalisation and development process, the cultural fabric is changing. Customs, traditions, beliefs, on one hand, food habits, art and craft, weaving, dyeing, and handloom artefacts, on the other, are undergoing a metamorphosis. It is imperative to explore, understand and document the process of changing diversity and relational inequalities. The series encompasses richness in art, craft, language, dance, music, folklore, food culture and beliefs, traditions and practices. It addresses the issues of development disparities, inequality, and segregation on the axes of caste, class, religion, region, ethnicity, and gender. The series publishes methodologically rigorous and theoretically sound, critical and comparative, empirical research peer-reviewed volumes related to non-codified healing practices, gender-based violence, migration induced vulnerabilities, child abuse, social identity-based work on a national, regional and local level, welcoming case studies, as well as comparative and applied research.

The series is of interest to the academicians and students in the discipline of sociology, anthropology, psychology, social work, history, philosophy, and public health, and to all of those interested in a wide-ranging overview of art, culture, and politics. It accepts monographs, edited volumes, and textbooks.

Ranbir Chander Sobti • Sudarshan Verma •
Dr. Vipin Sobti
Editors

Traditional Knowledge Systems in India: Sustainability, Ecology, and Resource Governance

Springer

ANTHROPOS INDIA FOUNDATION

Editors
Ranbir Chander Sobti
Biotechnology
Panjab University
Chandigarh, India

Sudarshan Verma
Department of Laws
Babasaheb Bhimrao Ambedkar University
Lucknow, India

Dr. Vipin Sobti
Chandigarh, India

ISSN 2662-6616 ISSN 2662-6624 (electronic)
People, Cultures and Societies: Exploring and Documenting Diversities
ISBN 978-981-92-1345-0 ISBN 978-981-92-1346-7 (eBook)
https://doi.org/10.1007/978-981-92-1346-7

This Springer imprint is published by the registered company Springer Nature Singapore Pte Ltd.
The registered company address is: 152 Beach Road, #21-01/04 Gateway East, Singapore 189721, Singapore

Preface

In an epoch characterized by unprecedented technological acceleration, ecological precarity, and the relentless pursuit of material advancement, the imperative to reclaim humanity's symbiotic relationship with nature has never been more urgent. The edited volume, ***Protection of Traditional Knowledge: Sustainable Practices and Resource Management***, emerges as an intellectual response to this global imperative—endeavouring to recontextualize the ancestral wisdom of indigenous and tribal communities as a viable paradigm for contemporary sustainability discourses.

Traditional knowledge—an intricate tapestry of experiential insights, ethical values, and environmental symbiosis—constitutes the living memory of civilizations that have flourished in consonance with their ecosystems. Far from being static or primitive, these knowledge systems represent dynamic repositories of empirical science, moral philosophy, and cultural adaptation, evolved meticulously through centuries of observation and experimentation. They articulate a worldview where nature is not a commodity to be conquered but a kinship network to be revered, nurtured, and coexisted with.

This compendium assembles a constellation of scholarly reflections, empirical explorations, and policy-oriented analyses that collectively interrogate the multifaceted dimensions of traditional knowledge—its conservation, validation, and integration into the framework of sustainable development. The volume traverses the disciplinary terrains of law, anthropology, geography, environmental science, and ethnobotany, thereby embodying a holistic, transdisciplinary ethos essential for addressing the complexities of ecological governance and cultural preservation in the twenty-first century.

The book is organized into two comprehensive sections.

Section I: "Tribal Indigenous Knowledge and Environmental Sustainability" illuminates the enduring wisdom embedded within tribal epistemologies that have long harmonized human livelihood with environmental stewardship. From indigenous water management in the Himalayas to forest governance in Myanmar, and from the biodiversity conservation frameworks under India's Biological Diversity Act, 2002, to the integration of Sustainable Development Goals (SDGs) into local traditions—this section underscores the transformative potential of traditional systems when recognized as legitimate forms of knowledge rather than folkloric residues. Chapters such as *"Preserving Tribal Wisdom: Integrating Western Himalayan Tribal Knowledge for Accelerating Sustainable Development Goals"* and *"Harnessing Indigenous Water Resources"* exemplify how tribal environmental ethics prefigure contemporary sustainability models long before their formal articulation in global policy forums.

Section II: "Tribal Indigenous Knowledge – Food Security and Health" delves into the biocultural continuum of sustenance and wellness among indigenous communities. The chapters traverse the diverse landscapes of traditional fermentation technologies in Manipur, nutraceutical and therapeutic uses of Himalayan flora, the ethnographic vitality of the Gaddi and Rung tribes, and the symbiosis between dietary heritage and holistic health. These contributions reveal that indigenous gastronomy and healing are not mere vestiges of cultural identity but integral mechanisms of ecological balance and human resilience.

Collectively, the contributions in this volume contest the epistemic hierarchies that have historically marginalized indigenous knowledge. They affirm that traditional ecological wisdom is neither anachronistic nor anecdotal—it is a sophisticated cognitive system with the capacity to inform contemporary science, law, and policy. The indigenous communities, who have long been custodians of biodiversity and sustainable resource management, must now be repositioned as equal partners in the global dialogue on environmental futures. Their cultural capital—codified through oral traditions, rituals, and collective practices—offers a moral compass to a world navigating the consequences of its own excesses.

At a time when climate catastrophes, biodiversity loss, and resource depletion threaten planetary stability, this book advocates for epistemological humility—an acknowledgment that sustainability cannot be achieved solely through technocratic interventions but through re-engagement with the moral and ecological intelligence of traditional societies. The chapters herein do not merely document indigenous wisdom; they reassert its contemporary relevance as a resilient framework for ethical coexistence, sustainable governance, and restorative justice.

The editors extend profound gratitude to all contributing scholars, reviewers, and institutional collaborators whose intellectual labour and commitment have rendered this volume possible. Their collective efforts reaffirm the conviction that **traditional knowledge is not a relic of the past, but a compass for the future**—a repository of living wisdom capable of steering humanity towards a more equitable, ecologically attuned, and ethically grounded civilization.

R.C.S. and V.S. acknowledge support rendered by their family members Er Aditi and Dr Aastha Sobti (daughters) and Er Vineet and Er Ankit (sons-in-law).

R.C.S. and S.V. thankful to the Indian National Science Academy and Panjab University, Chandigarh for providing his platform to continue his academic persuits as respectively as Honorary Scientist and Professor Emeritus.

R.C. Sobti
Sudarshan Verma
Vipin Sobti

Contents

List of Contributors

Irfan Ali Banka Department of Anthropology, Panjab University, Chandigarh, India

Khushnuma Bano Department of Law, SLS, Babasaheb Bhimrao Ambedkar (A Central) University, Lucknow, India

Khwairakpam Bembem Division of Transfer of Technology, Central Institute of Post-Harvest Engineering and Technology, Ludhiana, Punjab, India

Aribam Bijayasundari Devi Assistant Professor, Department of Anthropology, Dr. Harisingh Gour Vishwavidyalaya, Sagar, Madhya Pradesh, India

Kritika Bora Department of Geography, Kumaun University, Nainital, India

Diptendu Chatterjee Department of Anthropology, University of Calcutta, Kolkata, 700019, India

Nitesh Kumar Chaturvedi Atal Bihari Vajpayee, School of Legal Studies, CSJMU, Kanpur, India

Saamdu Chetri Shoolini University, Solan, India;
Dr. Vishwanath Karad MIT World Peace University, Pune, India

Nikita Das Research Scholar, Department of Anthropology, Dr. Harisingh Gour Vishwavidyalaya, Sagar, Madhya Pradesh, India

Pramod Tarachand Deshpande Department of History, Babasaheb Bhimrao Ambedkar University, Lucknow, India

Malay Dolai Department of Chemistry, Prabhat Kumar College, Contai, Purba Medinipur, 721404, West Bengal, India

Sharmistha Ganguly Department of Dravyaguna, Institute of Post Graduate Ayurvedic Education and Research, Kolkata, 700009, India

Sanjay Ghildiyal Department of History, Kumaun University, Nainital, India

Abhik Ghosh Department of Anthropology, Panjab University, Chandigarh, India

Sajal Ghosh Research Scholar, University Department of Geography, Ranchi University, Ranchi, Jharkhand, India

Rabiya Haritaksh Department of Anthropology, Panjab University, Chandigarh, India

Kamal Jaiswal Satellite Centre, Babasaheb Bhimrao Ambedkar University (A Central University), Amethi, Uttar Pradesh, India

Vibhash C. Jha Professor and Former Head, Department of Geography, Visva-Bharati University, Santiniketan, West Bengal, India

Basant Kumar Kanaujiya Department of History, Babasaheb Bhimrao Ambedkar University, Lucknow, India

Sonia Kaushal Department of Anthropology, Dr. Harisingh Gour Vishwavidyalaya (A Central University), Sagar, 470003, Madhya Pradesh, India

U. V. Kiran Satellite Centre, Babasaheb Bhimrao Ambedkar University (A Central University), Amethi, Uttar Pradesh, India

Subhamay Kisku Department of Anthropology, University of Calcutta, Kolkata, 700019, India

Pushp Lata Department of Zoology, University of Delhi, Delhi, 110007, India

Anjoy Majhi Department of Chemistry, Presidency University, 86/1 College Street, Kolkata, 700073, India

K. Murali Naik Babasaheb Bhimrao Ambedkar University (A Central University), Satellite Centre, Amethi, Uttar Pradesh, India

Shivanshi Nehria Department of Anthropology, Eliezer Joldan Memorial College, leh, Ladakh, India

Sourav Pakrashy Department of Chemistry, Presidency University, 86/1 College Street, Kolkata, 700073, India;
Department of Chemistry, Prabhat Kumar College, Purba Medinipur, Contai, 721404, West Bengal, India

Khwairakpam Priyanka Department of Horticulture, School of Agricultural Sciences and Technology, Babasaheb Bhimrao Ambedkar University, Lucknow, Uttar Pradesh, India

Muzamil Rashid Anthro International, Srinagar, India

V. M. Ravi Kumar Department of History, Babasaheb Bhimrao Ambedkar University, Lucknow, India

Rishav Rawal Department of History, Kumaun University, Nainital, India

Suman Rawat Dr. Vishwanath Karad MIT World Peace University, Pune, India

Shalvi Department of Anthropology, Panjab University, Chandigarh, India

Ningthoujam Samarendra Singh Division of Post-Harvest Management, Central Institute of Sub-Tropical Horticulture, Lucknow, Uttar Pradesh, India

Abhijit Sasmal Merck, KgAa, Bengaluru, 560087, India

Khwairakpam Sharmila Department of Human Development and Family Studies, School of Home Science, Babasaheb Bhimrao Ambedkar University, Lucknow, Uttar Pradesh, India

M. P. Sukthija DSLD- College of Horticultural Engineering and Food Technology, Devihosur, University of Horticultural Science, Bagalkot, India

Priya Verma Gurgaon, India

Sudarshan Verma Department of Law, School of Legal Studies, Babasaheb Bhimrao Ambedkar University, Lucknow, India

Chanchal Yadav Department of Zoology, University of Delhi, Delhi, 110007, India

Part I
Tribal Indigenous Knowledge and Environmental Sustainability

Chapter 1
Indigenous Knowledge and Community Centric Forest Management in Nepal

V. M. Ravi Kumar and Pramod Tarachand Deshpande

Abstract The natural resource conservation sense of tribal, ethnic and indigenous communities has emerged as an important component in the academic an activist discourse related to human ecology. Influenced by the concept of decentralised development and governance that gained ascendancy during the 1990s, the intrinsic potential of forest dependent communities to manage natural resources attracted the attention of public policies of several countries in South Asia. This chapter, focusing on historical and contemporary dimensions of participatory natural and forest management in Nepal, proposes that the domain of community emerged as an important component of public policy for implementation of decentralised forest management. It has been postulated in this chapter that the inherent potential of communities for management of forests was incorporated into community centric forest management to increase of forest cover and improvement in livelihood conditions of local communities which deepened upon forests for survival.

Keywords Nepal · Community forestry · Conservation · Participatory forestry · Communities

Introduction

The Himalayas constitute one of the important natural ecosystems that exercise a significant impact upon climatic conditions and livelihood systems of the people in South Asia. The Himalayan country of Nepal has been an abode of beautiful and valuable forests. These forests are extremely important for the Himalayan ecosystem at one level and source of critical importance to livelihoods of pastoral communities,

---

V. M. Ravi Kumar (✉) · P. T. Deshpande
Department of History, Babasaheb Bhimrao Ambedkar University, Lucknow, India
e-mail: vmravikumar@gmail.com
e-mail: pramod.deshpande95@gmail.com

R. C. Sobti et al. (eds.), *Traditional Knowledge Systems in India: Sustainability, Ecology, and Resource Governance*, People, Cultures and Societies: Exploring and Documenting Diversities,
https://doi.org/10.1007/978-981-92-1346-7_1

Dalits and several other tribal communities who dependent upon forests for survival. On account of this cardinal relevance, forests are being revered and venerated by local communities as spiritual entities. This chapter endeavours to capture the interplay of nature, communities and conservation in Nepal. By so doing, the conservation strategies of local forest dependent communities are highlighted.

This chapter intend to projects a general picture on the interaction between tribal/ethnic and indigenous communities with forest ecosystem in Nepal. The main purpose of the chapter is to describe the evolutionary trajectory of community centric forest management from historical to contemporary times. This endeavour is useful to locate the long term trend of the dynamic interaction between communities and forests and help us to formulate better policy systems for efficient management of forests. The concept of indigenous knowledge and community centric forest management in this chapter refers to the domain of historical dimension to forest use practices adopted by local communities for survival and impart of such practices on the policy of community oriented forest management implemented in Nepal during the 1990s. The chapter has been divided into five sections. The first part outlines literature review on indigenous forest conservation practices of communities in Nepal. The second part unveils composition of forests of Nepal. The third part documents the livelihoods aspects of forested communities in Nepal. The fourth part captures and efforts made by communities in management of local forests with their socio-cultural values and sometimes with the support of the government in Nepal.

A Brief Review of Literature

The phenomenon of community-based natural resource management acquired significant prominence as a preferred choice of public policy model from the early 1990s onwards in South Asia. It has undoubtedly been emerged as one of the important policy paradigms for management forests (Poffenberger, 2000). Enthusiastically supported by international donor agencies such World Bank, Asian Development Bank, Swedish Development Bank and others, the community centric participatory forest management has been able to gain an immense popularity as a public policy strategy. Its main objective was to promote management of degraded forests and at the same time, improving the livelihood status of tribal, ethnic and indigenous communities that depended upon forest for survival (Seeland and Schmitusen, 2004). Nearly after three decades of its implementation, it could be observed that community forest management has brought about mixed results. Undoubtedly, in terms of improvement in forest cover was reported, but its impact upon livelihood fall short of expected level. In view of this, community forest management in South Asia is certainly stood itself at crossroads. Nevertheless its efficacy as a policy to manage forests and people was never doubted. One of the core features of community forestry is that local communities were recognised as a creditable source for the managing forests in a sustainable way (Bagainski & Blaikie, 2007).

Nepal represents an ideal example of how the community forest management brought about certain significant changes in the quality of forests cover at one level and changes in the livelihoods of forest dependent communities at another level (Thwaites et al., 2019). The forests in Nepal are valuable sources for commercial timber, several varieties of non-timber products and equally important from climatic point of view by the way of exercising a significant impact upon water flow of ravines, local temperature, composition of glaciers and several other aspects (Stainton, 1972). On account of these important utilities, the forests of Nepal attracted the attention of the state systems in pre-colonial, British colonial and post-colonial periods (Negi, 1994). Broadly, two parallel forest resource use systems existed in the history of Nepal and to some extent continued in the contemporary period in Nepal. On the one hand, a system controlled by the state machinery aimed at extracting valuable timber and on the other hand, forests and their products are customarily used by local communities for subsistence needs (Rajnit, 2019). Needless to say that contrary to the state centric system, the local community centric forest use, this chapter proposes is not only sustainable but also useful for protection of forest resources.

On account of high commercial value, the forests of Nepal in the Hills, the Mountains and the Terai regions attracted the attention of several international agencies. While India during the colonial and postcolonial period was the principle source for imported timber, of late, several international agencies took keen interest. From the 1990s onwards, with the help of several international funding agencies, the government of Nepal initiated policies for community centric forest management (Nuberg et al., 2019). One of the consistent policy principles that were forcefully articulated is that those local communities are perceived as not destroyers of forests rather perceived as collaborators for efficient management of forests. It is this premise that constitutes the centrality of community oriented forest management that is formulated and implemented in Nepal and still continues (Ojha et al., 2025).

Forests of Nepal-a Profile

Some study suggested that etymological roots of the word 'Nepal' was originated from the Sanskrit 'nipalaya' which means 'at the foot of the mountains' or "abode at the foot", a reference to its location in relation to the Himalayas (Thierry, 2015: 22). On account of its geographical and natural location, Nepal is widely known for its rich eco-cultural diversities. The variation in topographical and climatic conditions has enriched the country with remarkable diversity in flora, fauna and intricate ecosystems. Such diversity is apparently reflected in cultural heritage and practices which are to some extent what cultural ecologists use the concept 'adoptive strategies' of local communities living in forest ecosystem. Although

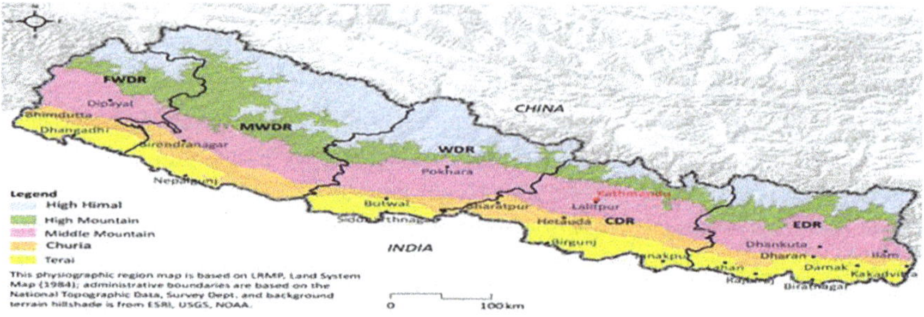

Fig. 1 Map of the five physiographic divisions of Nepal. *Source* Adapted from (DFRS, 2015, 2)

Nepal cover merely 0.1% of world's land area, it is a space nearly for 3.2% of the world's floral diversity (Ghimire et al., 2021). In addition to this, Nepal harbours 284 fascinating endemic species of flowering plant. Furthermore, Nepal's known flora comprises 6073 species of angiosperms; 26 gymnosperms; 534 pteridophytes; 1,10 bryophytes; 365 lichens; 1822 fungi and 1001 algae (DFRS, 2015). In geographical terms, Nepal lies between 260 22' N and 300 27' north latitude and 800 04' E and 880 12' east longitude (DFRS, 2015, 1). In terms of geology and geomorphology characteristics, Nepal comprises five physiographic regions such as Terai physiographic region, Churia region, Middle Mountains region, High Mountains region and High Himalayan region[1] (DFRS, Department of Forest Research and Survey (DFRS), 2015, 1–2). The factor of elevation from sea level is the main defining variable that determines the boundaries of these five physiographic regions as suggested by FAO working paper (Govil, 1999) (Fig. 1).

Vegetation Types and Eco-regions in Nepal

One of the working papers from Food and Agricultural Organisation (FAO) suggests that different researchers have classified the vegetation of Nepal into 118 ecosystems, six vegetation types based on altitudinal zones. For instance, Tropical, Sub-tropical, Temperate, etc.[2] However, Nepal recognised twelve eco-regions, as pointed out by the Conservation Science Programme WWF-US (1998), such as Trans-Himalayan, West Himalayan alpine shrub and East Himalayan broadleaf forest, etc.[3]

[1] See (DFRS Department of Forest Research and Survey (DFRS), 2015) for the brief description of each of these physiographic regions.

[2] See https://www.fao.org/4/ae154e/AE154E04.htm#P285_21659 for the brief description of vegetation in each of these zones.

[3] See for more details DFRS (Department of Forest Research and Survey (DFRS), 2015, p. 5).

Categories of Forests in Nepal

As per the FAO, Nepal's forest zones that situated outside protected areas are officially classified as national and private forest entities. The category of national forest comprises of several types such as Government managed Forest, Protected Forest, Community Forest, Leasehold Forest, Religious Forest and Collaborative Forest which are managed under supervision of the state. Further government holds the control over the national forest cover, while private forests managed by individual owners. Within national forests, except for government managed forests, local communities are bestowed only usufruct rights. Furthermore protected areas defined differently such as National Parks, Wildlife Sanctuaries, etc. (FAO, 2020, 18).

Forest Cover in Nepal

Nepal's forest zone extends across 5.96 million hectares, which was equals to 40.36% of the national territory. At the same time, other wooded land (OWL) accounts for 0.65 million hectares (4.38%) and other forest land covers 8.16 million hectares (55.26%). Within OWL the category shrub accounts for 0.12 million hectares (0.79%) and areas with tree canopy cover 5–10% occupies 0.53 million hectares, which was holds the 3.59% of the total area. In total both Forest and OWL together consist 6.61 million hectares, which was 44.74% of total area of the national territory (DFRS, Department of Forest Research and Survey (DFRS), 2015, 25). Nepal's total area of forest, about 82.68% (4.93 million hectares) is located outside protected areas, while 17.32% (1.03 million hectares) lies within them. Further within the protected areas, forest covers 0.79 million hectares in core zones and 0.24 million hectares in buffer areas (DFRS, Department of Forest Research and Survey (DFRS), 2015, ix). Further, out of the total area of forest cover, approximately 37.80% located in Middle Mountains physiographic region, 32.25% in High Mountains and High Himal, 23.04% in Churia and with small shares 6.90% in Terai. For other wooded land, the majority located in the High Mountains and High Himal physiographic regions 85.42%, followed by Middle Mountains 9.61%, the Churia, 3.50% and Terai 1.47% (Ibid, p. ix).

Further among the various type of the forest the Terai Mixed Hardwood (TMH) forest type has the largest share, covering 24.61% of the forest area, followed by the Upper Mixed Hardwood (UMH) (18.23%). The Shorea robusta and Pinus roxburghii forest types accounts 15.27% and 8.45%, respectively.[4] In total mixed forest types constitute around 60% of the total forest area (DFRS, 2015, 29).

As per the latest government draft plan document the extent of overall forest zone expanded from 44.74% (of the total national territories) in 2015 to 45.31% in

[4] See (DFRS, Department of Forest Research and Survey (DFRS), 2015) for details to know method they followed to land mapping area and total area of forest cover.

2022 and 46.08% in 2024 (National REDD + Strategy (2025–2034) Draft Document 2024, 8). Furthermore, it suggested that forest cover has expanded in the Terai, Siwalik, and Middle Mountain regions whereas it reduced in High Mountain and High Himal from 2000 to 2019. The area of other wooded land expanded in the Terai and Siwalik regions whereas it diminished in the Middle Mountain, High Mountain, and High Himal regions from 2000 to 2019. From 2019 to 2022, the forest area has increased in all physiographic regions whereas the area of other wooded land has decreased in all regions (National REDD + Strategy (2025–2034) Draft Document 2024, 9).

Forests of Nepal and Communities

The space of forest is the vital source of livelihood to forest depended communities. Forest Research Programme defines the concept of 'forest dependence' as being dependent on forest or tree resources for a significant part of livelihood, which cannot easily be substituted by other resources (Salim, 2001, 5). In the context of Nepal, it is a more vital as they rely on them for their sustenance. Agrarian sector, including animal husbandry played a significant part of rural livelihood and communities significantly rely on forests resources to meet their needs of fodder, fuelwood, farm equipment, timber and medicinal plants (Neha et al., 2022, 7). In Nepal, most of indigenous peoples are depended on forest sources. It would be worthwhile to grasp the nature of dependency that tribal and ethnic communities on forest ecosystem.

As per the 2021 census report of the Nepal, 35.08% of Nepal's population out of 29,164,578 peoples belongs to the native ethnic communities (Bhattachan, 2025, 271). Officially these communities are defined as "Indigenous Nationalities" means a tribe or community as mentioned in the Schedule having its own mother language and traditional rites and customs, distinct cultural identity, distinct social structure and written or unwritten history'[5]. This act enlisted a total of 59 communities (Adivasi/janjatis), identified as Indigenous Nationalities.[6] However, some groups eligible as indigenous communities under NFDIN Act criteria are still not formally included (Limbu and B., Limbu & Bhattachan, 2022). Recently, on July 11, 2024, the government of Nepal officially acknowledged the Humlo communities as Indigenous Nationalities of Nepal (Sunuwar, 2024).[7]

Most of the native ethnic groups are largely situated within their own territories that were inhibited by ancestors. In the mountain areas, communities such as

---

[5] Sec. 2 of the NFDIN Act, 2002.

[6] See for the enlisted communities names https://faolex.fao.org/docs/pdf/nep202214.pdf last accessed on 17th July 2025.

[7] See https://www.culturalsurvival.org/news/nepal-officially-recognizes-humlo-peoples-indigenous-nationality to know more about Humlo peoples.

Mustang, Manag, and Rasuwa, constitute about 75–95% of the population, however in the Hills and the Terai regions, they constitute less than 50% of population due to significant presence of non-indigenous and migratory communities (Limbu and B., Limbu & Bhattachan, 2022). The native ethnic communities in mountain regions remarkably sustain and continue with their traditional pattern of life regulated by socio-cultural patterns and norms. However, due to the impact of powerful groups who adopted the socio-cultural identities such as Gorkhalisation, Hinduisation, Sanskritisation, Nepalisation, and Westernisation, many less-powerful indigenous communities in the Hills and Terai gradually lost access to ancestral lands, resources, and eventually their traditional lifeway (Limbu & Bhattachan, 2022).

Tribal areas of Nepal are subjected to some of the following problems: discrimination in terms of development, social deprivation, and consequent subordination due to incursion and domination of hierarchy oriented value system practice by Hindu population a long period of time. These structural problems include the land dispossession in multiple ways which are as follows: demonization of indigenous traditional practices, forced evictions, and alienation because of aggressive development intervention by using legal mechanism such as laws, policies, rules and regulation, government programme and scheme (Bhattachan, 2025, 271). Based on socio-development indicators such as education, income, financial status, land ownership and other assets, the Nepal Federation of Indigenous Nationalities classified affected indigenous communities into five groups (Limbu & Bhattachan, 2022).[8]

The fact, however, is that most of the tribal and indigenous communities lived in forested areas and practiced diversified forms of traditional practices of forest use for survival. The active involvement of local communities and their collective action reflected in forest use patterns recognised as important in the conservation and sustainable utilisation of natural resources (Shrestha & McManus, 2008, 135). In fact, a few studies suggest that forests have been better protected by communities than by the state (Shrestha & McManus, 2008, 137). In Nepal, forest management pattern has gradually shifted from centralised governance by the state to community forestry (Chhetri et al., 2016, 33). The improvement of forest condition is attributed to the FUGs being focused on forest protection, not on utilization.

Forest as Livelihood

The forests of Nepal contribute in many ways to people's livelihood, including supplying timber for domestic use, fuel wood, grazing for livestock, medicinal herbs, non-timber forest products, agroforestry, dietary and nutrient supplements and enhance food security, particularly crop failure or lean season of the agriculture

[8] See (Limbu and B. Limbu & Bhattachan, 2022) Box no. 2; also see for Classification of indigenous peoples by population size, Census 2001, table no.1.

and several other aspects. Further, it provides the financial supports from selling forest and livestock products, generate income from the forest-based economic and labour activities (Fisher et al., 1997).[9]

Recently non-timber forest products (NTFPs) attracted significant attention on account a massive global demand (Ghimire et al., 2021, 7). NTFPs refers to the all forest resources other than timber products that used for people livelihood and other subsistence, such as medicinal herbs and aromatic plants, forest vegetables, fruit, bamboo products, wooden tools and utensils, fibre, fuelwood, fodder, etc. (Gurung et al., 2021, 447). As per the FAO, 'NTFPs as all goods of biological origin as well as services derived from forests, other wooded land and trees outside forests, and excluded woods in all its form' (Ghimire et al., 2021, 2).

In the context of Nepal, there has been a growing recognition of NTFPs for their vital role in rural welfare and livelihoods support, ecological conservation and economic potential could be observed. As a results, diverse disciplinary fields of academic and government institutions are seeking to improve rural economies, livelihoods and promote environmentally tenable methods (Ghimire et al., 2021, 7). It is estimated that nearly 90% of rural household income in Nepal coming from NTFPs-based economic activities (Gurung et al., 2021, 447). Furthermore, NTFPs are used for both supporting household consumption and generate tangible income source (Ibid, p. 455). Forest dependent communities in Nepal have historically dependent upon NTFPs as a crucial livelihood option, using the earnings from their gathering and selling to meet daily subsistence (Sherpa, 2025). Among the NTFPs bamboo product has largest potential, since it is important because of its social, economic, ecological, and cultural prospects[10] and Bamboo forests in Nepal are estimated to be about 63,000 hectares, accounting for around 4.6% of the nation's total land area. For centuries in Nepal, Bamboo was being used for diversified activities such as several varieties0 of handicrafts, furniture, as a source of food and diverse agricultural activities (Kafle et al., 2023, 2). The uses of bamboo in Nepal differ according to the type of species and regional variations as these processes significantly conditions the nature and quality of bamboos. Nevertheless, the significant presence of bamboo and its economic potential, the present use of bamboo in Nepal is underdeveloped due to lack of technological apparatus to tape it in a more efficient way.

The tribal and indigenous communities of Nepal, as mentioned above have critically been depended upon forests for survival. Their economic, cultural, social and spiritual life forms took shape in forest-based ecosystem. On account of this intimate relationship, tribal and indigenous communities evolved diversified forms of forest resource use to obtain sustained supply of resources for their needs. One

[9] Seehttps://www.fao.org/4/w7732e/w7732e04.htm#numbers%20of%20forest%20'dependent'%20peoples%20and%20types%20of%20people%20forest%20relationships last accessed on 19 July 2025.

[10] Seehttps://www.sciencedirect.com/science/article/pii/S2773139123000137#:~:text=We%20found%20that%20a%20total,of%20local%20communities%20in%20Nepal last accessed on 19 July 2025.

of the interesting examples of indigenous ecological knowledge of local communities as follows: several tribal communities in Nepal follow the tradition of planting minimum five trees in village common land for occasions such as birth or death. This practice evolved to make sure the existence of tree cover critical for survival of communities (Bagainski & Blaikie, 2007). A recent study on the dietary habits of Tharu tribes of the Terai region extensively mentions about how forest play a critical role influencing Tharu cuisine (Kumari & Ravi Kumar, 2024). It has further been pointed out that the forest ecosystem of the Terain region significantly shaped the belief system, cultural practices and social values that are followed by the Tharu tribes. It can thus be deduced conveniently that forest ecosystem played crucial role in formation of tribal and indigenous communities knowledge systems, practices and lifestyle patterns.

Forest Policies in Nepal: A History

The country of Nepal is basically a county consist predominantly of mountains, hills and forests. It is indeed an abode of thriving forests and dynamic communities such as tribes and indigenous communities dependent upon forest resources for survival. Out of total land area in Nepal, 29% is designated as forests and an extra 10.6% as shrub and diminished forests. While, the poor communities dependent upon forests for survival, the elite ruling class not only owned forests lands but also extracted them for profit. The forest space, thus, is a contested domain for divergent sections in Nepal. Historically the Shah dynasty and the Rana dynasty ruled Nepal for a long period of time and tapped the forest resources for commercial purpose. This extraction furthered by appointment of a British forester named J.V. Collier for management of valuable timber forests in Nepal on modern scientific lines (Negi, 1994).

The beginning of modern forest management was initiated with establishment of Department of Forests in 1942. In order to take forests away from the hands of the Rana ruling elite, the Private Forest Nationalisation Act of 1957 was promulgated. Further a forest policy for Nepal was issued in 1976 to streamline forest administration on a systematic footing. Things were further systematised with creation of the Master Plan for Forestry Sector in 1988. This plan, besides the role of forest department in managing forest, highlights the potential of local communities as partners in forest management. The Forest Act of 1993 made the community forestry as a legal category to be managed by communities under the guidance of the regulatory framework of the state. The forest regulations of 1995 proposed a systematic legal and administrative framework for creation of user groups for management of forest cover assigned to them by the government (Gautam et al., 2004). A visible trend in forest policy direction from centralised management to decentralised forest governance could be seen in Nepal.

Under the influence of neoliberal model of economic policies and governance strategies, community centric forest management became a favourable policy choice for management of the forests in Nepal. A trajectory of consistent expansion of community centric forest management could be seen during the period of 1993–2005. This trend is evident by the fact that by 2006, 14,000 community centric forest management committees in Nepal were created for management of managing 1,184,824 ha of forests involving an estimated number of 1,633,408 households are engaged management of forests mostly in Hill region. This remarkable expansion of community centric forest management is an axiomatic truth that local communities are capable of managing forest resources more efficiently.

Tribes, lower castes and other indigenous communities have significantly been dependent upon natural resources such as grazing, gathering from forests. Particularly, the livelihoods of tribes and lower caste dependent on forest-based eco-system. It exists in the form of cattle grazing, firewood, wood for construction of house, food such as mushrooms, etc. and finally several varieties of minor forest products collected and sold in towns. Some communities are being dependent upon shifting cultivation in forests. By all means, the forests in Nepal are very much resourceful to tribes and lower caste communities like Dalits. It is this critical dependency that resulted in evolution of diversified forest utilisation practices codified as cultural aspects and customary norms. The forest eco-system thus intertwined organically with the everyday lives of communities in Nepal. It is worth to explore some of the cultural practices that reflect an organic relationship between forests and communities in Nepal.

Communities and Conservation in Nepal and Some Trends

Nepal has an enigmatic history of relationship between communities with forest ecosystem. Several local level community centric customary forest utilisation practices existed in different parts of Nepal. Some of the examples are worth mentioning. The practice of *Kipat* a type of communal forms of land mostly found in east Nepal. Here local communities regulate common pool resources such as forests, grazing ground, water and other resources and make sure that resources are available to members of community by the way of imposing restrictions on their access to resources. Another traditional institution Riti-Thiti could be observed wherein forests and pastured lands are managed by local communities in different parts of Nepal. It is managed by communal councils which regulate socio-economic and cultural aspects of communities. In addition to this, the Khoria system which is a sort of shifting cultivation was being practiced communally. The trend to be noticed here is that Nepal has several local level communal institutional arrangements that exist for management of natural resources such as forests. These pre-conditions provide inspiration for introduction of community forest management in Nepal (Bagainski & Blaikie, 2007).

A systematic shape to legal and administrative formulations dealing with community forest management was introduced in Nepal from the early1990s onwards. Being predominantly a forest-based country; Nepal has implemented a vibrant community centric forest management policy in all most all parts. By 2006, the community forest management covered 14,000 community forest user group spreads across all over Nepal, managing 1,184,824 hectors of forests. From human participation point of view, it covers 1,633,408 households who were engaged in management of local forests. The massive quantitative expansion of the policy attests the expansionist trend of policy. It would be useful to evaluate the broad performativity of the programme by the way of focusing on some general trends. Overall trends on implementation process of the policy can be captured from several evaluative studies which attempted to capture the filed level process associated with implementation process of participatory forest management. This section attempts to highlights some of the dominant trends in terms of both positive and negative dimensions of the implementation of process of community centric forest management.

Several evaluative studies attempted to reflects upon the broader trends, dynamics and in fact shortcomings of participatory oriented community centric forest management in Nepal. Broadly, it has been contemplated that community forestry has been able to increase forest cover significantly and its impact on livelihood status of communities remained a question mark. It has been a general observation by many studies that community forestry did not address the issue of equity among intra community relations. Consequently, benefits of community forestry if not exclusively but mostly appropriated by elite sections that hold socio-economic and cultural power within communities (Adhikari et al., 2014). Some studies suggest that community oriented forest management has been able to sensitise citizens of Nepal on the importance of forest conservation. This sensitivity translated into active participation of local communities in forest management. The process, however fall short of expectation on account of lack of actual participation of wider stakeholders into execution process of forest management (Wagle & Ojha, 2002).

A report summoned by World Bank to evacuate the performance of the participatory community centric forest management in Nepal observed that the programme achieved good success in the middle hills region of Nepal and the same success could not be achieved in other parts on account of complex socio-economic and climatic conditions. Overall, the programme performed well in terms of rehabilitating degraded forests at one level and increasing forest cover at another level. The effective implementation of programme fall short of expected level due lack of inclusive participation of community in decision making process, ineffective gender participation and lack of flexibility in execution of rules and regulations (Nirmal Kumar, 2002).

Focusing on the Hills and the Terai regions, some evaluatory studies suggest that community centric forest management has brought about mixed results in terms impact. At the level of success, the policy execution process achieved remarkable success by the way of creating a massive network of forest stakeholder sections in all regions of Nepal. Additionally, several expels of increase of forest cover, livestock, improvement of ground water on account of increase of tree cover, careful conservation of valuable timber trees, etc. are reported. At the level of negative impact, it failed to address the needs of mobile tribal communities' stakes over forests, traditional dependency of women over forests and customary access of the rural folks for natural and forest resources. These problems aroused after implementation of community centric forest management in Nepal. Nevertheless, the community forest policy has performed well in some respects and failed do so in some other respect. By all means, one of the biggest successes of community forestry in Nepal is that it has been able to provide a platform to millions of tribal, lower castes and indigenous communities to participate in management of as partners.

Conclusion

The community centric forest management in Nepal represents a classic example of the complex relationship between communities and natural resources. Local communities traditionally evolved several institutional forms of collective action rooted in cultural believes and practices. Such traditional forms of forest management facilitated a harmonious relationship between communities and forests. The community centric forest management attempted to systematise such traditional communal management of forests with a policy framework. Undoubtedly this policy framework furthered the spirit of community oriented approach to forest management to protect and conserved forests along with due attention to the livelihoods needs of local communities.

References

Adhikari, S., Kingi, T., & Ganesh, S. (2014). Incentives for Community Participation in the Governance and Management of Common Property Resources: The Case of Community Forest Management in Nepal. *Forest Policy and Economics*, *44*(4), pp. 1–9.

Bagainski, O. S., & Blaikie, P. (2007). *Forests, people and power: The political ecology of reform in South Asia*. Earthscan.

Bhattachan, K. B. (2025). In D. Mamo, (Ed.), *Nepal, in the indigenous world 2025* (39th Ed, pp. 270–283). The International Work Group for Indigenous Affairs (IWGIA).

Chhetri, B., Patrick, A., & Yoshimoto, A. (2016). Forest dependence and inequality: An empirical study from community forests in Kaski, Nepal. *Formath Forest Resources & Mathematical Modeling*, *15*(4), 33–43.

Department of Forest Research and Survey (DFRS). (2015). *State of Nepal's Forests*. Department of Forest Research and Survey, Forest Resource Assessment (FRA), Nepal.

FAO. (2020). The Global Forest Resources assessments (FRA), Nepal.

Fisher, R. J., Srimongkontip, S., & Veer, C. (1997). Asia-Pacific Forestry Sector Outlook Study: People and Forests in Asia and the Pacific: Situation and Prospects. Working Paper No. APFSOS/WP/27. Rome: FAO.

Govil, K. (1999). *Forest Resources of Nepal*. Forestry Department, FAO, Working Paper. https://www.fao.org/4/ae154e/AE154E04.htm#P285_21659

Gautam, A. P., Shivakoti, G. P., & Webb, E. L. (2004). A review of forest policies, institutions, and changes in the resource condition in Nepal. *International Forestry Review*, *6*(2), 136–148.

Ghimire, P., Pudasaini, N., & Lamichhane, U. (2021). Status, prospect and challenges of NTFPs conservation in Nepal: A critical review. *Grassroots Journal of Natural Resources*, *4*(1), 1–16.

Gurung, L. J., Miller, K. K., Venn, S., & Bryan, B. A. (2021). Contributions of non-timber forest products to people in mountain ecosystems and impacts of recent climate change. *Ecosystem and People 17*(1), 447–463.

Kafle, A., Binfield, L., & Paudel, S. K. (2023). The commercialization of timber bamboo in Nepal: A SWOT-AHP analysis. *Advances in Bamboo Science*, 1–11.

Kumar, D. S. (2024). Nepal officially recognizes humlo peoples as an indigenous nationality. *Cultural Survival*, July 12, 2024. https://www.culturalsurvival.org/news/nepal-officially-recognizes-humlo-peoples-indigenous-nationality

Kumar, N. (2002). *The challenges of community participation in forest development in Nepal.* World Bank.

Kumari, K., & Ravi Kumar, V. M. (2024). Food as source of ethno-history: A study of tharu tribes of champaran district of North India. *Kurukul International Multidisciplinary Research Journal*, 26–33.

Limbu, S. K., & Bhattachan, K. B. (2022). Country technical notes on indigenous peoples' issues: Federal democratic Republic of Nepal. Technical Paper, Rome: IFAD and IWGIA. https://www.ifad.org/documents/48415603/49750747/nepal_ctn.pdf/af2bf9da-9035-9b20-4137-59473ff87703

Ministry of Forests and Environment, Nepal. (2024). *National REDD+ Strategy (2025–2034), Draft Document.* Ministry of Forests and Environment.

Negi, S. S. (1994). *Forests and forestry in Nepal.* Asish Publishing House.

Neha et al. (2022). Forest—People nexus in changing livelihood contexts: Evidence from community forests in Nepal.

Nuberg, I. K., Shrestha, K. K., & Bartlett, A. G. (2019). Pathways to forest wealth in Nepal. *Australian Forestry*, *82*(51), 106–122.

Ojha, H., Shrestha, K. K., & Paudel, N. S. (2025). *Restoring forests and improving livelihoods in Nepal: Four decades of community forestry*. Routledge.

Poffenberger, M., (Ed.). (2000). Communities and forest management in South Asia, WG-CIFM, DFID, and IUCN.

Rajnit, Y. (2019). History of forest management in Nepal: An analysis of political and economic perspective. *The Economic Journal of Nepal*, *42*(3–4), 12–28.

Salim, R. (2001). *Numbers of forest dependent poor people in Nepal.* DfID, Forestry Research Programme. https://assets.publishing.service.gov.uk/media/57a08d61e5274a31e0001812/ZF0163_-_forest_dependent_people_Nepal.pdf

Seeland, K., & Schmithüsen, F. (2004). *Indigenous knowledge, forest management, and forest policy in South Asia.* D.K. Print World.

Sherpa, C. (2025). Non-Timber Forest Products and Their Role in Rural Livelihoods: A Case Study from the Annapurna Conservation Area, Nepal. *Ethnobotany Research and Applications*, *30*, 1–15.

Shrestha, K., & McManus, P. (2008). The politics of community participation in natural resource management: Lessons from community forestry in Nepal. *Australian Forestry*, *71*(2), 135–146.

Stainton, J. D. A. (1972). *Forests of Nepal.* Hafner.

Thierry, B. (2015). *Regenerating forests and livelihoods in Nepal: A new lease on life and IFAD, by arrangement with CABI.*

Thwaites, R., Fisher, R., & Poudel, M. (2019). *Community forestry in Nepal: Adapting to a changing world* Routledge.

Wagle, M., & Ojha, H. (2002). Analysing participatory trends in Nepal's community forestry, policy trend report. https://www.iges.or.jp/policy-report/09_Nepal

Chapter 2
Ethnobotanical Wisdom of Tribal Communities

Ningthoujam Samarendra Singh, Khwairakpam Sharmila, Khwairakpam Bembem, and Khwairakpam Priyanka

Abstract The abstract provides an overview of the profound relationship between nature and human societies, particularly in the context of medicinal plant usage and conservation. It highlights the historical awareness of medicinal plant properties dating back thousands of years, underscoring the significance of traditional healing practices. Notably, many modern allopathic drugs trace their origins to traditional medicinal knowledge, emphasising the importance of ethnomedicinal exploration. The abstract also discusses the global reliance on plant-derived traditional medicines for primary healthcare, emphasising the accessibility and affordability of these remedies. Furthermore, it delves into the diverse cultural and geographical influences on medicinal plant usage, stressing the importance of documenting tribal medicinal practices for scientific and societal benefit. As a result, the urgent need arises to preserve ethnobotanical knowledge among tribal communities to advance medicinal understanding and foster sustainable resource management.

N. S. Singh (✉)
Division of Post-Harvest Management, Central Institute of Sub-Tropical Horticulture, Lucknow, Uttar Pradesh, India
e-mail: tairenki@gmail.com

K. Sharmila
Department of Human Development and Family Studies, School of Home Science, Babasaheb Bhimrao Ambedkar University, Lucknow, Uttar Pradesh, India
e-mail: khwairakpamsharmila@gmail.com

K. Bembem
Division of Transfer of Technology, Central Institute of Post-Harvest Engineering and Technology, Ludhiana, Punjab, India
e-mail: Bembemkhwai@gmail.com

K. Priyanka
Department of Horticulture, School of Agricultural Sciences and Technology, Babasaheb Bhimrao Ambedkar University, Lucknow, Uttar Pradesh, India
e-mail: khpriyanka5@gmail.com

R. C. Sobti et al. (eds.), *Traditional Knowledge Systems in India: Sustainability, Ecology, and Resource Governance*, People, Cultures and Societies: Exploring and Documenting Diversities,
https://doi.org/10.1007/978-981-92-1346-7_2

Keywords Traditional knowledge · Indigenous tribe · Botanical plant · Ethnomedicine · Bioactive compound

Introduction

With its vast resources, nature provides food, clothing, shelter, and antiques and serves as a source of medicine throughout history (Mahapatra et al. 2018). It was reported that human awareness of the medicinal properties of plants dates back to around 5000 years ago (Kabir et al., 2014). Importantly, many significant allopathic drugs like aspirin, atropine, ephedrine, and morphine have roots in traditional medicinal practices of indigenous peoples, highlighting the value of ethnomedicinal exploration (Mahapatra et al. 2018). This represents a mere 2.4% of the world's total geographical area, harbouring about 47,513 plant species, including microorganisms like viruses, bacteria, and fungi (Kumar et al., 2019). Of the estimated 270,000–422,000 flowering plant species globally, approximately 50,000–70,000 have been harnessed for medicinal use in both traditional and modern healthcare systems (Rahman et al., 2022), with around 53,000 plant species utilised for medicinal purposes worldwide (Malik et al., 2015). Plants hold a significant role in the healthcare practices of Himalayan remote communities, driven by their trust in herbal remedies and the lack of modern healthcare facilities (Rana et al., 2021). They serve as a rich source of diverse chemical compounds with medicinal and commercial value (Laldingliani et al., 2022). Recognising this, the World Health Organisation (WHO) notes that 80% of the world's population, especially in developing nations, relies on plant-derived traditional medicines for primary healthcare (Namsa et al., 2011; Sukumaran et al., 2021). These remedies are preferred due to their proximity, accessibility, and affordability (Roy Choudhury et al., 2015). Different cultural, geographic, and ethnic backgrounds influence medicinal plants' perspectives and usage patterns (Rahman et al., 2022), making adequate documentation of tribal medicinal practices valuable for scientists and the general population (Kabir et al., 2014). Medicinal plants, containing active constituents like alkaloids, glycosides, and essential oils in various parts, treat diverse ailments in humans and animals (Malik et al., 2015). Over time, the therapeutic potential of plant extracts and compounds has led to the development of popular modern medicines (Islam et al., 2014), which have been pharmacologically studied for various efficacies, including antibacterial, antifungal, and antidiabetic properties. Traditional medicines are typically prepared as decoctions, pastes, infusions, powders, or consumed fresh.

The traditional healthcare practices of India, such as Ayurveda, Siddha, and Unani, are globally renowned, with ancient scriptures like the Atharva Veda, Charaka Samhita, and Sushruta Samhita describing these practices (Rana et al., 2021). Evidence suggests that herbal medicine has been practised in China, India, Egypt, and Syria for over 5000 years (Khajuria et al., 2021). India, the second-largest producer of medicinal plants after China, has recorded around 20,000 medicinal plant species, of which traditional practitioners utilise 7000–7500 for treating various diseases (Rana et al., 2021). Indigenous people across different ethnic groups in India have long relied on plants for sustenance and healing (Panghal et al., 2010). With over 427 tribal communities in India, more than 130 major tribes reside in the North-East, where plants are extensively used for nutrition and medicine (Uniyal et al., 2006). India boasts over 17,209 flowering plant species, with more than 7918 possessing medicinal properties (Lingaraju et al., 2013). Traditional medicine draws from approximately 5000 plants in China, while about 1600 species serve medicinal purposes in north-west Amazonia (Singh & Lal, 2008). Ethnobotanical studies have unveiled numerous plants with significant medicinal properties previously unknown to the scientific community (Dutt et al., 2015). Thus, traditional medicine encompasses systematic codified systems like pharmacopoeias, non-codified folk medicine, spiritual or shamanistic practices, and allied health knowledge.

Globally, indigenous communities, comprising nearly 5% (370 million) of the total population, face significant poverty challenges, with approximately 15% categorized as 'extremely poor' (Priyadarshini & Abhilash, 2019). These communities, known by various names across different regions, have traditionally relied on plant resources, with indigenous tribal groups particularly adept at utilizing nature due to their closely intertwined lifestyle (Chase & Singh, 2016). The ethnomedicinal knowledge within these communities, passed down through diverse cultures and folklore, is notably rich and forms an integral part of indigenous cultures worldwide (Kichu et al., 2015). Typically, traditional knowledge is transmitted from generation to generation, primarily among elders (Bujurg) and traditional healers, highlighting the importance of experience-based wisdom (Tamang et al., 2023). Plants have historically served as the cornerstone of therapeutic practices, with traditional methods renowned for their minimal side effects, affordability, and widespread accessibility (Laldingliani et al., 2022). Economic benefits associated with developing indigenous medicine and using medicinal plants for treating various ailments have been well-documented (Kosalge & Fursule, 2009). Regarding healthcare preferences, a significant percentage of the population relies on their traditional clan healers, while others seek allopathic or homoeopathic treatments, indicating a diverse healthcare landscape (Kabir et al., 2014).

Ethnobotany examines how human societies, particularly indigenous and tribal communities, interact with surrounding flora (Lingaraju et al., 2013). Ethnomedicinal knowledge often serves as the source for new drugs or active compounds for critical ailments, recognised by the World Health Organisation as integral to healthcare strategies. This wisdom has spurred numerous drug development ventures, with pharmaceutical industries utilizing Phyto-resources to isolate active principles or for drug design (Lal et al., 2023). In India, over 50% of plant species in any ecosystem are utilised in ethnomedicine, with tribal populations possessing extensive traditional knowledge on biotic resource use (Pradhan & Badola, 2008). Necessary allopathic drugs like aspirin, ephedrine, and morphine have roots in indigenous medicinal practices (Kabir et al., 2014). Ethnomedicinal studies offer opportunities for new drug development, bridging traditional wisdom with modern science (Chander et al., 2016). Traditional societies, closely connected to nature, hold valuable knowledge of wild flora and fauna, often unfamiliar to urban populations (Dutt et al., 2015). Ethnobotanical research fosters interdisciplinary collaboration, benefiting fields like pharmacology, anthropology, and sociology (Namsa et al., 2011). Ethnomedicinal practices offer cost-effective and efficient healthcare solutions, particularly for low-income populations (Rahman et al., 2022). The rising use of medicinal plants stems from concerns over synthetic medicine side effects and high costs, driving interest in traditional herbal remedies (Tamang et al., 2023). Traditional herbal healing remains prevalent, especially in rural and forested areas, sustaining centuries-old practices (Panmei et al., 2019). However, traditional oral knowledge of medicinal plants faces threats, as seen among Gaddi shepherds in the northwestern Himalayas (Dutt et al., 2015).

About 80% of the population in many developing countries still heavily relies on these resources for their primary healthcare needs (Kumar et al., 2013). Recently, there has been a significant increase in the demand for medicinal plants in both local and global markets, leading to the overexploitation of these valuable plant genetic resources. Many of these tribes still rely wholly or partly on wild forest resources, exemplified by Manipur, situated in the Himalaya and Indo-Burma global biodiversity hotspot, showcasing rich flora and fauna diversity (Panmei et al., 2019). India, with an 8.6% tribal population, harbours an immense reservoir of indigenous knowledge, which, if properly recognised, adopted, and integrated, holds the potential to address various social challenges such as falling agricultural productivity, biodiversity loss, and water scarcity (Priyadarshini & Abhilash, 2019). The Himalayan region is renowned for its natural wealth, particularly medicinal and aromatic plants, which have been used by traditional healers for centuries to treat various ailments such as skin disorders, asthma, diabetes, snake bite, fever, pain, eye diseases, diarrhoea, indigestion, jaundice, burn, wound, liver disorder, etc. (Kumar et al., 2019). These socio-economic changes, along with environmental factors, pose a threat to the existence of many highly sought-after medicinal plants worldwide (Kumar et al., 2013). Active participation of knowledge holders in village-level activities can significantly enhance the promotion, learning, and conservation of traditional practices and related resources (Singh et al., 2010). Knowledge holders across different age groups and social systems possess numerous traditional practices that offer promising solutions to contemporary challenges. Promoting traditional knowledge-based products can also contribute to resource conservation and the subsistence survival of communities. Hence, it is imperative to emphasise the preservation of ethnobotanical knowledge among tribal communities, as it facilitates advancements in understanding the medicinal value of plants and contributes to the conservation of forests and their economic utilisation. This book endeavours to shed light on the diverse applications of ethnobotanical practices among tribes in treating ailments and for various other purposes, such as cultural practices, rituals, and sustainable resource management. By documenting and promoting this invaluable traditional knowledge, we can enhance healthcare practices and foster sustainable management of natural resources for the benefit of both present and future generations.

Ethnobotanical Study of Tribal Communities

Understanding the traditional lifestyle of ethnic peoples relies heavily on scientific investigation and documentation of indigenous plant knowledge (Sukumaran et al., 2021). Despite remarkable advances in allopathic medicine, herbal practices still hold significant sway in managing and treating various ailments in remote and rural areas of India (Malik et al., 2015). Notably, ethnic and tribal groups residing in geographically distant regions possess distinct dialects, cultures, and subsistence, yet they share common knowledge about particular plant species (Lingaraju et al., 2013). For instance, the usage of *Passiflora subpeltata* against jaundice remains consistent among the Jenu Kuruba, Kadu Kuruba, and Mullu Kuruba tribes of the study area. The biodiversity, ethnic, religious, and linguistic diversity of a region or country directly reflects its biocultural diversity, highlighting the interconnectedness between nature and culture (Priyadarshini & Abhilash, 2019). An Ethnobotanical Survey of Medicinal Plant Species used by Tribal Communities Around Katghora Tehsil of Korba District of Chhattisgarh state has been conducted (Kunjam et al., 2021). Medicinal plants have served as sources of medicine in many indigenous communities worldwide (Namsa et al., 2011). The World Health Organisation estimates that approximately 80% of the world's population relies on traditional medicine for primary health care needs (Lepcha et al., 2019). Medicinal plants have been crucial in curing diseases, especially in developing countries (Laldingliani et al., 2022). In India, out of over 427 tribal communities, the northeastern states, including Sikkim, boast over 130 major tribes, comprising about 300 sub-tribes or groups (Mahapatra et al. 2018). India has the second-largest tribal population globally, after Africa (Kosalge & Fursule, 2009). India is inhabited by 16 major tribes distributed across 11 districts (Chase & Singh, 2016). These districts encompass numerous villages with distinct languages, cultural practices, and traditional knowledge, contributing to a unique assemblage of plants, animals, and ecosystems. Unfortunately, this region is recognized as one of the world's biodiversity hotspots, and its bioresources are facing rampant exploitation.

India's ethnic and rural communities possess extensive knowledge about medicinal plants in their surroundings, often passed down orally through generations (Tamang et al., 2023). Rich in bioactive compounds with diverse pharmacological properties, these plants serve as both an economic resource and a primary healthcare source for these communities. Living in remote areas with limited access to modern healthcare facilities, they heavily rely on ethnobotanical practices for daily medical needs (Islam et al., 2014). Represented by over 145 tribal communities, the northeastern part of India, including the eastern Himalaya, boasts rich biodiversity and cultural heritage (Panmei et al., 2019). Similarly, Bangladesh is home to approximately 150 tribes, many of which, particularly smaller communities with populations below 500, are at risk of disappearing due to declining populations and habitat loss (Kabir et al., 2014). Despite their valuable knowledge, many individuals are hesitant to share it, as it provides them recognition in society, resulting in oral transmission of knowledge across generations, albeit with some loss (Uniyal et al., 2006). An ethnobotanical study in the Madhupur forest area of

Bangladesh documented 78 medicinal plant species from 45 families, treating 77 different ailments (Islam et al., 2014). Leaves were the most commonly used part, with decoction as the primary form of preparation, and Fabaceae had the highest number of species (10), followed by Lamiaceae (5), Apocynaceae (4), Rutaceae (4), Solanaceae (4), and Combretaceae (3).

An ethnoveterinary study among tribal communities in Thakht-e-Sulaiman Hills, Pakistan, was conducted by Ahmad et al. (2021), who found that leaves were predominantly used to prepare orally administered decoctions. Although skeletal, musculoskeletal, and gastrointestinal ailments were commonly treated, ritual uses and dermatological illnesses had the highest informant consensus factor (ICF). Salix tetrasperma, Berberis calliobotrys, and Litsea monopetala showed the highest fidelity level (FL). Furthermore, traditional medicinal knowledge among Buner's tribal communities in Pakistan documented 282 plant species from 94 families, including one fungal taxon, two ferns, and five gymnosperms (Rahman et al., 2022). Herbs dominated (57.80%), leaves were the most used part (41%), and most recipes were prepared in boiling water (55.67%). The top medicinal plant families by use reports (UR) were Fabaceae (306), Lamiaceae (256), and Rosaceae (210). The largest disease categories were gastrointestinal (1020 UR) and cardiovascular (972 UR), with key species being *Berberis lyceum*, *Azadirachta indica*, *Prunus dulcis*, *Skimmia laureola*, and *Mentha longifolia*. Additionally, a study on traditional uses of medicinal plants against malaria by tribal communities in the Lesser Himalayas, Pakistan, recorded 84 plant species from 69 genera and 50 families (Shah et al., 2014). Asteraceae was the most commonly used botanical family (11.9%), followed by Lamiaceae (5.9%), Solanaceae and Verbenaceae (4.7% each), and Violaceae (3.5%). Furthermore, Wali et al. (2022) found that *Mentha longifolia* had the highest use reports (5), while *Artemisia maritima* had the highest use value (1.79) for the treatment of gastrointestinal disorders treated by tribal communities using various plants in Diamir district, Western Himalayas, Pakistan (Table 2.1).

Ethnomedicine of Plant Origin in the Indian Context

India is renowned for its ancient indigenous knowledge and practices, particularly through medicinal systems like Ayurveda, Siddha, and Unani (Laldingliani et al., 2022). Despite the vast diversity of traditions among over 427 tribal communities, ethnomedicines face criticism due to regional variations and socio-economic challenges. Approximately 6000 species are used in traditional and herbal medicine in India, meeting about 75% of the needs of developing countries, with 3,000 plants officially recognised for their medicinal value (Mao & Roy, 2017). It is estimated that 70–80% of rural Indians rely on non-allopathic medicines (Ayurveda, Siddha, Unani, and Homoeopathy) for healthcare, with herbs as a significant component (Roy Choudhury et al., 2015). For instance, the Bhangalis are particularly rich in traditional knowledge, especially regarding medicinal plant use (Uniyal et al., 2006) (Fig. 2.1).

Table 2.1 Various ethnobotanical plants used by tribal communities

S. No.	Scientific name	Local name	Family	Parts used	Purpose /usage/treatment	References
1	*Abroma augusta*	Ulot-kombol	Sterculiaceae	Leaf and root	Menstrual disorders	(Islam et al., 2014)
2	*Abelmoschus esculentus*	Kasthuri vendai	Malvaceae	Leaf and flower	Skin diseases	(Francis Xavier et al., 2015)
3	*Abies pindrow*	Saru	Pinaceae	Bark	Allergy	(Rana et al., 2021)
4	*Abrus precatorius*	Gunj	Fabaceae	Seed	Mouth ulcer	(Kosalge & Fursule, 2009)
5	*Abutilon indicum*	Malvārū	Malvaceae	Leaves	Cough, chest pain and vomiting	(Chander et al., 2014)
6	*Adenia trilobata*	Tenik tepang	Passifloraceae	Leaves	Relieve pain	(Kichu et al., 2015)
7	*Andrographis lineata*	Siriyanangai	Acanthaceae	Leaves	Diabetes	(Ignacimuthu et al., 2006)
8	*Acalypha indica*	Kuppi gida	Euphorbiaceae	Leaves	Scabies and ringworms	(Lingaraju et al., 2013)
9	*Achyranthes bidentata*	Puthkanda	Amaranthaceae	Wood	Kidney ailments	(Dutt et al., 2015)
10	*Aerva sanguinolenta*	Iodine pata	Amaranthaceae	Leaf	To stop bleeding from external cuts and wounds	(Kabir et al., 2014)
11	*Aconitum ferox Wal*	Bikh	Ranunculaceae	Tubers	Cough, throat infection, stomach disorder, and vomiting	(Lepcha et al., 2019)
12	*Acorus calamus*	Themeprii	Araceae	Rhizome	Fever, diarrhoea and dysentery	(Chase & Singh, 2016)

(continued)

Table 2.1 (continued)

S. No.	Scientific name	Local name	Family	Parts used	Purpose /usage/treatment	References
13	*Aconitum heterophyllum*	Patish	Ranunculaceae	Roots	Stomach ache and fever	(Uniyal et al., 2006)
14	*Acmella paniculata*	Geuteuheu	Asteraceae	Leaf eaten fresh	Toothache	(Panmei et al., 2019)
15	*Andrographis paniculata*	Kairata	Acanthaceae	Whole plant	Small worms and stomach troubles	(Shil et al., 2014)
16	*Angelica glauca*	Churra	Apiaceae	Roots	Stomach disorder	(Singh & Lal, 2008)
17	*Anethum graveolens*	Shatapushpa	Apiaceae	Seeds	Diabetic	(Kumar et al., 2019)
18	*Amaranthus viridis*	Lalisaag	Amaranthaceae	Roots and leaves	Laxatives and urinary disorders.	(Tamang et al., 2023)
19	*Artemisia nilagirica*	Merangma	Asteraceae	Leaves	Wounds, cuts, scabies, and inflammations	(Namsa et al., 2011)
20	*Asclepias curassavica*	Parija	Asclepiadaceae	Root	Blood coagulation	(Shil et al., 2014)
21	*Bambusa vulgaris*	Moongil	Poaceae	Leaf	Stomachache	(Sukumaran et al., 2021)
22	*Barleria prionitis*	Vjradanti	Acanthaceae	Whole plant	Diabetic	(Kumar et al., 2019)
23	*Barringtonia asiatica*	Tapōngońg	Barringtoniaceae	Leaves	Body ache and paralysis	(Chander et al., 2014)
24	*Basella alba*	Latsungen	Basellaceae	Leaves	Gastritis and laxatives	(Kichu et al., 2015)
25	*Bidens pilosa*	Tenyiitsiithu	Asteraceae	Leaf	Heals old wounds	(Chase & Singh, 2016)
26	*Benincasa hispida*	Maipawl	Cucurbitaceae	Decoction	Anti-obesity, anti-ulcer, anti-inflammatory, and anxiolytic	(Laldingliani et al., 2022)
27	*Berberis asiatica*	Chunchri	Berberidaceae	Roots	Jaundice	(Uniyal et al., 2006)
28	*Betula utilis*	Bhojpatra	Betulaceae	Bark	Cold	(Rana et al., 2021)

(continued)

Table 2.1 (continued)

S. No.	Scientific name	Local name	Family	Parts used	Purpose /usage/treatment	References
29	*Bombax ceiba*	Simal	Malvaceae	Roots, gums, barks, leaves	Fungal infection and sexual impotence	(Tamang et al., 2023)
30	*Blechnum orientale*	Sikiomamoidu	Blechnaceae	Pinnae	Abscess	(Shil et al., 2014)
31	*Blumeopsis flava*	Tingmandibae	Asteraceae	Steam juice	Body ache	(Panmei et al., 2019)
31	*Bryophyllum pinnatum*	Gorol	Begoniaceae	Leaf	Allergy	(Roy Choudhury et al., 2015)
32	*Boehmeria macrophylla*	Muithlimsu	Urticaceae	Root	Leucorrhoea	(Shil et al., 2014)
33	*Boerhaavia diffusa*	Satodi	Nyctaginaceae	Root	Jaundice	(Kosalge & Fursule, 2009)
34	*Caltha palustris*	Guttalu	Ranunculaceae	Leaf	Remove the dead skin from hands and feet	(Dutt et al., 2015)
35	*Capsicum chinense*	Tasu	Solanaceae	Raw	Dysentery	(Panmei et al., 2019)
36	*Carica papaya*	Kumita	Caricaceae	Fruits and sap	Gastritis	(Kichu et al., 2015)
37	*Caryota urens*	Ulathi	Arecaceae	Flower	Hair growth	(Sukumaran et al., 2021)
38	*Cassia tora*	Powadya	Caesalpinaceae	Seeds	Diarrhoea	(Kosalge & Fursule, 2009)
39	*Carum carvi*	Kashmirajiraka	Apiaceae	Seeds	Diabetic	(Kumar et al., 2019)
40	*Chaerophyllum villosum*	Shigu-Zera	Apiaceae	Leaf	Stomach disorder	(Singh & Lal, 2008)
41	*Chrysanthemum indicum*	Guldaudi	Asteraceae	Leaf	Boils	(Rana et al., 2021)

(continued)

Table 2.1 (continued)

S. No.	Scientific name	Local name	Family	Parts used	Purpose /usage/treatment	References
42	*Cissus adnata*	Nrungi	Vitaceae	Leaf	Urinary trouble	(Panmei et al., 2019)
43	*Cissus quadrangularis*	Harkuchla	Lauraceae	Bark	Asthma	(Roy Choudhury et al., 2015)
44	*Citrus aurantiifolia*	Champara	Rutaceae	Juice	Antimicrobial, anti-obesity, and anthelminthic	(Laldingliani et al., 2022)
45	*Citrus reticulata*	Suntala	Rutaceae	Flowers and fruits	Diarrhoea, vomiting and cough	(Tamang et al., 2023)
46	*Centella asiatica*	Thankuni	Apiaceae	Leaf, stem	Stomach disorders	(Kabir et al., 2014)
47	*Clerodendrum cordatum*	Bhati	Rutaceae	Leaf	Sone, Jaundice	(Roy Choudhury et al., 2015)
48	*Cleome viscosa*	Kulching	Capparaceae	Leaves	Ear ache	(Chander et al., 2014)
49	*Clitoria ternatea*	Gokarna	Papilionaceae	Leaf	Inflammation	(Kosalge & Fursule, 2009)
50	*Costus speciosus*	Khwiinothi	Costaceae	Rhizome	Weak heart and inflammation	(Chase & Singh, 2016)
51	*Cochlospermum religiosum*	Kenkechi	Bixaceae	Root	Urinary infection	(Shil et al., 2014)
52	*Curcuma longa*	Holud	Zingiberaceae	Rhizome	Itches and ringworm.	(Kabir et al., 2014)
53	*Crataeva adansonii*	Varno	Capparidaceae	Bark	To cure wounds	(Kosalge & Fursule, 2009)
54	*Crotalaria spectabilis*	Skamayomakakho	Leguminosae	Root and leaf	Health tonic and ringworm	(Shil et al., 2014)
55	*Cyathula tomentosa*	-	Amaranthaceae	Leaves	Antidote to snake bite and skin diseases	(Malik et al., 2015)
56	*Dendrocnide sinuata*	Urticaceae	Zaklojawa	Stem	Fresh cuts and wounds	(Kichu et al., 2015)

(continued)

Table 2.1 (continued)

S. No.	Scientific name	Local name	Family	Parts used	Purpose /usage/treatment	References
57	*Desmos longiflorus*	Khorjun	Annonaceae	Bark	Toothache	(Shil et al., 2014)
58	*Dendrobium densiflorum*	Ghogaysunakhari	Orchidaceae	Leaves	Bone fracture	(Tamang et al., 2023)
59	*Dolichos lablab*	Ciiba	Fabaceae	Leaf, root, and pod	Treats tuberculosis	(Chase & Singh, 2016)
60	*Diospyros peregrina*	Gab	Ebenaceae	Bark	Dysentery and cholera	(Islam et al., 2014)
61	*Dryopteris filix*	Nachav	Dryopteridaceae	Whole plant	Snake and insect bites	(Kichu et al., 2015)
62	*Ehretia acuminata*	Dusukchaoma	Boraginaceae	Leaf	Bones fracture	(Shil et al., 2014)
63	*Elaeocarpus ganitrus*	Rudraksh	Elaeocarpaceae	Shoot	Pneumonia	(Rana et al., 2021)
64	*Engelhardtia spicata*	Mowa	Juglandaceae	Bark	Diarrhoea	(Malik et al., 2015)
65	*Ephedra gerardiana*	Tse	Ephederaceae	Shoot	Rheumatism	(Singh & Lal, 2008)
66	*Erigeron borealis*	Mathok-lugmik	Asteraceae	Leaf	Stomach disorder	(Singh & Lal, 2008)
67	*Erythrina variegata*	Madar	Fabaceae	Leaf	Menstruation	(Islam et al., 2014)
68	*Fagopyrum esculentum*	Khuru	Polygonaceae	Roots, leaves and grains.	Rheumatic pain and liver problems	(Tamang et al., 2023)
69	*Flacourtia indica*	Lukluki	Flacourtiaceae	Leaf	Carbuncle	(Shil et al., 2014)
70	*Flemingia fruticulosa*	Shinkani	Fabaceae	Root and leaf	Haematuria	(Rana et al., 2021)
71	*Fragaria vesca*	Ban achoo	Rosaceae	Flower	Diuretic and laxative	(Dutt et al., 2015)
72	*Garcinia pedunculata*	Rehiachi	Clusiaceae	Fruit	Haemorrhoids	(Panmei et al., 2019)
73	*Geranium wallichianum*	Sucha phulli	Geraniaceae	Root	Gum bleeding	(Dutt et al., 2015)

(continued)

Table 2.1 (continued)

S. No.	Scientific name	Local name	Family	Parts used	Purpose /usage/treatment	References
74	*Goniothalamus sesquipedalis*	Damnagra	Annonaceae	Root	Leucorrhoea	(Shil et al., 2014)
75	*Gynura bicolor*	Vilhunha	Asteraceae	Leaf paste is applied to wounds	Stops bleeding and promotes blood clot	(Chase & Singh, 2016)
76	*Hodgsonia macrocarpa*	Assa	Cucurbitaceae	Leaves and seeds	Laxative	(Kichu et al., 2015)
77	*Hippophae tibetana*	Chha-Tuan	Elaeagnaceae	Flower	Sexual dysfunction	(Singh & Lal, 2008)
78	*Houttuynia cordata*	Musanderi	Saururaceae	Plant juice	Joint pain, muscle pain, and blood dysentery	(Sarma & Devi, 2017)
79	*Houttuynia cordata*	Nekir name	Saururaceae		Measles, gonorrhoea and skin troubles	(Khongsai et al., 2011)
80	*Iris ensata*	Thema	Iridaceae	Shoot	Stomach disorder	(Singh & Lal, 2008)
81	*Ipomoea batatas*	Hereukumbe	Convolvulaceae	Leaf	Body ache	(Panmei et al., 2019)
82	*Jatropha curcas*	Kangala	Euphorbiaceae	Leaf	Tooth ache	(Kabir et al., 2014)
83	*Justicia adhatoda*	Vasa	Acanthaceae	Roots	Diabetic	(Kumar et al., 2019)
84	*Juniperus macropoda*	Pashtun	Cupressaceae	Seeds	Anti-inflammatory	(Khajuria et al., 2021)
85	*Kaempferia rotunda*	Bhui champa	Zingiberaceae	Tubers	To avoid swelling	(Lepcha et al., 2019)
86	*Leea aequata*	Kin heut	Leeaceae	Leaves	Dysentery and infertility	(Chander et al., 2014)
87	*Leonurus sibiricus*	Kempu tumbe	Lamiaceae	Whole plant	Leucorrhoea	(Lingaraju et al., 2013)
88	*Leucas cephalotes*	Goma	Lamiaceae	-	Snake bite	(Panghal et al., 2010)
89	*Litsea monopetala*	Kutmero	Lauraceae	Bark	Pneumonia	(Tamang et al., 2023)

(continued)

Table 2.1 (continued)

S. No.	Scientific name	Local name	Family	Parts used	Purpose /usage/treatment	References
90	*Lyonia ovalifolia*	Anyar	Ericaceae	Seed	Wounds and Boils	(Malik et al., 2015)
91	*Macropanax undulatus*	Semza	Araliaceae	Leaves	Cold and high fever	(Kichu et al., 2015)
92	*Mentha spicata*	Tephinitso	Lamiaceae	Leaf	Stomachache, uterus infection and throat infection	(Chase & Singh, 2016)
93	*Michelia champaca*	Champa	Magnoliaceae	Seed	Jaundice	(Shil et al., 2014)
94	*Moringa oleifera*	Sajna	Moringaceae	Leaf, stem	Jaundice	(Kabir et al., 2014)
95	*Mussaenda roxburghi*	Khurmumu	Rubiaceae	Bark	Diarrhoea	(Shil et al., 2014)
96	*Myricaria germanica*	Humbu	Tamaricaceae	Leaf	Rheumatism	(Singh & Lal, 2008)
97	*Nerium indicum*	Lal kaner	Apocynaceae	Wood	Treat skin diseases	(Dutt et al., 2015)
98	*Oenothera rosea*	Rose evening	Onagraceae	Leaves	Anti-obesity	(Khajuria et al., 2021)
99	*Oroxylum indicum*	Bhatgilla	Bignoniaceae	Barks	Tuberculosis and diarrhoea	(Khongsai et al., 2011)
100	*Oxyria digyna*	Chukru	Polygonacea	Root	Cough	(Rana et al., 2021)
101	*Oxytropis lapponica*	Chhushin Darm	Fabaceae	Wood	Rheumatism	(Singh & Lal, 2008)
102	*Picrorhiza kurroa*	Karu	Plantaginaceae	Root	Cough	(Rana et al., 2021)
103	*Pedicularis pectinata*	Langna-Marfo	Scrophulariaceae	Flower	Stomach disorder	(Singh & Lal, 2008)
104	*Phyllanthus reticulatus*	Khaukora	Euphorbiaceae	Stem	Diarrhoea	(Kabir et al., 2014)
105	*Polygonum molle*	Thomcha	Polygonaceae	Leaves	Diarrhoea and dysentery.	(Lepcha et al., 2019)
106	*Polygonum tomentosum*	Bis kathali	Polygonaceae	Leaves	Ringworm	(Islam et al., 2014)

(continued)

Table 2.1 (continued)

S. No.	Scientific name	Local name	Family	Parts used	Purpose /usage/treatment	References
107	*Punica granatum*	Anaa	Lythraceae	Leaf	Heat stroke	(Rana et al., 2021)
108	*Rhus chinensis*	Bhakimlo	Anacardiaceae	Fruits	Dysentery and stomach disorder	(Tamang et al., 2023)
109	*Scindapsus officinalis*	Duleiro	Araceae	Leaf	Leprosy	(Shil et al., 2014)
110	*Scaevola taccada*	Tuful	Goodeniaceae	Leaves	Body ache and paralysis	(Chander et al., 2014)
111	*Seidenfia rheedii*	Kairagai illai	Orchidaceae	Flower	Cholera	(Francis Xavier et al., 2015)
112	*Shorea robusta*	Shal	Dipterocarpaceae	Leaf	Dysentery, gonorrhoea, and weak digestion	(Islam et al., 2014)
113	*Smilax macrophylla*	Kumira lota	Smilacaceae	Stem	Weakness, tremor in hands or legs	(Kabir et al., 2014)
114	*Stellaria media*	Losdhi	Caryophyllaceae	Seeds	Skin infection	(Khajuria et al., 2021)
115	*Solanum indicum*	Ao longkok	Solanaceae	Fruits and leaves	Anthelminthic	(Kichu et al., 2015)
116	*Solanum trilobatum*	Thoodhuvalai	Solanaceae	Leaves	Asthma	(Ignacimuthu et al., 2006)
117	*Sonchus brachyotus*	Blonghamchen	Asteraceae	Leaf	Rheumatism	(Shil et al., 2014)
118	*Tacca leontopetaloid*	Sasana-bhala	Taccaceae	Root	To cure boils	(Kosalge & Fursule, 2009)
119	*Taxus wallichian*	Poshtul	Taxaceae	Bark	Anti-cancer	(Khajuria et al., 2021)
120	*Terminalia chebula*	Haida	Combretaceae	Seed	Asthma	(Malik et al., 2015)
121	*Thysanolaena latifolia*	Pasham	Poaceae	Roots	Mouthwash	(Lepcha et al., 2019)

(continued)

Table 2.1 (continued)

S. No.	Scientific name	Local name	Family	Parts used	Purpose /usage/treatment	References
122	*Tridax procumbens*	Muriampachilai	Compositae	Leaf	Cuts	(Sukumaran et al., 2021)
123	*Tupistra nutans*	Nakima	Asparagaceae	Roots and spikes	Urinary disease, food poisoning, and gastrointestinal problems	(Tamang et al., 2023)
124	*Urtica dioica*	Zaklutasula	Urticaceae	Leaves	Dog and snake bites	(Kichu et al., 2015)
125	*Urtica dioica*	Shocha	Urticaceae	Whole plant	High blood pressure, cough, cold, and a stone problem	(Lepcha et al., 2019)
126	*Vitex negundo*	Notchi	Verbenaceae	Leaf	Fever	(Francis Xavier et al., 2015)
127	*Viola canescens*	Vanaksha	Violaceae	Root, flower and leaf	Urinary pain	(Rana et al., 2021)
128	*Viola pilosa*	Vanaksa	Violaceae	Flowers	Fever, cough and cold	(Uniyal et al., 2006)
129	*Withania somnifera*	Ashwagandha	Solanaceae	Leaf and root	Epilepsy	(Islam et al., 2014)
130	*Zanthoxylum acanthopodium*	Narik	Rutaceae	Fruits	Stomachache, gastritis, toothache, and cold	(Lepcha et al., 2019)
131	*Zanthoxylum armatum*	Tim bur	Rutaceae	Fruits	Stomach disorder	(Tamang et al., 2023)
132	*Zanthoxylum rhetsa*	Ongret	Rutaceae	Seeds	Poison fish	(Kichu et al., 2015)
133	*Zingiber rubens*	Brintaukru	Zingiberaceae	Rhizome	Urinary infection	(Shil et al., 2014)
134	*Ziziphus mauritiana*	Boroi	Rhamnaceae	Root and leaves	Dysentery, diarrhoea, and ulcer	(Islam et al., 2014)

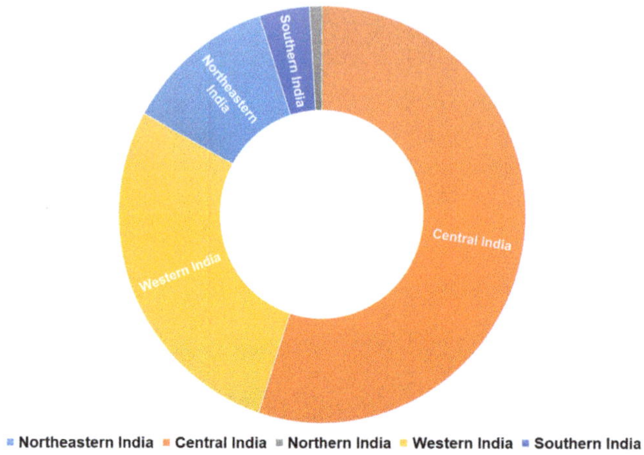

Fig. 2.1 Geographical distribution of tribes in India, i.e., Northeastern India, Central India, Western India, Southern India and Northern India, as per sources of Sivakumar et al. (2013)

Northeastern India: The North East (NE) India is one of the richest reservoirs of plant diversity in the country (Roy Choudhury et al., 2015), lying within the Indo-Burma biodiversity hotspot (Laldingliani et al., 2022). This region, encompassing Arunachal Pradesh, Assam, Manipur, Meghalaya, Mizoram, Nagaland, Tripura, and Sikkim, harbours nearly 50% of India's flowering plants and features a rich diversity of species (Mao & Roy, 2017). More than 250 tribes speak with 200 dialects, where agriculture is their principal occupation, and they possess extensive traditional knowledge of medicinal plants. Traditional remedies are crucial in primary healthcare for the Reang people in Tripura (Shil et al., 2014). They preserve a rich heritage of medicinal plant knowledge and have traditional physicians who utilise these plants. For instance, the Deb barma tribal healer uses Andrographis paniculata to treat malaria (Kabir et al., 2014).

Similarly, the tribals of Sikkim rely heavily on herbal medicine passed down through generations, with 490 medicinal plants thriving in the region due to its diverse altitude and climate. (Panda & Misra, 2010). Similarly, the Zeliangrong people continue to depend on medicinal plants for primary healthcare, some displaying established pharmacological activities (Panmei et al., 2019). In a recent study conducted in Pakyong subdivision of East Sikkim during 2016–17, 71 informants from Lepcha, Bhotia, and Nepali tribes provided information on 60 valuable plant species used for local healthcare needs, including veterinary use (Tamang et al., 2023). Herbaceous plants were most commonly utilised (54%), followed by trees (33%), shrubs (8%), and climbers (5%). Among the documented medicinal plants in Pakyong subdivision, species like *Astilbe rivularis* and *Artemisia indica* are commonly used as first aid for cuts and wounds, potentially harbouring novel phytoconstituents. A study on the Apatami, Mongpa, Sinpho, and Tangsa tribes identified 28 medicinal plant species, primarily herbs (Khongsai et al., 2011). The study documented 145 medicinal plants utilised by the Zeliangrong tribes, primarily Angiospermic and mostly found in the wild (Panmei et al., 2019). These plants treat approximately 59 health ailments, with the highest Informant Consensus Factor (ICF) observed for digestive disorders (0.75), involving 44 species. Additionally, established medicinal plants like *Ageratum conizoides*, *Blumeopsis flava*, *Clerodendrum glandulosum*, *Gynura cusimbua*, *Hedyotis scandens*, and *Paederia foetida* were frequently used. In the Deb Barma clan, most households (14 out of 20) were led by elderly males and 6 by females, who were still active members (Kabir et al., 2014). The traditional healer utilised 44 medicinal plants to treat various ailments. The Reang tribe in Tripura, India, employed 125 medicinal plant species belonging to 59 families to address 42 ailments (Shil et al., 2014). Leaves were the primary plant part used, often administered orally. Fever and gastrointestinal diseases had the highest Informant Consensus Factor (FIC) of 0.79, followed by dermatological problems (FIC 0.78), wherein the general health problems and inflammation/pain shared a consensus factor of 0.77.

Eastern India: Traditional medicinal plant usage as a febrifuge is observed among the tribes of Purulia district, West Bengal, India (Islam et al., 2014). Similarly, Nicobarese tribes on Car Nicobar Island practice ethnomedicine and healthcare traditions (Chander et al., 2014). Notably, the gastrointestinal system exhibits the highest Informant Consensus Factor (ICF) of 0.68, with the Euphorbiaceae family. *Morinda citrifolia*, *Tabernaemontana crispa*, and *Colubrina asiatica* have the highest use values (UV). Shrubs constitute the most common growth form (28%), with leaves being the primary plant part used in medicine preparation (Ayyanar & Ignacimuthu, 2011). Additionally, antimicrobial and antimalarial properties of six medicinal plants traditionally used by Nicobarese tribes were studied, including *Alstonia macrophylla*, *Claoxylon indicum*, *Dillenia andamanica*, *Jasminum syringifolium*, *Miliusia andamanica*, and *Pedilanthus tithymaloides* (Chander et al., 2016).

Northern India: The literature highlights the prevalent use of surveyed plants in various anti-diabetic formulations and patent drugs prescribed by Ayurvedic practitioners in India (Kumar et al., 2019). These species range from low sub-tropical foothills to high elevations of temperate mountains in the northwestern Himalayas (Dutt et al., 2015). Among plant parts, leaves are favoured by the nomadic Gaddi tribe in Jammu and Kashmir, India. Indigenous medicinal plant knowledge of the Saperas community in Khetawas, Jhajjar District, Haryana, India, revealed the use of 57 species from 35 families, primarily treating snakebites (Panghal et al., 2010). Notably, more medicinal plants were claimed by men, with the Fabaceae family having the highest representation. Along the migratory route of Gaddi Shepherds, 190 plant species from 70 families are used to treat over 80 ailments (Dutt et al., 2015). Leaves are predominantly utilised, and older shepherds possess more traditional medicinal knowledge than younger ones. Specifically, 56 plant species are used for gastrointestinal and liver disorders, while diabetes and snake/scorpion bites are treated using only two species each. Moreover, Singh and Lal (2008) studied the new or lesser-known medicinal uses of plant bioresources, validating traditional knowledge used by tribal communities to cure common ailments in the Lahaul-Spiti region of the western Himalaya. This area, inhabited by Lahaulis and Bodhs, is located in the cold arid zone of Himachal Pradesh, India, where observations on prevalent ailments like rheumatism, stomach issues, liver disorders, and sexual disorders were recorded during ethnobotanical explorations from 2002 to 2006. Similarly, an ethnomedicinal study of wild plants used by tribal communities in the Temperate Forests of Himalayan Kashmir, India, identified 82 medicinal plant species from 52 families (Khajuria et al., 2021). *Artemisia absinthium* had the highest Relative Frequency of Citation value (0.89), while *Dioscorea deltoidea* had the lowest (0.03).

Additionally, indigenous medicinal knowledge among the local traditional healers, tribal, and non-tribal people of the Churah subdivision in Chamba district, Himachal Pradesh, was investigated (Rana et al., 2021). New ethnomedicinal claims were also documented from the Gujjar and Bakerwals tribes in Rajouri and Poonch districts of Jammu and Kashmir, India (A Shah et al., 2015). Furthermore, the use of medicinal plants for treating various ailments by the indigenous people of the Churah subdivision in Chamba, Himachal Pradesh, was documented (Rana et al., 2021). Additionally, the roots of Saussurea costus were reported to treat dysentery and stomach ache in the Spiti area (Singh & Lal, 2008). Ethnomedicinal investigations of medicinal plants in the Chakrata region (Uttarakhand) used by the Jaunsari tribe for diabetes identified key species such as Pteridium aquilinum, Juglans regia, Corylus jacquemontii, Urtica dioica, Diplazium maximum, and Angelica glauca (Kumar et al., 2019). Notably, 32 plant species were reported for ethnomedicinal uses, followed by food plants (22 species), household purposes (16 species), edible fruits (15 species), and fodder plants (14 species). The informants' agreement was highest for food and fruit plants, slightly lower for fodder and household uses, and least for ethnomedicinal uses. Fidelity values ranged from 8% to 100% across all use categories. Similarly, ethnobotanical knowledge among the semi-pastoral Gujjar tribe in the high-altitude Adhwari of the Churah subdivision in Chamba district identified 58 plant species from 45 genera and 24 families for treating various diseases (Panghal et al., 2010). The plants were most commonly used for stomach disorders (29 species), rheumatism (18 species), liver problems (15 species), and sexual ailments (9 species) (Singh & Lal, 2008). Leaves were predominantly used in herbal preparations (20 species), followed by flowers (12 species) and roots (11 species). Most formulations were powder, with some prepared as juice or decoctions.

Southern India: In fact, Ayyanar and Ignacimuthu (2011) conducted an ethnobotanical survey of medicinal plants used by Kani tribals in the Tirunelveli hills of the Western Ghats, India. The author found that the leaf is the most commonly used plant part, with medicines typically prepared as a paste and administered orally. *Gymnemasylvestre* had the highest use value (2.00), followed by *Melia azedarach*, *Murraya koenigii*, *Syzygium cumini*, and *Terminalia chebula* (1.83). Similarly, ethnobotanical investigations among Paliyar tribes in Madurai District of Tamil Nadu, India, revealed frequent use of species from the Acanthaceae and Asclepiadaceae,

Euphorbiaceae and Solanaceae families to treat diseases (Ignacimuthu et al., 2006). Furthermore, Lingaraju et al. (2013) conducted an ethnopharmacological survey of traditional medicinal plants in tribal areas of Kodagu district, Karnataka, India. They found that 126 plant species from 60 families are used as ethnic drugs to treat various ailments, including diabetes, skin disorders, diarrhoea, wound healing, intestinal worms, snake bites, and jaundice. Indeed, investigating the ethnomedicinal claims among tribals of Satpuda Hills in India revealed their dependence on medicinal plants for healthcare (Kosalge & Fursule, 2009).

The tribals regularly used 52 plant species from 36 families to treat various diseases ranging from skin disorders to urinary problems. Moreover, Francis Xavier et al. (2015) observed traditional phytotherapy among Malayali tribes in the Eastern Ghats of Tamil Nadu, South India. They found that pastes are the most common preparation method, with plants used to treat various diseases grouped under 19 categories. *Asystasia chelonoides* and *Piper nigrum* are found to be the most commonly used medicinal plant species. Additionally, an ethnobotanical study of medicinal plants used by the Kani tribes of Pechiparai Hills, Western Ghats, India, reported the use of 138 species from 60 families, with Leguminosae being the most speciose family (Sukumaran et al., 2021). Leaf is the most commonly used part, often prepared as a leaf extract and administered orally. *Aristolochia indica* had the highest use value, while the ailment category 'fever' had the highest informant consensus factor.

Western India: Traditional ethnomedicinal knowledge is preserved within the Pawra Tribe of Satpura Hills, Maharashtra, India, with about 80 unique or lesser-known uses of 79 plant species documented (Jagtap et al., 2009). These traditional uses predominantly treat asthma, inflammation, lactation issues, menstrual problems, poisonous bites, skin conditions, stomach and toothaches. Similarly, the Tharu tribal community in Uttar Pradesh, India, practices traditional dermatological healthcare management using medicinal plants (Kumar et al., 2013). In this community, information was gathered from 230 informants across 46 villages in the Terai region, revealing 44 medicinal plants across 34 families for treating a variety of ailments such as pain, coughs, colds, gastrointestinal issues, cuts, wounds, diabetes, malaria, heart disorders, and paralysis (Kabir et al., 2014). This highlights the deep-rooted tradition of using medicinal plants within indigenous communities for healthcare practices.

Central Indian: Medicinal plants traditionally used by tribes in Madhya Pradesh, India, show significant anti-mycobacterial activity against tuberculosis-related symptoms (Gupta et al., 2018). Notably, eleven plants, including *Alstonia scholaris*, *Glycyrrhiza glabra*, *Holarrhena antidysentrica*, *Eulophia nuda*, *Cocculus hirsutus*, *Pueraria tuberosa*, *Curcuma caesia*, *Sphaeranthus indicus*, and *Plumbago zeylanica,* demonstrate promising anti-mycobacterial properties, validating their traditional uses. Another study highlights medicinal plants employed by Majhgawan Block tribes, i.e., Kol, Gond, and Mawasi tribes, where *Holarrhena antidysenterica* is employed to address diarrhoea, with utilizing 50%, 33%, and 25% (R Singh & Sharma, 2011).

Bioactive Compounds in Ethno-Botanical Plant

This ethnobotanical plant harbours diverse potent bioactive compounds suitable for treating various diseases. The robust antimalarial activity observed in the crude extracts of *M. andamanica*, *P. tithymaloides*, *J. syringifolium*, and *D. andamanica* presents a promising avenue for further investigation. These effects, combined with their in vitro antimicrobial activity, are likely attributable to alkaloids, flavonoids, triterpenes, sterols, tannins, and saponins in the methanol extracts of these plants (Chander et al., 2016). Additionally, the traditional usage of *Moringa oleifera* among the tribals of the Kolli Hills region for enhancing sexual vigour may be supported by its rich content of flavonoids and phenolic compounds such as kaempferol, quercetin, rutin, and caffeoylquinic acids (Francis Xavier et al., 2015). These antioxidants are recognised for mitigating oxidative stress-related testicular impairments in animal tissues. Similarly, the presence of diverse bioactive compounds in the roots of *Cissampelos pareira* and *Stephania glabra*, including pelosine, l-curine, hayatinine, gindarine, and cycleanine, underscores their potential therapeutic value in traditional medicine systems like Indian Ayurveda and Chinese medicine. Moreover, Achyranthes aspera and Bauhinia variegata show promising bioactivity, with compounds like Saponin A, saponin B, and achyranthine for treating skin eruptions, and alanine, aspartic acid, glycine, serine, oxaloacetic acid, and palmitic acid for blood purification in skin diseases (Singh & Singh, 2009). These findings emphasise the rich potential of traditional medicinal plants in addressing various health concerns.

Usage of Ethnobotanical Plant Parts by Tribal Communities

Traditionally, these tribal groups have a rich tradition of utilising wild plants for various purposes, including medicine, food, fodder, fuel, cultural practices, and other miscellaneous uses (Panghal et al., 2010). Moreover, the exponential growth in the human population has escalated the demand for novel plant resources, posing a threat to natural resources (Laldingliani et al., 2022). Other factors contributing to this decline include the use of synthetic drugs for immediate relief, accessibility to modern medicine, and the reluctance of younger generations to pursue ancestral professions as traditional healers (Islam et al., 2014). 97 medicinal plant species, representing 52 families and 83 genera, have been documented for treating various ailments such as fever, cough, cold, digestive disorders, constipation, and menstrual disorders (Malik et al., 2015). Studies have shown that the majority of plant species used for human health care by these communities are trees (43%), followed by herbs (25%), shrubs (18%), and climbers (14%) (Sarma & Devi, 2017). Additionally, Francis Xavier et al. (2015) stated that the herbs constitute the largest growth habit (48 species, 56%), followed by trees (13 species, 15%), and shrubs (9 species, 11%). Additionally, leaves (46%) are the most frequently used plant part, followed by the whole plant (10%), seeds and fruits (10%), and roots (8%). Furthermore, leaves were the most frequently utilized plant part for medicinal preparation, comprising 50% of the species studied (69 species), followed by roots (8%; 11 species), fruits (7%; 9 species), aerial parts (6%; 8 species), barks (5%; 7 species), flowers (4%; 6 species), seeds (4%; 5 species), bulbs and stems (3% each; 4 species), and latex and rhizomes (2% each; 3 species) (Sukumaran et al., 2021) (Fig. 2.2).

Interestingly, Islam et al. (2014) reported that various plant parts, such as leaves, seeds, roots, tubers, fruits, stems, flowers, and rhizomes, were frequently utilised for treating various ailments. Notably, leaves emerged as the most commonly employed plant part, accounting for 34% of usage, followed by seeds and fruits (20%), roots, root-bark, and rhizomes (16%), bark (15%), whole plants (6%), stems (4%), flowers (4%), and heartwood (1%). Whereas, in herbal formulations, leaves (39%) are the most preferred, followed by fruits and seeds (18%), roots (16%), whole plants (13%), stem bark (11%), and latex (3%) (Lingaraju et al., 2013). Among drug formulations, paste (39.06%) and decoctions (34.37%) were commonly used, with juice (15.62%) and raw preparations (10.93%) less prevalent. Oral administrations (77%) were generally preferred for most diseases, while external applications (23%) were prescribed for skin diseases, snake bites, and wound healing. The widespread use of leaves is attributed to their easy availability in the area (Panghal et al., 2010). The most common methods of drug preparation included infusion (23%), powder (16%), decoction (10%), and paste (10%), while remedies were less commonly prepared as pellets (9%), juice (6%), band (2%), and fumes (2%). In some cases, tribal people of this region frequently add sugar, salt, orange peel, banana, or lemon to enhance the palatability of a preparation (Islam et al., 2014). Herbs (66 species) were predominantly used for medicinal purposes, and the most common method of remedy preparation was decoction/tea (27.8%) (Shah et al., 2015) (Fig. 2.3).

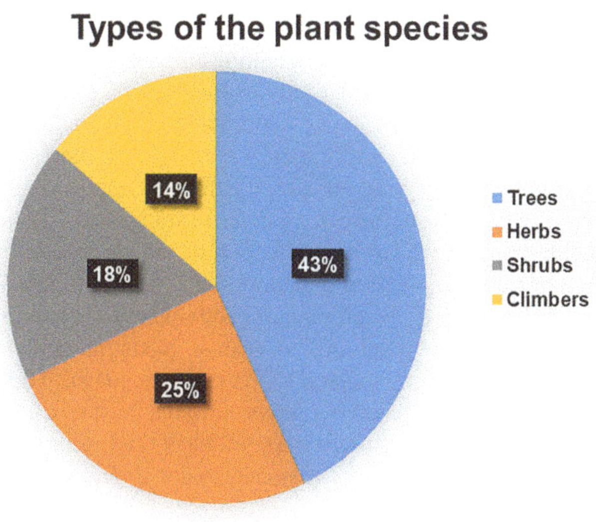

Fig. 2.2 Illustrating the utilisation of different plant types: Trees, herbs, shrubs, and climbers

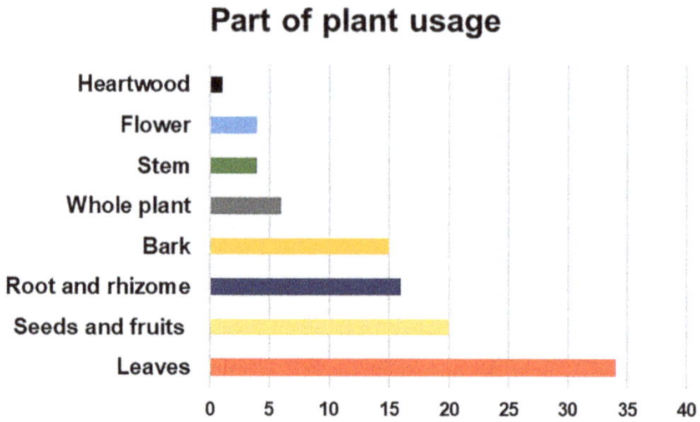

Fig. 2.3 Different parts of the plant utilised for ethnobotanical applications

Herbs are predominantly utilised for treating ailments due to their local abundance, with fresh plant materials commonly used in remedy preparation (Shil et al., 2014). Dosage varies based on age and physical condition, with children typically receiving smaller doses than adults. Treatment duration depends on the ailment, and patients not showing improvement after herbal treatment are referred to nearby modern health centres for further treatment by physicians.

Challenges for the Ethnobotany of Tribal Communities

Urbanisation and outmigration are accelerating the loss of traditional medicinal knowledge, underscoring the urgent need for preservation efforts (Tamang et al., 2023). It can be challenging, as many healers initially hesitate to share openly, which is typically passed orally within families only through substantial payment, maintaining confidentiality and preserving one's local significance (Roy Choudhury et al., 2015). Nonetheless, changes in traditional culture and the rise of modern technologies have led to the decline of indigenous traditional knowledge concerning medicinal plants and therapies among various local communities (Kumar et al., 2019). Due to developmental activities, deforestation, population growth, and indiscriminate exploitation, native medicinal plants are rapidly dwindling, endangering some species (Kichu et al., 2015). Thus, there is an urgent need to focus on preserving them through sustainable management. Enhancing medicinal plants' effectiveness, efficacy, and rational use is crucial, particularly by integrating them into national, regional, and local health policies and programs.

Conclusion

The current research underlined the continued significance of ethnobotanicals in aiding the health of tribal communities. Traditional healing practices utilising medicinal plants are prevalent for treating various ailments. Further exploration is needed to develop conservation strategies and ensure sustainable use of medicinal plants. Additionally, detailed investigations will facilitate future validation through phytochemical and pharmacogenetic studies, enhancing the acceptance of plant-based remedies in healthcare systems worldwide. Identifying active ingredients in local plants may lead to new drug development, bridging traditional knowledge with modern pharmaceutical advancements for the benefit of humanity.

SUMMARY

In this chapter summary, the ethnobotanical plants utilised by different tribal communities are discussed in terms of their applications in medicine, food, fodder, fuel, cultural practices, health, and other purposes. These plants are used in various forms, including decoctions, pastes, infusions, and powders, and they are consumed fresh. Ethnobotanical plants harbour diverse potent bioactive compounds suitable for treating various diseases. In India, over 50% of plant species in any ecosystem are used in ethnomedicine, with tribal populations possessing extensive traditional knowledge on biotic resource use. Approximately 6000 species are employed in traditional and herbal medicine in India, fulfilling about 75% of the medicinal needs of developing countries, with 3000 plants officially recognised for their medicinal value. Traditional knowledge is typically transmitted from generation to generation, primarily among elders and traditional healers, underscoring the importance of experience-based wisdom. If properly recognised, adopted, and integrated, Indigenous knowledge holds the potential to address various social challenges such as declining agricultural productivity, biodiversity loss, and water scarcity.

References

Ahmad, K., Ahmad, M., Huber, F. K., & Weckerle, C. S. (2021). Traditional medicinal knowledge and practices among the tribal communities of Thakht-e-Sulaiman hills, Pakistan. *BMC Complementary Medicine and Therapies*, *21*, 1–9. https://doi.org/10.1186/s12906-021-03403-1

Ayyanar, M., & Ignacimuthu, S. (2011). Ethnobotanical survey of medicinal plants commonly used by Kani tribals in Tirunelveli hills of Western Ghats, India. *Journal of Ethnopharmacology*, *134*, 851–864. https://doi.org/10.1016/j.jep.2011.01.029

Chander, M. P., Kartick, C., Gangadhar, J., & Vijayachari, P. (2014). Ethno medicine and healthcare practices among Nicobarese of Car Nicobar—An indigenous tribe of Andaman and Nicobar Islands. *Journal of Ethnopharmacology*, *158*, 18–24. https://doi.org/10.1016/j.jep.2014.09.046

Chander, M. P., Pillai, C. R., Sunish, I. P., & Vijayachari, P. (2016). Antimicrobial and antimalarial properties of medicinal plants used by the indigenous tribes of Andaman and Nicobar Islands, India. *Microbial Pathogenesis*, *96*, 85–88. https://doi.org/10.1016/j.micpath.2016.04.017

Chase, P., & Singh, O. P. (2016). Ethnomedicinal plants used by Angami Tribe of Nagaland, India. *Indian Journal of Tropical Biodiversity*, *21*, 29–42.

Dutt, H. C., Bhagat, N., & Pandita, S. (2015). Oral traditional knowledge on medicinal plants in jeopardy among Gaddi shepherds in hills of northwestern Himalaya, J&K, India. *Journal of Ethnopharmacology*, *168*, 337–348. https://doi.org/10.1016/j.jep.2015.03.076

Francis Xavier, T., Kannan, M., & Auxilia, A. (2015). Observation on the traditional phytotherapy among the Malayali tribes in Eastern Ghats of Tamil Nadu, South India. *Journal of Ethnopharmacology*, *165*, 198–214. https://doi.org/10.1016/j.jep.2015.02.045

Gupta, V. K., Kaushik, A., Chauhan, D. S., et al. (2018). Anti-mycobacterial activity of some medicinal plants used traditionally by tribes from Madhya Pradesh, India for treating tuberculosis related symptoms. *Journal of Ethnopharmacology*, *227*, 113–120. https://doi.org/10.1016/j.jep.2018.08.031

Ignacimuthu, S., Ayyanar, M., & Sivaraman, K. S. (2006). Ethnobotanical investigations among tribes in Madurai District of Tamil Nadu (India). *Journal of Ethnobiology and Ethnomedicine*, *2*, https://doi.org/10.1186/1746-4269-2-25

Islam, M. K., Saha, S., Mahmud, I., et al. (2014). An ethnobotanical study of medicinal plants used by tribal and native people of Madhupur forest area, Bangladesh. *Journal of Ethnopharmacology*, *151*, 921–930. https://doi.org/10.1016/j.jep.2013.11.056

Jagtap, S. D., Deokule, S. S., Pawar, P. K., & Harsulkar, A. M. (2009). Traditional ethnomedicinal knowledge confined to the Pawra tribe of. *Ethnobotanical Leaflets*, *13*, 98–115.

Kabir, M. H., Hasan, N., Rahman, M. M., et al. (2014). A survey of medicinal plants used by the Deb barma clan of the Tripura tribe of Moulvibazar district, Bangladesh. *Journal of Ethnobiology and Ethnomedicine*, *10*, 19.

Khajuria, A. K., Manhas, R. K., Kumar, H., & Bisht, N. S. (2021). Ethnobotanical study of traditionally used medicinal plants of Pauri district of Uttarakhand, India. *Journal of Ethnopharmacology*, *276*, 114204. https://doi.org/10.1016/j.jep.2021.114204

Khongsai, M., Saikia, S. P., & Kayang, H. (2011). Ethnomedicinal plants used by different tribes of Arunachal Pradesh. *Indian Journal of Traditional Knowledge*, *10*, 541–546.

Kichu, M., Malewska, T., Akter, K., et al. (2015). An ethnobotanical study of medicinal plants of Chungtia village, Nagaland, India. *Journal of Ethnopharmacology*, *166*, 5–17. https://doi.org/10.1016/j.jep.2015.02.053

Kosalge, S. B., & Fursule, R. A. (2009). Investigation of ethnomedicinal claims of some plants used by tribals of Satpuda Hills in India. *Journal of Ethnopharmacology*, *121*, 456–461. https://doi.org/10.1016/j.jep.2008.11.017

Kumar, A., Aswal, S., Chauhan, A., et al. (2019). Ethnomedicinal investigation of medicinal plants of Chakrata region (Uttarakhand) used in the traditional medicine for diabetes by Jaunsari tribe. *Natural Products and Bioprospecting*, *9*, 175–200. https://doi.org/10.1007/s13659-019-0202-5

Kumar, A., Pandey, V. C., Singh, A. G., & Tewari, D. D. (2013). Traditional uses of medicinal plants for dermatological healthcare management practices by the Tharu tribal community of Uttar Pradesh, India. *Genetic Resources and Crop Evolution*, *60*, 203–224. https://doi.org/10.1007/s10722-012-9826-6

Kunjam, S., Thakur, G. S., & Jadhav, S. K. (2021). Ethnobotanical survey of medicinal plant species used by tribal communities around Katghora Tehsil of Korba District. *Advances in Plant Sciences*, *34*, 43–50.

Lal, M., Chandraker, S. K., & Shukla, R. (2023). Quantitative ethnobotanical study of therapeutic plants of Amarkantak hills in Achanakmar-Amarkantak Biosphere Reserve, Central India. *Acta Ecologica Sinica, 43*, 139–153. https://doi.org/10.1016/j.chnaes.2022.03.002

Laldingliani, T. B. C., Thangjam, N. M., Zomuanawma, R., et al. (2022). Ethnomedicinal study of medicinal plants used by Mizo tribes in Champhai district of Mizoram, India. *Journal of Ethnobiology and Ethnomedicine, 18*. https://doi.org/10.1186/s13002-022-00520-0

Lepcha, T. T., Pradhan, P., Gaira, K. S., et al. (2019). Ethnomedicinal use of plants by Bhutia tribe in Sikkim Himalaya. *Proceedings of the 1st Himalayan Research Consortium, 1*, 71–78.

Lingaraju, D. P., Sudarshana, M. S., & Rajashekar, N. (2013). Ethnopharmacological survey of traditional medicinal plants in tribal areas of Kodagu district, Karnataka, India. *Journal of Pharmacy Research, 6*, 284–297. https://doi.org/10.1016/j.jopr.2013.02.012

Mahapatra, A. D., Bhowmik, P., Banerjee, A., et al. (2018) Ethnomedicinal wisdom: An approach for antiviral drug development. In *New look to phytomedicine: Advances in herbal products as novel drug leads*, 35–61. https://doi.org/10.1016/B978-0-12-814619-4.00003-3

Malik, Z. A., Bhat, J. A., Ballabha, R., et al. (2015). Ethnomedicinal plants traditionally used in health care practices by inhabitants of Western Himalaya. *Journal of Ethnopharmacology, 172*, 133–144. https://doi.org/10.1016/j.jep.2015.06.002

Mao, A. A., & Roy, D. K. (2017). Ethnobotanical studies in North East India: A review. In *Indian Ethnobotany: Emerging Trends* (pp. 99–112).

Namsa, N. D., Mandal, M., Tangjang, S., & Mandal, S. C. (2011). Ethnobotany of the Monpa ethnic group at Arunachal Pradesh, India. *Journal of Ethnobiology and Ethnomedicine, 7*, 1–14. https://doi.org/10.1186/1746-4269-7-31

Panda, A. K., & Misra, S. (2010). Health traditions of Sikkim Himalaya. *Journal of Ayurveda and Integrative Medicine, 1*, 183–189. https://doi.org/10.4103/0975-9476.72617

Panghal, M., Arya, V., Yadav, S., et al. (2010). Indigenous knowledge of medicinal plants used by Saperas community of Khetawas, Jhajjar District, Haryana, India. *Journal of Ethnobiology and Ethnomedicine, 6*, 1–11. https://doi.org/10.1186/1746-4269-6-4

Panmei, R., Gajurel, P. R., & Singh, B. (2019). Ethnobotany of medicinal plants used by the Zeliangrong ethnic group of Manipur, northeast India. *Journal of Ethnopharmacology, 235*, 164–182. https://doi.org/10.1016/j.jep.2019.02.009

Pradhan, B. K., & Badola, H. K. (2008). Ethnomedicinal plant use by Lepcha tribe of Dzongu valley, bordering Khangchendzonga Biosphere Reserve, in North Sikkim, India. *Journal of Ethnobiology and Ethnomedicine, 4*. https://doi.org/10.1186/1746-4269-4-22

Priyadarshini, P., & Abhilash, P. C. (2019). Promoting tribal communities and indigenous knowledge as potential solutions for the sustainable development of India. *Environment and Development, 32*. https://doi.org/10.1016/j.envdev.2019.100459

Rahman, S. U., Ullah, Z., Ali, A., et al. (2022). Traditional knowledge of medicinal flora among tribal communities of Buner Pakistan. *Phytomedicine Plus, 2*, 100277. https://doi.org/10.1016/j.phyplu.2022.100277

Rana, D., Bhatt, A., Lal, B., et al. (2021). *Use of medicinal plants for treating different ailments by the indigenous people of Churah subdivision of district Chamba, Himachal Pradesh, India.* Springer.

Roy Choudhury, P., Dutta Choudhury, M., Ningthoujam, S. S., et al. (2015). Ethnomedicinal plants used by traditional healers of North Tripura district, Tripura, North East India. *Journal of Ethnopharmacology, 166*, 135–148. https://doi.org/10.1016/j.jep.2015.03.026

Sarma, J., & Devi, A. (2017). Ethnomedicinal wisdom of Garo community from Erstwhile Sonitpur District of Assam, Northeast India. *Pleione, 11*, 429. https://doi.org/10.26679/pleione.11.2.2017.429-439

Shah, A., Bharati, K. A., Ahmad, J., & Sharma, M. P. (2015). New ethnomedicinal claims from Gujjar and Bakerwals tribes of Rajouri and Poonch districts of Jammu and Kashmir, India. *Journal of Ethnopharmacology, 166*, 119–128. https://doi.org/10.1016/j.jep.2015.01.056

Shah, G. M., Abbasi, A. M., Khan, N., et al. (2014). Traditional uses of medicinal plants against malarial disease by the tribal communities of Lesser Himalayas-Pakistan. *Journal of Ethnopharmacology, 155*, 450–462. https://doi.org/10.1016/j.jep.2014.05.047

Shil, S., Dutta Choudhury, M., & Das, S. (2014). Indigenous knowledge of medicinal plants used by the Reang tribe of Tripura state of India. *Journal of Ethnopharmacology, 152*, 135–141. https://doi.org/10.1016/j.jep.2013.12.037

Singh, A., & Singh, P. K. (2009). An ethnobotanical study of medicinal plants in Chandauli District of Uttar Pradesh, India. *Journal of Ethnopharmacology, 121*, 324–329. https://doi.org/10.1016/j.jep.2008.10.018

Singh, K. N., & Lal, B. (2008). Ethnomedicines used against four common ailments by the tribal communities of Lahaul-Spiti in western Himalaya. *Journal of Ethnopharmacology, 115*, 147–159. https://doi.org/10.1016/j.jep.2007.09.017

Singh, R., & Sharma, A. (2011). Medicinal plants used for diarrhoea by tribals from Majhgawan block of district Satna, Madhya Pradesh, India. *Studies on Ethno-Medicine, 5*, 205–208. https://doi.org/10.1080/09735070.2011.11886410

Singh, R. K., Pretty, J., & Pilgrim, S. (2010). Traditional knowledge and biocultural diversity: Learning from tribal communities for sustainable development in northeast India. *Journal of Environmental Planning and Management, 53*, 511–533. https://doi.org/10.1080/09640561003722343

Sivakumar, P. S., Anantharaman, M., Thirugnanavel, A., et al. (2013). *Traditional tuber crops food of Northeastern India* (pp. 1–36)

Sukumaran, S., Sujin, R. M., Geetha, V. S., & Jeeva, S. (2021). Ethnobotanical study of medicinal plants used by the Kani tribes of Pechiparai Hills, Western Ghats, India. *Acta Ecologica Sinica, 41*, 365–376. https://doi.org/10.1016/J.CHNAES.2020.04.005

Tamang, S., Singh, A., Bussmann, R. W., et al. (2023). Ethno-medicinal plants of tribal people: A case study in Pakyong subdivision of East Sikkim, India. *Acta Ecologica Sinica, 43*, 34–46. https://doi.org/10.1016/j.chnaes.2021.08.013

Uniyal, S. K., Singh, K. N., Jamwal, P., & Lal, B. (2006). Traditional use of medicinal plants among the tribal communities of Chhota Bhangal, Western Himalaya. *Journal of Ethnobiology and Ethnomedicine, 2*, 1–8. https://doi.org/10.1186/1746-4269-2-14

Wali, R., Khan, M. F., Mahmood, A., et al. (2022). Ethnomedicinal appraisal of plants used for the treatment of gastrointestinal complaints by tribal communities living in Diamir district, Western Himalayas, Pakistan. *PLoS One, 17*, 1–21. https://doi.org/10.1371/journal.pone.0269445

Chapter 3
Ethno-botany and Ethno-pharmacology of Indravaruni: An Important Medicinal Plant of Tribal Culture and Its Action Against SARS-CoV-2

Sourav Pakrashy, Sharmistha Ganguly, Abhijit Sasmal, Malay Dolai, and Anjoy Majhi

Abstract The SARS-CoV-2 or COVID-19 pandemic witnessed from late 2019 to 2021tested health-care structures, creating the infection a global pandemic that led to a community healthiness emergency. Other comorbidities such as pneumonia, diabetes, stress, asthma, etc. developed hastily as a major comorbidity for SARS-CoV-2severity, accumulating the risk of undesirable results if one got sick with the SARS-CoV-2 virus. The SARS-CoV-2pandemic has encouraged wide spread investigation into prospective treatments, including the study of natural products as repurposing medications. They have high assistance, with diverse biological activity, and exceptional variety. This brief review proposes to highlight the active properties of the plants Indravaruni or *Citrulluscolocynthis (L.) Schrad* and its metabolites that are recognized to offer healing aids against cancer, leprosy, diabetes, jaundice, bronchitis, asthma, etc., and can be cast as possible adjuvant

A. Majhi (✉)
Department of Chemistry, Presidency University, 86/1 College Street, Kolkata, 700073, India
e-mail: anjoy.chem@presiuniv.ac.in

S. Pakrashy
Department of Chemistry, Presidency University, 86/1 College Street, Kolkata, 700073, India; Department of Chemistry, Prabhat Kumar College, Contai, Purba Medinipur, 721404, West Bengal, India

S. Ganguly
Department of Dravyaguna, Institute of Post Graduate Ayurvedic Education and Research, Kolkata, 700009, India

A. Sasmal
Merck, KgAa, Bengaluru, 560087, India

M. Dolai
Department of Chemistry, Prabhat Kumar College, Contai, Purba Medinipur, 721404, West Bengal, India

R. C. Sobti et al. (eds.), *Traditional Knowledge Systems in India: Sustainability, Ecology, and Resource Governance*, People, Cultures and Societies: Exploring and Documenting Diversities,
https://doi.org/10.1007/978-981-92-1346-7_3

treatment in severe acute respiratory syndrome (SARS) coronavirus (CoV) 2 infection. Determining the knowledge of the phyto constituents which have various medical properties and may reveal anti-viral effects when put to use, such that it can help in infection prevention. The medicinal plants consist of numerous secondary metabolites and potent ethno-pharmacological shreds of evidence, the chosen medicinal plant constitutes several active secondary metabolites, which are extensively used by Ayurvedic practitioners to treat a wide number of diseases. Herein, we give those active secondary metabolites that were used with the help of molecular docking to understand their fit affinity with the Coronavirus-2019-nCoV - protein (6VWW). The described literature was obtained from PubMed, Science Direct, Google Scholar, Web of Science, Scopus, Clinical Trials.gov, and Wiley online library with a focus on peer-reviewed English-language publications that exclusively addressed different types of diseases and SARS-CoV-2collaboration, clinical implications, and the healing treatments of traditional medicine with natural products. Indravaruni or *Citrullus colocynthis (L.) Schrad* has a variety of ethnomedicinal and ethnopharmacological uses against cancer, bronchitis, asthma, etc. which may also be used for the COVID-19 virus and its strains.

Keywords Indravaruni · Ethno-botany · Ethno-pharmacology · Traditional medicine · Sars-cov-2

Introduction

Indravaruni a botanically active medicinal herb known as *Citrullus colocynthis (L.) Schrad* belongs to the family Cucurbitaceae. They are normally recognized as bitter apples or colocynth (Abdel-Hassan et al., 2000; Nighantu et al., 2015). It is a perpetual prostrate sprawling herb originating around sandy manors wilds of Northwest Punjab, Central India, Sind, and Southern parts of India. They have also been seen ethnically in West-Asia, Arabia, Tropical-Africa, and also along a region of the Mediterranean (Pravin et al., 2013). The stems are in angular orientation and tendrils of a bifid manner; it has sinuately pinnatifid leaves, monoecious flowers, axillary, single, and colour in yellow; globose pepo, 3.50–10.50 cm in overall diameter, smooth, green blotchy yellow marks fruits, pulp is mostly spongy, more or less bitter, and seeds of its are numerous with white to light brown colour. Among the key chemical ingredients mainly secondary metabolites are citrollic acid, Colocynthoside A citrulonol, Colocynthoside B, methyl lengenol, citrulluin, Cucurbitacin E and inositol (Sastry, 2005). The used parts that are medicinally active are fruits, roots, and leaves. The Ayurvedic pharmacopeia of India specified the usage of this fruit in case of jaundice, the root for liver and spleen diseases, and the leaves have cutaneous fondness andalopecia (Rohilla et al., 2017). Conventionally it is cast to remedy or avoid several diseases like diabetes disease, malaria disease, constipation disease, rheumatism disease, jaundice, etc.

Indravaruni is recognised to help control blood sugar levels. Studies advocate it can act as an insulin-sensitizer, aiding progress of glucose easiness and decrease blood-sugar levels in individuals with diabetes. The plant encompasses secondary metabolites that have anti-oxidant properties, which assist to counteract free radicals in the body and defend cells from oxidative damage, this contributes to overall health and may help in avoiding diseases connected to oxidative stress, such as cancers, heart diseases, etc. It has anti-inflammatory properties, which make it beneficial for handling disorders like arthritis. It can decrease inflammation and pain in the joints. Indravaruni is recognized to have hepato-protective properties, meaning it aids in guarding the liver against harm caused by toxins, alcohol, and other unsafe affluence, it is habitually used in liver ailments like jaundice. Indravaruni is known to endorse healthy digestion. It sometimes service conditions like indigestion, constipation, and bloating, it has slight laxative properties that support systematic bowel activities. *Citrullus colocynthis (L.) Schrad* has a wide range of secondary metabolites that shows antimicrobial properties against a wide range of microorganisms such different types of bacteria and fungi. It has been conventionally used against skin infections, gastrointestinal infections, and wounds since it can be applied externally to inhibit infections or castoff in therapeutic formulations to prevent internal infections, as its antimicrobial and anti-inflammatory synergic properties make it advantageous for wound remedy. It can be applied as a paste or ointment to accelerate the healing process and prevent infection in cuts or wounds. *Citrullus colocynthis (L.) Schrad* has possible anticancer properties due to the occurrence of cucurbitacins as secondary metabolites, which may constrain the development of cancer cells. Scientific researches has specified that cucurbitacins have cytotoxic effects against several types of cancer. *Citrullus colocynthis (L.) Schrad* has been used as a pain reliever in traditional medicine. The secondary metabolites may help improve pain, principally in circumstances like headaches, back pain, etc.

The fruit of *Citrullus colocynthis (L.) Schrad* is characteristically used in its dehydrated form as it is extremely toxic when fresh. It is habitually pulverised into powder form or used in extracts and solutions. The seeds are sometimes also used in medicinal preparations, paste made from the plant's pulp or seeds can be applied externally to treat wounds or skin conditions.

Out breaking of the new Corona Virus Disease in December 2019 (COVID-19) they have resulted in a global pandemic and Public health emergency (Dey et al., 2020). Till now no approved anti-viral drug has been declared for the disease. Neither does it have any current standard operative protocol for care. Medical practitioners should be careful about the toxicity and usage of unfamiliar substances being tested or reused to treat COVID-19. The coronavirus protein macromolecular structure has encouraged the finding of anti-SARS-CoV-2 ingredients through *in silico* analysis (Anand et al., 2003). Herbal remedies or herbal medicines presented by Mother Nature have long been the home remedy for treating illnesses in patients in the oral tradition.

Nature always has an answer or hint of an answer for any problem and medicinal plants for ages have shown their versatility and produce valuable medicine for mankind, as in, the first-ever anti-malarial compound discovered in the extract of Cinchona bark. In today's world the novel coronavirus SARS-CoV-2 disease "COVID-19" was first found in China in late 2019 at Wuhan (Wu et al., 2020) and is rapidly spreading to other countries of the world. These belong to the family namely Coronaviridae and are RNA viruses, non-segmented and of positive sense and shifted in humans (Richman et al., 2016) and Orthocoronaviridae, a sub-family, there are five variants of coronaviruses which is causing concerns till now; Alpha, Beta, Gamma, Delta and Omicron coronavirus variants respectively. It has become a severe concern in this twenty-first century (Dey et al., 2020). The genome of COVID-19 or SARS-CoV-2 shows a complex group organization encrypting different structural besides Non-structural proteins (Nsps) (Kim et al., 2020). The viral genome's major part (replicase ORF1ab encompassing Nsps) is construed into two overlaying poly-proteins pp1a and pp1ab namely (Amoretti et al., 2002). Codes made by polypeptides are nearly 306 amino acids long, the main protease (M^{pro}) (Jin et al., 2020) which produces the conserved sites of polypeptides making a total of 16 functional viral Nsps showing ample enzymatic activities, viral replication mainly. Along with Nsps the genome of SARS-CoV-2 also has the structural protein S (spike) gene which encodes genes, the N gene which is the nucleo-capsid protein, and the viral envelop protein which is the E gene (Khan et al., 2020). Viral spike glycoproteins have a high affinity for the human ACE2 receptor (angiotensin-converting enzyme 2) this helps the virus to fuse with integral membrane protein to move inside our human cells (Hussain et al., 2020). The renin-angiotensin system (RAS) is the main playground of it as the blood pressure homeostasis and functioning of the heart are maintained by it (Oudit et al., 2003).

Its path gnomonic symptoms are mild to high dry and harsh cough, high fever, and breathlessness (dyspnea) (Wu et al., 2020). It's known to cause lymphopenia in severe conditions along with disseminated intravascular coagulation, hypercytokinemia, kidney failure, acute respiratory distress syndrome (ARDS), and death in the majority of cases (Thomas-Rüddel et al., 2020). The secondary metabolites of Indian Medicinal plants which are components of natural sources have several varieties in structure. Thus, computationally screened (In silico) with the respective binding target. Identify novel natural anti-SARS-CoV-2 active principles against the SARS-CoV-2 drug target PDB-ID—6VWW.Our review aims to provide an extensive assortment of valuable data for exploration and it will lead to the development of new structural leads along with natural treatment against COVID-19 or SARS-COV-2 in the coming days (Deora & Venkatraman, 2024).

Medicinal Plants in Tribal Culture

India's tribal societies uphold a strong and amusing quantity of medicinal plant knowledge fashioned through generations of empirical opinion and divine assembly to nature. Peer-reviewed investigation displays consistent configurations of plant usage, including Azadirachta indica, Justicia adhatoda, Centella asiatica, Vitex negundo, Rauvolfia serpentina, among others, across tribal cultures in West Bengal, Madhya Pradesh, Assam, Tamil Nadu, and the Himalayas (Datta et al., 2014).

However, this arrangement faces growing intimidations from conservational dilapidation, cultural alteration, and lack of defensive legal backgrounds. Conserving and responsibly assimilating this gorgeous legacy weights careful certification, irrepressible community corporations, and scientific authentication of key ethnomedicinal kinds. Such efforts can support both biodiversity management and comprehensive, culturally entrenched healthcare approaches (Rajpoot & Kareti, 2025).

Indravaruni, scientifically known as *Citrullus colocynthis* (family Cucurbitaceae), is an unpleasant perennial vine widely encountered across India's parched and semi-arid regions. In tribal and folk healthcare systems, this plant holds noteworthy therapeutic significance for circumstances like constipation, abdominal pain, rheumatism, diabetes, and menstrual disorders. Traditional therapists have specifically cherished its roots, fruit pulp, seeds, and leaves for assorted therapies. Contemporary ethnobotanical surveys authorize the plant's continued use amongst tribal inhabitants for gastrointestinal ailments, metabolic syndromes (notably diabetes), and microbial contaminations. For example, an ethnobotanical assessment in Kashmir Himalayan communities found *C. colocynthis* commonly used for diabetes and blood purification (Cheng et al., 2023).

Molecular Docking of Secondary Metabolites of Indravaruni

The in-silico assortment of the compounds explicitly docking analysis of proteins with diverse ligands was executed using UCSF Chimera, AutoDockVina (Trott & Olson, 2010) to comprehend the binding attraction between all compounds and the protein target of SARS-CoV-2.

The protein NSP15 Endoribonuclease from SARS CoV-2 was taken from the RCSB Protein databank (PDB) (http://www.rcsb.org/pdb), PDB-ID: 6VWW in PDB format. Before docking studies, water molecules other solvent molecules, and the co-crystallized residues (if any), were removed using UCSF Chimera. Preparation of the protein structure was done with DockPrep tool by assigning the H-atoms, charges, and steps for energy minimization (Krivov et al., 2009). AM1-BCC method assigned the charges which is quick and efficient and ANTECHAMBER algorithm (Wang et al., 2001) computed the charges. Six-hundred (600) iterations with 0.020 Armstrong step size for the energy minimization were performed using ten(10) update intervals and then again saved in PDB format. Every step was done in UCSF Chimera. The protein energy minimization was further done with SwissPDB viewer (Guex & Peitsch, 1997).

All *in-silico* interaction assays with ligands were secondary metabolites of Indigenous medicinal plant and the structures which were present in PubChem were retrieved from there in SDF format while others were directly drawn on chemdraw, these drawn structures were copied to Chem3D pro where their energy minimization was carried using MM2 calculations (not for float structures), after that, kept in SDF format. Now before carrying out the in-silico-docking between the secondary metabolites (ligands) and proteins, hydrogen addition of the ligands and addition of charge using Gasteiger algorithm (Gasteiger & Jochum, 1979), energy minimization was performed using one thousand (1000) iteration with nearly 0.020 Armstrong step size and ten(10) update interval in the Chimera software and then again saved in PDB format, which works on the principle of chemoinformatics of electronegativity equilibration then files were saved in PDB format. A grid box which assigns the binding region was chosen in such a way that it would cover the whole protein for the hydrophobic surface of the concave region of protein to fit in properly the hydrophobic surface of the ligand giving the best binding score. For each ligand and each protein three runs on UCSF Chimera were done and the best score that is least free energy value in Kcal/mol among them was chosen. Then for visualising in different formats we used the software Discovery studio and UCSF Chimera.

The reason behind considering the whole protein apart from its binding site is to understand the inhibitory property of the phytoconstituents (Cho et al., 2021).

Ethnobotany of Indravaruni

Ethnobotany is a multi-disciplinary field that supports us to comprehend the dynamic assembly amongst humans and plants, it conglomerates aspects of botany, anthropology, and ecology. It highlights the prominence of conserving traditional information, subsidiary biodiversity, and controlling contemporary scientific research. By distinguishing the value of plants in cultural backgrounds, ethnobotany donates to both the conservation of cultural inheritance and the examination of new possibilities for medical, agricultural, and ecological progressions. Numerous native and local populations have deep, traditional information about the herbs and plants in their surroundings, which has been conceded down through generations. Ethnobotany plays a serious role in maintaining this knowledge, which is often in jeopardy of being lost due to modernization and ecological changes. Ethnobotanists also scrutinize maintainable ways of using plant possessions, as many customary rehearses are concentrated on maintaining balance with nature. Ethnobotanical studies can escort scientists to prospective new therapeutic combinations or lead compounds. Like the discovery of taxol (a chemotheraputic drug) derived from the bark of the Pacific yew tree. Various floras like ginseng, ginger, and licorice are used in Chinese medicine for their remedial properties. Ethnobotanists studies the efficacies and uses of plant parts, serving reservation to traditional understanding while also exploring potential modern solicitations. Indravaruni or *Citrullus colocynthis (L.) Schrad*plays a significant role in the ethnobotany of numerous cultures, predominantly in the tropics, where it is appreciated for its therapeutic properties, nutritional benefits, and traditional importance. Its presentations in handling conditions like diabetes, liver disorders, and digestive issues have been well documented in traditional systems of medicine like Ayurveda and Unani. While the plant continues to be used in various forms today, its rich ethnobotanical heritage emphasizes the importance of preserving both its knowledge and the plant itself for future generations.

Botanical Background:

Taxonomic classification

Bentham and Hooker's system of classification:
Kingdom—Plantae
Sub kingdom- Tracheobionta
Super division- Spermatophyta
Division—Magnoliophyta
Class—Magnoliopsida
Sub- class—Dilleniidae
Order—Cucurbitales
Family—Cucurbitaceae
Genus—*Citrullus*
Species epithet- *C. colocynthis*

APG IV system of classification:

Superregnum: Eukaryota
Regnum: Plantae
Cladus: Angiosperms
Cladus: Eudicots
Cladus: Core eudicots
Cladus: Rosids
Cladus: Eurosids I
Ordo: Cucurbitales
Familia: Cucurbitaceae
Tribus: Benincaseae
Genus: *Citrullus*
Species: *Citrulluscolocynthis*

- **Synonyms**: *Citrulluscolocynthis* var. *insipidus* Pangalo; *Citrullus colocynthis* subsp. *insipidus* (Pangalo) Fursa; *Citrullus colocynthis* var. *stenotomus* Pangalo; *Citrullus colocynthis* subsp. *stenotomus* (Pangalo) Fursa; *Citrullus colocynthoides* Pangalo; *Citrullus pseudocolocynthis* M. Roem.; *Colocynthis officinalis* Schrad.; *Colocynthis vulgaris* Schrad.; *Cucumis colocynthis* L. (Plantlist)
- *Cucumiscolocynthis* L., Sp. pl. 2:1011 (1753) (Scientificlib.com)

- **Description**: The plant resembles a watermelon. Its and annual herbaceous plant with roughly pubescent, the leaves and long petiolate with alternate phyllotaxy. The stems are rough, tough, vine-like, spreads on the ground (Pravin et al., 2013). The leaves are pubescent with undulations, obtuse, triangular in shape with green upside and pale yellow underneath, flowers yellow, axillary, fruits are globular, orange to yellow, with hard coriaceous covering and pulpy white inside, bitter in taste, seed are white to brown, ovate and compressed. Fruits are considered to be extremely toxic (Rahimi et al., 2012).

Uses: The ripened fruits (approx. 100 g) are given twice a day to cattle to cure indigestion, colic pain, cough and intestinal worms (Qureshi et al., 2010). The methanolic extract of ripe and deseeded fruit shows antimycobacterial action against *Mycobacteriumtuberculosis* H37Rv thereby making the patients free from the disease like tuberculosis which affects the lung tissues and forcing the patient to cough up (Mehta et al., 2013). The fruits are bitter and used against asthma (Mariod et al., 2017). The fruit pulp reduces cough and respiratory congestion (Kīrtikara et al., 2001). The leaves are used against asthma (Joshi et al., 2021; Uma & Sekar, 2014). The fruits are rich sources of alkaloids, resins and pectins were made to consume with mixture with anison before sleeping to cure asthma which was followed by traditional Iranian medicine (Jalali et al., 2020).

Ethnopharmacology of Indravaruni

Ethnopharmacology is the education of how dissimilar philosophies and civilisations use plants, animals, fungi, and other natural materials in medication and healing practices. It combines elements of anthropology, pharmacology, and botany to investigate the relationship between humans and their natural environment, particularly how native and local information systems have contributed to the development of medicinal substances.

Ethnopharmacologists explore how various indigenous and local communities have used plants, herbs, minerals, and animal products for centuries to treat illnesses, prevent diseases, and promote health, they are concerned about the identification of bioactive compounds in traditional medications and their pharmacological effects. Scientists isolate compounds from plants or other sources, test them for potential medicinal benefits, and explore their mechanisms of action in the body. Many modern pharmaceutical drugs are derived from compounds initially acknowledged through ethnopharmacological studies. Like aspirin was developed from the active ingredient in willow bark, and the anti-cancer drug paclitaxel (Taxol) was derived from the Pacific yew tree. Ethnopharmacology continues to play an enthusiastic role in the search for new therapeutic means. Ethnopharmacology encompasses a diversity of disciplines, comprising anthropology (to comprehend cultural background), pharmacology (to study the therapeutic properties of secondary metabolites), chemistry (to isolate and analyse bioactive secondary metabolites), and ecology (to apprehend the association between humans and their environment).

Other Indian Name:—Indarun/Indrayan
Part used: Fruit, root and leaf
Ethnopharmacology:

Phytochemicals		Parts of plant	References
Cucurbitacins	Cucurbitacins E	Whole Fruit	(Hameed et al., 2020)
	Cucurbitacins B	Whole Fruit	(Hameed et al., 2020)
	Cucurbitacins I	Whole Fruit	(Hameed et al., 2020)
	Cucurbitacins A	Fruit	(Adam et al., 2001)
	Cucurbitacins C	Fruit	(Adam et al., 2001)
	Cucurbitacins D	Fruit	(Adam et al., 2001)
	Cucurbitacins J	Fruit	(Adam et al., 2001)
	Cucurbitacins K	Fruit	(Yoshikawa et al., 2007)
	Cucurbitacins L	Fruit	(Yoshikawa et al., 2007)
	Colosynthosides A	Fruit	(Yoshikawa et al., 2007)
	Colosynthosides B	Fruit	(Yoshikawa et al., 2007)
Glycosides, flavonoids and Phenolic acids	2-O-β-D-glucopyranosyl-Cucurbitacin I	Fruit	(Yoshikawa et al., 2007)
	2-O-β-D-glucopyranosyl-Cucurbitacin L	Fruit	(Delazar et al., 2006)
	Isosaponarin	Fruit, Seeds	(Delazar et al., 2006)
	Isovitexin	Fruit, Seeds	(Gurudeeban et al., 2011)
	Isoorientin 3-o-methylether	Fruit, Seeds	(Gurudeeban et al., 2011)
	Catechin	Fruit pulp	(Gurudeeban et al., 2011)
	Myricetin	Fruit pulp	(Hussain et al., 2013)
	Quercetin	Fruits	(Hussain et al., 2013)
	Kaempferol	Fruit pulp	(Hussain et al., 2013)
	Gallic acid	Fruit pulp	(Hussain et al., 2013)
	p-Hydroxybenzoicacid	Fruit pulp	(Hussain et al., 2013)
	Chlorogenic acid	Fruit pulp	(Hussain et al., 2013)
	Caffeic acid	Fruit pulp	(Hussain et al., 2013)
	Vanillic acid	Fruit pulp	(Hussain et al., 2013)
	p-Coumeric acid	Fruit pulp	(Hussain et al., 2013)
	Sinapic acid	Fruit pulp	(Hussain et al., 2013)
	Ferulic acid	Fruit pulp	(Hussain et al., 2013)

(continued)

(continued)

Alkaloids	Choline	Whole fruit, Pulp, and Fruit pulp	(Hussain et al., 2013)
Volatiles and Terpenoids	4-(1-methyl)ethoxy-1butanol;5-methoxy,2methyl,2-pentanol;1cyclopentyl,2-propene-1-ol and 2-furanmethanol, tetrahydro-5-methyl- cis and trans isomers, 3,4-dimethyl,2-hexanone; 2- methyl,4-heptanoneand3-Methyl,2-heptanone,1-propoxypentaneand2,3-epoxymethylPropionate	Fruit pulp	(Hameed et al., 2020)
Fatty acids	Palmitic acid, Stearic acid, Linoleic acid, oleic acids	Fruit seed oil	(Gurudeeban et al., 2011)
	Myristic, palmitic, stearic, oleic, linoleicandlinolenicacids	Seeds	(Sebbagh et al., 2009)
Tocopherol and carotenes	α-tocopherol, γ-tocopherol, β-carotene	Fruit seed oil	(Gurudeeban et al., 2011)

Pharmacological Effects of *Citrulluscolocynthis* Fruits.

Pharmacological effects	Extracts/ Compounds	Dose/Concentration	In vitro/ In vivo	References
Antidiabetic	Petroleum ether extract of fruit	Dose: 300 mg/kg	In vivo	(Jayaraman et al., 2009)
	Hydro-ethanol Extract of fruit	Dose: 300 g/kg	In vivo	(Dallak, 2011)
	Aqueous extract of fruits	300 mg/kg	In vivo	(Abdel-Hassan et al., 2000)
	Alkaloid rich fraction of fruit	50 mg/kg	In vivo	(Abdel-Hassan et al., 2000)
	Glycosides rich fraction of fruit	50 mg/kg	In vivo	(Abdel-Hassan et al., 2000)
	Saponin rich fraction of fruit	50 mg/kg	In vivo	(Abdel-Hassan et al., 2000)
	Saponin rich fraction of fruit	20 mg/kg	In vivo	(Abdel-Hassan et al., 2000)
	Fruit powder	100mgcapsule/ three times a day	In vivo	(Huseini et al., 2009)
	Aqueous extract of seeds	300 mg/ kg	In vivo	(Al-Ghaithi et al., 2004)
Hypolipidemic	Seed powder	300 mg seed powder/day	In vivo	(Rahbar & Nabipour, 2010)
	Aqueous ethanol extract	1.2 g/kg/day	In vivo	(Daradka et al., 2007)

(continued)

(continued)

Antimicrobial	Water extract, ethanol extract	5 mg/mL	In vitro	(Najafi et al., 2010)
	Ethanol extract of fruit (18mmInhibitionzoneagainst *Staphylococcus aureus*)	500 mg/mL	In vitro	(Hameed et al., 2020)
	Ethanol extract of fruit (15 mm Inhibition zone against *Bacillus cereus*)	500 mg/mL	In vitro	(Hameed et al., 2020)
	Ethanol extract of fruit (9 mm Inhibition zone against *Klebsiella pneumonia*)	500 mg/mL	In vitro	(Hameed et al., 2020)
	Cucurbitacin E (18 mm Inhibition zone against *Staphylococcus aureus*)	500 mg/mL	In vitro	(Hameed et al., 2020)
	Cucurbitacin E (9 mm Inhibition zone against *Klebsiella pneumonia*)	500 mg/mL	In vitro	(Hameed et al., 2020)
	Cucurbitacin E (16 mm Inhibition zone against *Bacillus cereus*)	500 mg/mL	In vitro	(Hameed et al., 2020)
	Cucurbitacin B (16 mm Inhibition zone against *Staphylococcus aureus*)	500 mg/mL	In vitro	(Hameed et al., 2020)
	Cucurbitacin B (8 mm Inhibition zone against *Klebsiella pneumonia*)	500 mg/mL	In vitro	(Hameed et al., 2020)
	Cucurbitacin B (10 mm Inhibition zone against *Bacillus cereus*)	500 mg/mL	In vitro	(Hameed et al., 2020)
	Cucurbitacin I (16 mm Inhibition zone against *Staphylococcus aureus*)	500 mg/mL	In vitro	(Hameed et al., 2020)
	Cucurbitacin I (8 mm Inhibition zone against *Klebsiella pneumonia*)	500 mg/mL	In vitro	(Hameed et al., 2020)
	Cucurbitacin I (10 mm Inhibition zone against *Bacillus cereus*)	500 mg/mL	In vitro	(Hameed et al., 2020)
Anti-inflammatory	Water extract	4 mg/ kg	In vivo	(Marzouk et al., 2010)
	Aqueous ethanol extract gel (3%)	2 g of gel	In vitro	(Aly & Ahmad Naddaf, 2006)
Anti-oxidant	Hydro-ethanol extract of fruits	300 g/ kg	In vivo	(Dallak, 2011)
	Methanol extract (88% DPPH radical Scavenging activity was found against ascorbic acid)	Concentration 2500 μg/mL	In vitro	(Kumar et al., 2008)
	Methanol extract (61.4% Nitric oxide scavenging activity was found against ascorbic acid)	Concentration 2500 μg/mL	In vitro	(Kumar et al., 2008)
	Aqueous extract	$IC_{50} = 0.021$ mg/mL	In vitro	(Marzouk et al., 2010)
	Cucurbitacin glycosides combination	IC_{50} 145 μM	In vitro	(Tannin-Spitz et al., 2007)

(continued)

(continued)

Cytotoxic	Ethanol extract of fruits (DPPH radical scavenging activity)	$IC_{50} = 7.14$ µg/mL	In vitro	(A.I. Hussain et al., 2013)
	Ethanol extract of fruits (76.5% inhibition of linoleic acids peroxidation was observed)	Concentration: 5 mg/100 mL	In vitro	(A.I. Hussain et al., 2013)
	Cucurbitacin B, Cucurbitacin E (1:1)	20 µM	In vitro	(Tannin-Spitz et al., 2007)
	Alkaloid rich fraction	$IC_{50} = 3.30$ µg/mL	In vitro	(Mukherjee & Patil, 2012)
	Alkaloid rich fraction of fruits	IC_{50} against MCF-7 = 17.20 µg/mL; against HepG- 2 = 12.54 µg/mL	In vitro	(Mukherjee & Patil, 2012)
	Cucurbitacin B isolated from fruits (inhibited cellular proliferation of human laryngeal cancer cell line (Hep-2)	1-100 µM	In vitro	(Liu et al., 2010)
	Cucurbitacin B isolated from fruits (Tested on human GBM cell lines in liquid culture, cucurbitacin B inhibited 50% growth at 0.1 µM)	$LD_{50} = 0.1$ µM	In vitro	(Yin et al., 2008)
Insecticidal	n- hexane extract	LC_{50}23,065 ppm	In vitro	(Torkey et al., 2009)
	Methylenechloride extract	LC_{50}19,497 ppm	In vitro	(Torkey et al., 2009)
	Chloroform extract	LC_{50}17,328 ppm	In vitro	(Torkey et al., 2009)
	Ethanol extract	LC_{50}11,003 ppm	In vitro	(Torkey et al., 2009)
Antiallergic	Methanol extract	Dose—200 mg/ Kg	In vivo	(Yoshikawa et al., 2007)
	Cucurbitacin E 2-O-β-D glucopyranoside	Dose—200 mg/kg	In vivo	(Yoshikawa et al., 2007)

Indravaruni or *Citrulluscolocynthis (L.) Schrad*is having a wide range of ethno medicinal uses which are cancer, leprosy, diabetes, jaundice, bronchitis, asthma, epilepsy, cough, common cold, wound, joint pain, toothache, mastitis, and medicinal uses also in gastroenteritis and gastrointestinal disorders such as constipation, abortification, dysentery, colic pain, indigestion, hair growth promotion and fever from different microbial infections. This fruit is also used in sciatica and gout treatment. The fruits and roots were rubbed together with simultaneous addition of water and then applied to pimples and boils. The root has been used in the breast inflammation, joints pain externally. It was used in ophthalmia and in uterine pains. The root is pasted and applied on enlarged abdomen of children. Leaves of this plant are used for the treatment of jaundice, asthma and epilepsy (Jeeb et al., 2016).

The names and docking results of the secondary metabolites of Indravaruni (Fig. 1) are provided in Table 1 from our previous paper (Pakrashy et al., 2022).

Active principles: Secondary Metabolites of Indravaruni

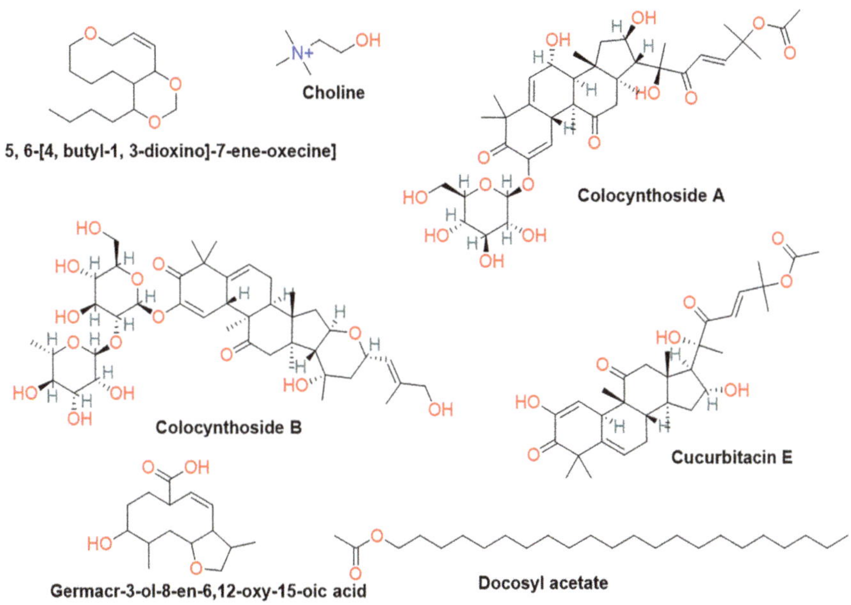

Fig. 1 Structures of some secondary active-metabolites of Indravaruni

Table 1 Names and docking results of some secondary active-metabolites of **Indravaruni** (***Citrulluscolocynthis*** **(L.) Schrad.**) Against 6VWW protein

Name and Structure	PubChem ID	Docking Score 6VWW
1). 5, 6-[4, butyl-1, 3-dioxino]-7-ene-oxecine]	–	−5.5
2). Choline	305	−3.8
3). Colocynthoside A	16,216,752	−8.3
4). Colocynthoside B	16,216,649	−8.1
5). Cucurbitacin E	5,281,319	−8.4
6). Docosyl acetate	69,969	−3.9
7). Germacr-3-ol-8-en-6,12-oxy-15-oic acid	-	−6.3

Conclusion

Indravaruni or *Citrulluscolocynthis (L.) Schrad*h as a variety of ethnomedicinal and ethnopharmacological uses against cancer, leprosy, diabetes, jaundice, bronchitis, asthma, epilepsy, cough, common cold, wound, joint pain, toothache, mastitis, and medicinal uses also in gastroenteritis and gastrointestinal disorders such as constipation, abortification, dysentery, colic pain, indigestion, hair growth promotion and fever from different microbial infections. This can also be used for the COVID-19 virus and its strains.

Acknowledgments This research was supported by the Science and Engineering Research Board (SERB), India. AM acknowledges SERB for research grant ref: EEQ/ 2019/000194. The authors are also thankful to the Presidency University, Kolkata for providing FRPDF research grant, laboratory and NMR facilities.

Authors' Statement SP, SG and AS collected the data and prepared the draft. MD analysed the data, reviewed and edited the manuscript. AM supervised, arranged funds, analysed the data, reviewed and edited the manuscript.

Competing interest The authors declare that there are no conflicts of interest.

Abbreviations

ACE2	Angiotensin-converting enzyme 2
ARDS	Acute respiratory distress syndrome
COVID-19	Corona Virus Disease in December 2019
MM2	Molecular mechanics of 2nd order
Mpro	Main protease
Nsps	Non-structural proteins
PDB	Protein Data Bank
Pp	Poly-proteins
RAS	Renin-angiotensin system

References

Abdel-Hassan, I. A., Abdel-Barry, J. A., & Mohammeda, S. T. (2000). The hypoglycaemic and antihyperglycaemic effect of Citrullus colocynthis fruit aqueous extract in normal and alloxan diabetic rabbits. *Journal of Ethnopharmacology*, *71*, 325–330.

Adam, S. E. I., Al-Yahya, M. A., & Al-Farhan, A. H. (2001). Response of Najdi sheep to oral administration of Citrullus colocynthis fruits, Nerium oleander leaves or their mixture. *Small Ruminant Research*, *40*, 239–244.

Al-Ghaithi, F., El-Ridi, M. R., Adeghate, E., & Amiri, M. H. (2004). Biochemical effects of Citrullus colocynthis in normal and diabetic rats. *Molecular and Cellular Biochemistry*, *261*, 143–149.

Aly, A. M., & Ahmad Naddaf, A. N., Anti-inflammatory activities of Colocynth topical gel (2006).

Amoretti, M., Amsler, C., Bonomi, G., Bouchta, A., Bowe, P., Carraro, C., Cesar, C. L., Chaliton, M., Collier, M. J. T., Doser, M., Filippini, V., Fine, F. S., Fontana, A., Fujiwara, M. C., Funakoshi, R., Genova, P., Hangst, J. S., Hayano, R. S., Holzscheiter, M. H., … Van Der Werf, D. P. (2002). Production and detection of cold antihydrogen atoms. *Nature*, *419*, 456–459. https://doi.org/10.1038/nature01096

Anand, K., Ziebuhr, J., Wadhwani, P., Mesters, J. R., & Hilgenfeld, R. (2003). Coronavirus main proteinase (3CLpro) Structure: Basis for design of anti-SARS drugs. *Science*, *80-.). 300*, 1763–1767. https://doi.org/10.1126/science.1085658

Cheng, X., Qin, M., Chen, R., Jia, Y., Zhu, Q., Chen, G., Wang, A., Ling, B., & Rong, W. (2023). Citrullus colocynthis (L.) Schrad.: A promising pharmaceutical resource for multiple diseases. *Molecules*, *28*, 6221.

Cho, E., Rosa, M., Anjum, R., Mehmood, S., Soban, M., Mujtaba, M., Bux, K., Moin, S. T., Tanweer, M., Dantu, S., Pandini, A., Yin, J., Ma, H., Ramanathan, A., Islam, B., Mey, A. S. J. S., Bhowmik, D., & Haider, S., (2021). Dynamic Pro fi ling of β - Coronavirus 3CL M.

Dallak, M. (2011). In vivo, hypolipidemic and antioxidant effects of Citrullus colocynthis pulp extract in alloxan-induced diabetic rats. *African Journal of Biotechnology*, *10*, 9898–9903.

Daradka, H., Almasad, M. M., WSh, Q., El-Banna, N. M., & Samara, O. H. (2007). Hypolipidaemic effects of Citrullus colocynthis L. in rabbits. *Pakistan Journal of Biological Sciences*, *10*, 2768–2771.

Datta, T., Patra, A. K., & Dastidar, S. G. (2014). Medicinal plants used by tribal population of Coochbehar district, West Bengal, India–an ethnobotanical survey. *Asian Pacific Journal of Tropical Biomedicine*, *4*, S478–S482.

Delazar, A., Kosari, A. R., Nazemieh, H., Modaresi, M., Gibbons, S., Nahar, L., & Sarker, S. D. (2006). Flavone C-glycosides and cucurbitacin glycosides from Citrullus colocynthis.

Deora, N., & Venkatraman, K. (2024). Potential use of plant-based therapeutics for the management of SARS-COV2 infection in Diabetes mellitus-A review. *Journal of Herbal Medicine*, 100923.

Dey, S. K., Rahman, M. M., Siddiqi, U. R., & Howlader, A. (2020). Analyzing the epidemiological outbreak of COVID-19: A visual exploratory data analysis approach. *Journal of Medical Virology*, *92*, 632–638. https://doi.org/10.1002/jmv.25743

Gasteiger, J., & Jochum, C. (1979). An Algorithm for the Perception of Synthetically Important Rings. *Journal of Chemical Information and Computer Sciences*, *19*, 43–48. https://doi.org/10.1021/ci60017a011

Guex, N., & Peitsch, M. C. (1997). SWISS-MODEL and the Swiss-PdbViewer: An environment for comparative protein modeling. *Electrophoresis*, *18*, 2714–2723. https://doi.org/10.1002/elps.1150181505

Gurudeeban, S., Ramanathan, T., & Satyavani, K. (2011). Characterization of volatile compounds from bitter apple (Citrullus colocynthis) using GC-MS. *International Journal of Chemical and Analytical Science*, *2*, 108–110.

Hameed, B., Ali, Q., Hafeez, M. M., & Malik, A. (2020). Antibacterial and antifungal activity of fruit, seed and root extracts of Citrullus colocynthis plant. *Biological and Clinical Sciences Research Journal*, *33*, 1–5.

Huseini, H. F., Darvishzadeh, F., Heshmat, R., Jafariazar, Z., Raza, M., & Larijani, B. (2009). The clinical investigation of Citrullus colocynthis (L.) schrad fruit in treatment of Type II diabetic patients: A randomized, double blind, placebo-controlled clinical trial, Phyther. Res. *An International Journal Devoted to Pharmacology and Toxicological Evaluation of Natural Product Derivatives*, *23*, 1186–1189.

Hussain, A. I., Rathore, H. A., Sattar, M. Z. A., Chatha, S. A. S., Ud Din Ahmad, F., Ahmad, A., & Johns, E. J. (2013). Phenolic profile and antioxidant activity of various extracts from Citrullus colocynthis (L.) from the Pakistani flora, Ind. *Crops and Products*, *45*, 416–422. https://doi.org/10.1016/j.indcrop.2013.01.002

Hussain, M., Jabeen, N., Raza, F., Shabbir, S., Baig, A. A., Amanullah, A., & Aziz, B. (2020). Structural variations in human ACE2 may influence its binding with SARS-CoV-2 spike protein. *Journal of Medical Virology*, *92*, 1580–1586. https://doi.org/10.1002/jmv.25832

Jalali, A., Vanani, A. R., & Shirani, M. (2020). Ethnobotanical approaches of traditional medicinal plants used in the management of asthma in Iran. *Jundishapur Journal of Natural Pharmaceutical Products*, *15*.

Jayaraman, R., Shivakumar, A., Anitha, T., Joshi, V. D., & Palei, N. N. (2009). Antidiabetic effect of petroleum ether extract of Citrullus colocynthis fruits against streptozotocin-induced hyperglycemic rats. *Romanian Journal of Biology - Plant Biology*, *4*, 127–134.

Jeeb, S. A. A. A., Mohammed, M. S., Fathelrahman, A. E., & Wadah, J. (2016). Bioactivity-guided isolation of two sesquiterpenes with potential anti-fungal activity from Citrullus colocynthis L.(Schard). *Journal of Pharmacognosy and Phytochemistry*, *5*, 422–425.

Jin, Z., Du, X., Xu, Y., Deng, Y., Liu, M., & Zhao, Y., Structure of M pro from SARS-CoV-2 and discovery of its inhibitors, (2020).

Joshi, R., Mishra, P., Kesh, R., & Patni, V. I., Citrullus colocynthis: A Potential Source of Alternative Medicine. 2021: pp. 43–66.

Khan, M. F., Khan, M. A., Khan, Z. A., Ahamad, T., & Ansari, W. A. (2020).Identification of dietary molecules as therapeutic agents to Combat COVID-19 Using Molecular Docking Studies. https://doi.org/10.21203/rs.3.rs-19560/v1.

Kim, Y., Jedrzejczak, R., Maltseva, N. I., Wilamowski, M., Endres, M., Godzik, A., Michalska, K., & Joachimiak, A. (2020). Crystal structure of Nsp15 endoribonuclease NendoU from SARS-CoV-2. *Protein Science*, *29*, 1596–1605. https://doi.org/10.1002/pro.3873

Kīrtikara, K. R., Basu, B. D., An, I. C. S., Blatter, E., Caius, J. F., & Mhaskar, K. S. (2001). I ndian medicinal plants, with illustrations.

Krivov, G. G., Shapovalov, M. V., & Dunbrack, R. L. (2009). Improved prediction of protein side-chain conformations with SCWRL4. *Proteins: Structure, Function, and Bioinformatics*, *77*, 778–795. https://doi.org/10.1002/prot.22488

Kumar, S., Kumar, D., Saroha, K., Singh, N., & Vashishta, B. (2008). Antioxidant and free radical scavenging potential of *Citrullus colocynthis* (L.) Schrad. methanolic fruit extract. *Acta Pharmaceutica*, *58*, 215–220.

Liu, T., Peng, H., Zhang, M., Deng, Y., & Wu, Z. (2010). Cucurbitacin B, a small molecule inhibitor of the Stat3 signaling pathway, enhances the chemosensitivity of laryngeal squamous cell carcinoma cells to cisplatin. *European Journal of Pharmacology*, *641*, 15–22.

Mariod, A. A., Mirghani, M. E. S., & Hussein, I. H. (2017). *Unconventional oilseeds and oil sources*. Academic Press.

Marzouk, B., Marzouk, Z., Haloui, E., Fenina, N., Bouraoui, A., & Aouni, M. (2010). Screening of analgesic and anti-inflammatory activities of Citrullus colocynthis from southern Tunisia. *Journal of Ethnopharmacology*, *128*, 15–19.

Mehta, A., Srivastva, G., Kachhwaha, S., Sharma, M., & Kothari, S. L. (2013). Antimycobacterial activity of *Citrullus colocynthis* (L.) Schrad. against drug sensitive and drug resistant Mycobacterium tuberculosis and MOTT clinical isolates. *Journal of Ethnopharmacology*, *149*, 195–200.

Mukherjee, A., & Patil, S. D. (2012). Effects of alkaloid rich extract of Citrullus colocynthis fruit on Artemia salina and human cancerous (MCF-7 and HEPG-2) cells. *Journal of PharmaSciTech*, *1*, 15–19.

Najafi, S., Sanadgol, N., Nejad, B. S., Beiragi, M. A., & Sanadgol, E. (2010). Phytochemical screening and antibacterial activity of Citrullus colocynthis (Linn.) Schrad against Staphylococcus aureus. *Journal of Medicinal Plants Research*, *4*, 2321–2325.

Nighantu, B. (2015). In K. C. Chunekar & D. G. S. Pandey (Eds.), *Commentary*. Chaukhambha Bharati Academy.

Oudit, G. Y., Crackower, M. A., Backx, P. H., & Penninger, J. M. (2003). The role of ACE2 in cardiovascular physiology. *Trends in Cardiovascular Medicine*, *13*, 93–101. https://doi.org/10.1016/S1050-1738(02)00233-5

Pakrashy, S., Mandal, P. K., Dey, S. K., Choudhury, S. M., Alasmary, F. A., Almalki, A. S., Islam, M. A., & Dolai, M. (2022). Design of a structurally novel multipotent drug candidate by the scaffold architecture technique for ACE-II, NSP15, and MproProtein inhibition: Identification and isolation of a natural product to prevent the severity of future variants of Covid 19 and a. *ACS Omega*, *7*, 33408–33422. https://doi.org/10.1021/acsomega.2c04051

Pravin, B., Tushar, D., Vijay, P., & Kishanchnad, K. (2013). Review on Citrullus colocynthis. *International Journal of Research in Pharmacy and Chemistry*, *3*, 46–53.

Qureshi, R., Bhatti, G. R., & Memon, R. A. (2010). Ethnomedicinal uses of herbs from northern part of Nara desert, Pakistan. *Pakistan Journal of Botany*, *42*, 839–851.

Rahbar, A. R., & Nabipour, I. (2010). The hypolipidemic effect of *Citrullus colocynthis* on patients with hyperlipidemia. *Pakistan Journal of Biological Sciences*, *13*, 1202–1207.

Rahimi, R., Amin, G., & Ardekani, M. R. S. (2012). A review on citrullus colocynthis schrad.: From traditional iranian medicine to modern phytotherapy. *Journal of Alternative and Complementary Medicine*, *18*, 551–554. https://doi.org/10.1089/acm.2011.0297

Rajpoot, V. S., & Kareti, S. R. (2025). Ethnobotanical survey of medicinal plants used for the treatment of malaria by tribal communities of Anuppur district, Central India: Ethnobotanical medicinal plants for malaria. *Indian Journal of Traditional Knowledge*, *24*, 158–167.

Richman, D. D., Whitley, R. J., & Hayden, F. G. (2016). Introduction. *Clinical Virology*. https://doi.org/10.1128/9781555819439.ch1

Rohilla, L., Singh, D. C., & Yadav, C. (2017). Review on Indravaruni (Citrullus colocynthis Schrad): An ethnomedicinally rich plant wsr to its therapeutic uses. *International Journal of Ayurveda and Pharma Research*.

Sastry, J. L. N. (2005). Illustrated Dravyaguna Vijnana Chaukhambha Orientalia.

Sebbagh, N., Cruciani-Guglielmacci, C., Ouali, F., Berthault, M.-F., Rouch, C., Sari, D. C., & Magnan, C. (2009). Comparative effects of Citrullus colocynthis, sunflower and olive oil-enriched diet in streptozotocin-induced diabetes in rats. *Diabetes and Metabolism*, *35*, 178–184.

Tannin-Spitz, T., Grossman, S., Dovrat, S., Gottlieb, H. E., & Bergman, M. (2007). Growth inhibitory activity of cucurbitacin glucosides isolated from Citrullus colocynthis on human breast cancer cells. *Biochemical Pharmacology*, *73*, 56–67.

Thomas-Rüddel, D., Winning, J., Dickmann, P., Ouart, D., Kortgen, A., Janssens, U., & Bauer, M. (2020). Coronavirus disease 2019 (COVID-19): Update for anesthesiologists and intensivists March 2020„Coronavirus disease 2019 "(COVID-19): Update für Anästhesisten und Intensivmediziner März 2020, Anaesthesist. https://doi.org/10.1007/s00101-020-00760-3

Torkey, H. M., Abou-Yousef, H. M., Abdel Azeiz, A. Z., & Hoda, E. A. F. (2009). Insecticidal effect of cucurbitacin E glycoside isolated from Citrullus colocynthis against Aphis craccivora. *Australian Journal of Basic and Applied Sciences*, *3*, 4060–4066.

Trott, O., & Olson, A. J. (2010). AutoDock Vina: Improving the speed and accuracy of docking with a new scoring function, efficient optimization, and multithreading. *Journal of Computational Chemistry*, *31*, 455–461.

Uma, C., & Sekar, K. G. (2014). Phytochemical analysis of a folklore medicinal plant *Citrullus colocynthis* L (bitter apple). *Journal of Pharmacognosy and Phytochemistry*, *2*, 195–202.

Wang, J., Wang, W., Kollman, P. A., & Case, D. A. (2001). Case, antechamber, an accessory software package for molecular mechanical calculations. *Journal of the American Chemical Society*, *222*, U403.

Wu, C., Liu, Y., Yang, Y., Zhang, P., Zhong, W., Wang, Y., Xu, Y., Li, M., Li, X., Zheng, M., Chen, L., & Li, H. (2020). *Journal Pre-proof.*

Wu, F., Zhao, S., Yu, B., Chen, Y. M., Wang, W., Song, Z. G., Hu, Y., Tao, Z. W., Tian, J. H., Pei, Y. Y., Yuan, M. L., Zhang, Y. L., Dai, F. H., Liu, Y., Wang, Q. M., Zheng, J. J., Xu, L., Holmes, E. C., & Zhang, Y. Z. (2020). A new coronavirus associated with human respiratory disease in China. *Nature*, *579*, 265–269. https://doi.org/10.1038/s41586-020-2008-3

Yin, D., Wakimoto, N., Xing, H., Lu, D., Huynh, T., Wang, X., Black, K. L., & Koeffler, H. P. (2008). Cucurbitacin B markedly inhibits growth and rapidly affects the cytoskeleton in glioblastoma multiforme. *International Journal of Cancer*, *123*, 1364–1375.

Yoshikawa, M., Morikawa, T., Kobayashi, H., Nakamura, A., Matsuhira, K., Nakamura, S., & Matsuda, H. (2007). Bioactive saponins and glycosides. XXVII. Structures of new cucurbitane-type triterpene glycosides and antiallergic constituents from *Citrullus colocynthis. Chemical and Pharmaceutical Bulletin*, *55*, 428–434.

Chapter 4
Role of Bodies for the Conservation of Bio-diversity Under Biological Diversity Act, 2002

Sudarshan Verma and Nitesh Kumar Chaturvedi

Abstract Biological Diversity is essential part of agriculture and food production and processing, because it makes an essential contribution for feeding the millions of the population of the every nation of the world. Biological Diversity is multi-disciplinary area, which involve multifaceted actions & activities. Preservation of Biological Diversity is a common concern of human being and humanity. In this chapter author has also discussed about Community Forest Resource Rights, which is a recognized right under Sections Enactment Related to the Conservation of Biodiversity in India and Community Forest Resource Rights Under Forest Dwellers Act, 2006 of Forest Dwellers (Recognition of Forest Rights) Act, 2006, which aims to empower local communities with rights over forest land and resources and promote their participation in the conservation and sustainable management of forests. In this chapter, researcher has explored and analyses the role of three tiered management structure of bodies under the Biological Diversity Act. Three tiered management structure of bodies at National, State and local level has discussed in Biological Diversity Act, 2002. Constitution, Composition and Functions of Bodies has elaborated with number of provisions of above said Act, in which setting up National Biodiversity Authority, State Biodiversity Board and Biodiversity Management Committee.

Keywords Biological Diversity · Community Forest Resource · National Biodiversity Authority · State Biodiversity Board and Biodiversity Management Committee

---

S. Verma (✉)
Department of Law, School of Legal Studies, Babasaheb Bhimrao Ambedkar University, Lucknow, India
e-mail: drsudarshanvermabbau@gmail.com

N. K. Chaturvedi
Atal Bihari Vajpayee, School of Legal Studies, CSJMU, Kanpur, India

R. C. Sobti et al. (eds.), *Traditional Knowledge Systems in India: Sustainability, Ecology, and Resource Governance*, People, Cultures and Societies: Exploring and Documenting Diversities,
https://doi.org/10.1007/978-981-92-1346-7_4

Introduction

Preservation of Biological Diversity is a common concern of human being and humanity. All forms of Biological Diversity including species, ecosystems and genetic resources are covered by the Convention on Biological Diversity, 1992. Biological Diversity is essential part of agriculture and food production and processing, because it makes an essential contribution for feeding the millions of the population of the every nation of the world. Biological Diversity is multi-disciplinary area, which involve multifaceted actions & activities. The subject matter of Biological Diversity Act, 2002 is Conservation of Biodiversity and according to provision of above said Act, stakeholders of conservation of Biological Diversity are Government of India, Provincial Government, Bodies of Local-self Government and Biodiversity Management Committees and industry.

Unity in diversity is one of the most amazing features in Indian society from ancient time. India has the second-largest tribal population in the world. According to the 6th schedule of Indian Constitution, provision of the 6th schedule provides the administration of tribal areas. As per the Census of 2011, the tribal population constitutes 8.61% of the total population in India. Scheduled Tribes (ST) constitute approximately 8.61% of population of India numbering around 10.4 crores citizens. There are over 730 Scheduled Tribes communities notified under Article 342 of the Constitution of India and this diversity provides ample scope for ethno-biological studies. State of Madhya Pradesh is first position in population of tribal community, while second and third position has state of Maharashtra and Orissa in India. The tribal communities throughout the territory of India have rich traditions, cultures and heritage with unique lifestyles, ecosystem and customs. (The Constitution of India, 1950 (Bare Act)).

Sustainable management of natural resources has become an important aspect of survival of life on earth in modern days, where the ecological degradation and loss of Biological Diversity pose a serious threat to sustainable development. In recent times, indiscriminate human being are in-discriminatively exploiting the natural resources and over population coupled with the continuous grazing of (under) productive cattle population pose the greatest threat to species and ecosystems. Many valuable species have reduced to an alarming number and at times disappeared from the scene. Natural hazards owing to their adverse impacts, also contribute to the loss of biodiversity, due to the climate change, natural calamities and manmade disasters.

International Instrument on Conservation of Biodiversity

Convention on Biological Diversity, 1992

Firstly at International level in the year of 1984, contracting parties are starting an effort to conserve and sustainable use of the Biodiversity with the help of a global convention after that in the year of 1987 under the United Nations Environment

Program recognized the need to protect Biological Diversity. The secretariat of UNEP prepared an initial draft related to the Convention on Biological Diversity in the year of 1991. On May 22, 1992 contracting parties of the Convention on Biological Diversity was adopted the above Convention in Nairobi, so international day for Biological Diversity is celebrated on 22thMay of every year (Singh & Bahari, 2024).

The United Nations Conference on Environment and Development (UNCED), also known as the 'Earth Summit' was held at Rio-de-Janeiro in June 1992. In this Conference the sovereign rights of the State over their biological resources on earth reaffirmed. United Nations Framework Convention on Climate Change entered into force on March 21, 1994. 195 States contracting parties that have ratified the convention are called parties to the Convention. The UNFCCC is a "Rio-Convention", one of three adopted at the Earth Summit, 1992 was held at Rio-de-Janeiro. The intergovernmental Panel on Climate Change is the international body for assessing the science related to climate change. It was set up in 1988 by the World Meteorological Organization and the UNEP to provide policymakers with regular assessment of the scientific basis of Climate Change, its impacts and future risks, and options for adaption and mitigation.

The 'Earth Summit' had many enormous achievements, in which particularly the Rio-de-Janeiro Declaration and its 27 universal principles, the United Nations Framework Convention on Climate Change, the Convention on Biological Diversity and the Declaration on the principles of forest management. Convention on Biological Diversity came into existence on December 29, 1993. Convention on Biological Diversity has been ratified by 196 nations over the World. The Secretariat of the Convention on Biological Diversity (SCBD) is situated in Montreal, Canada.

Aim of this Convention is to provide a holistic approach for the conservation of Biological Diversity, sustainable use of its components and fair and equitable sharing of the benefits arising out of the use of biological resources and Indigenous Knowledge or Traditional Knowledge. In the light of the two core issues i.e., environment and development, the Convention on Biological Diversity, 1992 is historic because it is the first of its kind to address the sustainable use and utilization of biological resources and the conservation of Biological Diversity on earth from a particular to a comprehensive perspective (Convention on Biological diversity, 1992).

According to Article 8 & 9 of the CBD, each contracting party shall be taking part into the promotion of in-situ & ex-situ conservation of Biological Diversity. In-situ conservation means the conservation of species in their natural habitats, is considered the most appropriate way of conserving Biological Diversity on earth, while ex-situ conservation is the preservation of components of Biological Diversity outside their natural habitats. According to the Article 8(j) of the above said Convention, each contracting Party of the Convention shall take as far as possible and appropriate action on subject to its implementation in national legislation, respect, preserve and maintain knowledge, innovations and practices of indigenous and local communities embodying traditional lifestyles relevant for the conservation and sustainable use of Biological Diversity and also promote their wider application with the approval and involvement of the holders of such knowledge, innovations and practices and encourage the equitable sharing of the benefits arising from the utilization of such knowledge, innovations and practices (Shastri, 2022).

Declaration on the Rights of Indigenous Peoples

The United Nations Declaration on the Rights of Indigenous Peoples (UNDRIP) has discussed the provision for the Conservation of Biological Diversity, in which Article 1 of the above said Declaration states that the Indigenous Peoples have the right to the full enjoyment, as a collective or as individuals, of all human rights and fundamental freedoms as recognized in the Charter of the United Nations, the Universal Declaration of Human Rights and other international human rights law.

Article 29(1) of the Declaration empowered to the Indigenous peoples of the society have the right to the conservation and protection of the environment and the productive capacity of their lands or territories and resources. Parties of the above said Declaration shall establish and implement assistance programmes for indigenous peoples for such conservation and protection without any discrimination amongst them. Article 29(2) of the Declaration States shall take effective measures to ensure that no storage or disposal of hazardous materials shall take place in the lands or territories of indigenous peoples without their free, prior and informed consent and Article 29(3) of the above Declaration said that the-the States shall also take effective measures to ensure, as needed, that programmes for monitoring, maintaining and restoring the health of indigenous peoples, as developed and implemented by the peoples affected by such material, are duly implemented.

Enactment Related to the Conservation of Biodiversity in India

India being signatory to the Convention on Biological Diversity found it necessary to give immediate effect to the above said Convention in 1994 and after that with the help Article 253of the Constitution of India for the Conservation of Biological Diversity. In the light of above Convention, Government of India being the signatory State was introduced Biological Diversity Bill in the Parliament. Firstly the Bill was introduced in Lower-House of the Parliament on May 15, 2000. Report of the Standing Committee on Ministry of Science and Technology and Environment and Forests, Government of India was laid the above Bill on the table of Lower-House of Parliament on December 04, 2001. It was passed by the Lower-House of the Parliament on December 02, 2002 and by Upper-House of Parliament on December 11, 2002.President of India was given the assent on the above Bill on February 05, 2003 under Article 111 of Constitution of India. Biological Diversity Act shall came into force on October 01, 2003by notification in the Official Gazette of India Article 253, The Constitution of India, 1950a (Bare Act).

Definition of Biological Diversity and Biological Resources

Biological Diversity Act, 2002 have 12 chapters, in which 65 sections and number of subsection. For the purpose proper implementation of the above Act, rules were also be notify by the Ministry of Environment and Forests, Government of India. In exercise of the powers conferred by section 62 of the above said Act, series of rules has been called as Biological Diversity Rules, 2004. Section 2(b) of The Biological Diversity Act, 2002 describe the term 'Biological Diversity or Biodiversity', which means the variability among living organisms from all sources and the ecological complexes of which they are part and includes diversity within species or between species and of eco-systems and Section 2(c) of The Biological Diversity Act, 2002 describe the term 'Biological Resources', which include plants, animals, micro-organisms or parts of their genetic material and derivatives (excluding value added products), with actual or potential use or value for humanity, but does not include human genetic material.

Constitution, Composition and Functions of Bodies Under Biological Diversity Act, 2002

Three tiered management structure of bodies at National, State and local level has discussed in Biological Diversity Act, 2002. Constitution, Composition and Functions of Bodies has elaborated with number of provisions of above said Act, in which setting up National Biodiversity Authority, State Biodiversity Board & Biodiversity Management Committee.

Section 2(j) of The Biological Diversity Act, 2002 provides the definition of 'National Biodiversity Authority', which means the National Biodiversity Authority established under section 8 of the above said Act. Section 2(n) of The Biological Diversity Act, 2002 provides the definition of 'State Biodiversity Board, which means the State Biodiversity Board established under Section 22 of the Act and rule 2(c) of the Biological Diversity Rules, 2004 define the term 'Biodiversity Management Committee', which means a Biodiversity Management Committee established by a local body under sub-section(1) of Section 41 of The Biological Diversity Act, 2002.

Constitution and Composition of National Biodiversity Authority

Section 8 of the Act provides the rule for establishment of National Biodiversity Authority. According to section 8(1) of the Act Union Government of India may, by notification in the Official Gazette, appoint a body, which is called the National Biodiversity Authority. The National Biodiversity Authority shall be a corporate body, having perpetual succession and a common seal, with power to acquire, hold and dispose of property with both movable and immovable. Authority has power to make a contract and shall by the same name sue and be sued. Section 8(3) of the Act provide the rule for establishment head office of the National Biodiversity Authority shall be at Chennai and with the previous approval of the Union Government establish offices at other places in the territory of India.

According to the Section 8(4) the National Biodiversity Authority shall consist of 25 members, which of the following members are as follows-

i. A Chairperson to be nominated by Union Government, who have adequate knowledge, expertise and experience in the conservation and sustainable use of biological diversity, to be appointed by the Union Government.
ii. 16 members to be appointed by the Government of India, representing from the various Ministries i.e., Agricultural Research and Education, Agriculture and Farmers Welfare, Ayurveda, Unani, Siddha, Sowa Rigpa, Yoga and Naturopathy and Homoeopathy, Biotechnology, Environment and Climate Change, Forests and Wildlife, Indian Council of Forestry Research and Education, Earth Sciences, Panchayati Raj, Science and Technology, Scientific and Industrial Research, Tribal Affairs.
iii. Members are appointed as representatives on rotational basis from different State Biodiversity Boards.
iv. 05 non-official members to be appointed from amongst experts including legal experts and scientists having special knowledge of, or experience in, matters relating to conservation of biological diversity, sustainable use of biological resources and [fair and equitable] sharing of benefits arising out of the use of biological resources, representatives of industry, conservers, creators and knowledge-holders of biological resources.

Functions and Powers of National Biodiversity Authority

According to the section 18 of Biological Diversity Act, 2002 the National Biodiversity Authority shall with the prior approval of the Government of India; make a regulation/policy to provide for access to biological resources and traditional knowledge associated thereto and also determines the just, fair and equitable sharing of benefits for all. It is the duty of the Authority to regulate any activity referred to in Sections 3, 4 and 6 of aforementioned Act by granting or rejecting approvals. Rule 12 of the Biological Diversity Rules, 2004 has discussed the common functions of the Authority. The National Biodiversity Authority may perform their action, which are as follows:

i. Authority will advise to the Government of India on the matters relating to the conservation of biodiversity, sustainable use of its components and just and equitable sharing of benefits arising out of the use of biological resources for all.
ii. Authority will advise to the State Governments in the selection of areas of biodiversity importance to be notified under sub-section (1) of section 37 of said Act as biodiversity heritage sites and measures for the management of such heritage sites.
iii. Authority will advise to the State Biodiversity Boards on any matter relating to the implementation of the above aforementioned Act,
iv. Authority will also perform such other functions as may be necessary to carry out the provisions of the said Act.

The National Biodiversity Authority may on behalf of the Government of India, take any measures necessary to oppose the grant of intellectual property rights in any country outside India on any biological resource which is found in or brought from India, including those deposited in repositories outside India, or traditional knowledge associated thereto accessed (Biological Diversity Act, 2002).

Constitution and Composition of State Biodiversity Board

According to the Section 22, State Biodiversity Board shall be established State Government for the purpose of Biological Diversity Act. State Government may notify the constitution of State Biodiversity Board in official Gazette from such date. State Biodiversity Board shall not be constituted for a Union Territory so in this condition National Biodiversity Board shall exercise the powers and perform the functions as State Biodiversity Board for the Union Territory. State Biodiversity Board shall be corporate body having perpetual succession and a common seal. SBB has power to acquire, hold and dispose of both the movable and immovable property. Head office of the State Biodiversity Board shall be at any place, which is declared and specify via a notification by the Provincial Government in the Official Gazette. States Biodiversity Board will be make a contract and sue or sued by the above said name.

Composition of the State Biodiversity Board will be formed with13 members, which are as follows:

i. A Chairperson, who has adequate Knowledge, expertise and experience in the conservation and sustainable use of Biological Diversity
ii. Maximum 07 members of States Biodiversity Board will be appointed by the State Government from the concerned Departments i.e., Panchayati Raj and Tribal Affairs.

iii. Maximum 05 non-official members of the Board will be appointed from the field of Law and science having special Knowledge, expertise and experience in the conservation and sustainable use of Biological Diversity.

Power and Functions of State Biodiversity Board

According to section 23 of the Biological Diversity Act, functions of the State Biodiversity Board are as follows:

i. Board will be advise to the Provincial Government on matters relating to the conservation of biodiversity, sustainable use of its components and fair and equitable sharing of benefits arising out of the utilisation of biological resources or traditional knowledge associated thereto, in conformity with the regulations or guidelines if any, issued by the Central Government or the National Biodiversity Authority;
ii. Board will regulate any activity, which is referred in section 7 of the Act by granting or rejecting approvals;
iii. Board will also perform such other functions as may be necessary to carry out in the light of the provisions of said Act or as may be prescribed by the Provincial Government.
iv. Board will also be determine the just and equitable allocation of benefits provided under the regulations made in this behalf by the National Biodiversity Authority while giving the approvals.

Constitution and Composition of Biodiversity Management Committee

According to the Section 41 of Biological Diversity Act, 2002 of every local self government unit at the Village Panchayat level and Nagar Panchayat or Municipal Committee/Corporation level shall constitute a Biodiversity Management Committee

within its locality for the purpose of promoting conservation, sustainable utilization of biological diversity including preservation of habitats, conservation of landraces, folk varieties, farmers' varieties, and cultivars, domesticated stocks and breeds of animals, living things in water bodies and microorganisms and chronicling of knowledge relating to Biological Diversity. Provincial Government may also constitute BMC at the intermediate or District Panchayat level for achieving the objectives of the Act. The Biodiversity Management Committee shall be constituted in prescribed composition by the Provincial Government and above said Committee have not less than seven and not exceeding to the eleven members in any condition. Rule 22 of the Biological

Diversity Rules, 2004 has also specified the Constitution of Biodiversity Management Committees. The BMC shall consist of a Chairperson and not more than six persons nominated by the local self government unit, of whom not less than one third should be women and not less than 18% should belong to the community of SC/ST.

Power and Functions of Biodiversity Management Committee

The most important function of the BMC is to prepare Biodiversity register in consultation with local people. The Register shall contain comprehensive information on availability and knowledge of local biological resources, their medicinal or any other use or any other traditional knowledge associated with that person, who name has registered and other functions of the BMC are to advice on any matter referred to it by the SBB or NBA for granting approval, to maintain data about the local aids and practitioners using the biological resources. The NBA shall take steps to specify the form of the People's Biodiversity Registers, and the particulars it shall contain and the format for electronic database.

The National Biodiversity Authority and the State Biodiversity Boards shall provide guidance and technical support to the Biodiversity Management Committees for preparing People's Biodiversity registers of those persons, who are contributing in Biological Diversity. Registers shall be maintained and validated by the BMC. The BMC shall also maintain a register giving information about the details of the access to biological resources and traditional knowledge granted, details of the collection fee imposed and details of the benefits derived and the mode of their sharing. According to the section 41(3) of the above said Act, BMC may charges fees to any person, who are accessing or collecting any biological resource for commercial purposes within the territorial jurisdiction of particular local self government unit.

Community Forest Resource Rights Under Forest Dwellers Act, 2006

Community Forest Resource Rights (CFRR) refer to the legal recognition and protection of traditional forest rights of local communities including indigenous and tribal communities, over forests and forest resources in their customary rules by the provision of Scheduled Tribes and other Traditional Forest Dwellers (Recognition of Forest Rights) Act, 2006. Under Section 2(a) of Forest Dwellers (Recognition of Forest Rights) Act, 2006, 'community forest resource' means customary common forest land within the traditional or customary boundaries of the village or seasonal use of landscape in the case of pastoral communities, including reserved forests, protected forests and protected areas such as Sanctuaries and National Parks to which the community had traditional access over the long year.

Community Forest Resource (CFR) is a recognized right under Sections 3, 5 of above said Act, which aims to empower local communities with rights over forest land and resources and promote their participation in the conservation and sustainable management of forests. Section 3 of the Scheduled Tribes and other Traditional Forest Dwellers (Recognition of Forest Rights) Act, 2006 recognizes and vests the forest and occupational rights in forest land in Forest Dwelling Scheduled Tribes and Other Traditional Forest Dwellers who have been residing in such forests area for generations. Under section 5 of the above said Act, the holders of any forest right, Gram Panchayat or Village Panchayat level institutions in areas have certain responsibility assigned, which will be performed by these rights holder as follows-

(a) To protect the wild life, forest and Biological Diversity.
(b) To ensure that adjoining catchments area, water sources and other ecological sensitive areas are adequately protected.
(c) To ensure that the habitat of forest dwelling Scheduled Tribes and other traditional forest dwellers is preserved from any form of destructive practices affecting their cultural and natural heritage.
(d) to ensure that the decisions taken in the Gram-Sabha to regulate access to community forest resources and stop any activity which adversely affects the wild animals, forest and the biodiversity are complied with.

So, the Gram Panchayat/Sabha or Village Level institution is the authority to initiate the process for determining the nature and extent of individual forest rights or community forest rights or both that may be given to FDST and OTFD. Under the Community Forest Resource Rights Forest Dwelling Scheduled Tribes and Other Traditional Forest Dwellers have certain benefits, which are as follows-

i. Source of income and livelihood of forest dwellers.
ii. Sustainable management of forests
iii. Protection of Traditional knowledge and Cultural Rights of Indigenous Person.
iv. Empowerment and participation of Indigenous Person.
v. Promote social justice and equitable rights amongst the member of Indigenous Communities.
vi. Conservation of Biological Diversity and climate change mitigation (The Community Forest Resource Rights under Forest Dwellers Act., 2006).

Conclusion

Major issues faced by the tribal community at present time are about the lack of knowledge and awareness about their own rights, which is conferred by the number of enactments and Constitutional of India. Members of tribal community have been worshiping of nature considering as God from time immemorial and this was one of the main reason they are protecting nature or environment. This is the need of the hour properly implement the objectives of Biological Diversity Act, 2002 and other related laws to specify their rights. In this alarming condition conservation or preservation of Biological Diversity rich prosperity is very important for the earth. Most of these plant and animal varieties have been reduced to a greater extent and even some endemic and endangered species disappeared from the Earth. Over the last two decades from the enactment of Biological Diversity Act, 2022 there has been a wind of change with greater recognition and commitments to reconciliation and respect for rights holders of indigenous person in India.

References

Singh, G., & Bahri, A. (2024). Environmental law (eastern book of company, 3rd Edn.)

Shastri, S. C., Environmental law (eastern book of company, Lucknow 7th Edn, 2022 reprint 2024). Available at https://tribal.nic.in/ST/Statistics8518.pdf. (last visited on June 10, 2024).

The Constitutional of India. (1950).

The Biological Diversity Act. (2002).

The Community Forest Resource Rights under Forest Dwellers Act. (2006).

Available at https://www.un.org/en/observances/biological-diversity-day/convention (last visited on June 15, 2024).

The Constitution of India, 1950a art.253. Legislation for giving effect to international agreements. Notwithstanding anything in the foregoing provisions of this Chapter, Parliament has power to make any law for the whole or any part of the territory of India for implementing any treaty, agreement or convention with any other country or countries or any decision made at any international conference, association or other body.

Chapter 5
Harnessing Indigenous Water Resources: Integrating Traditional Water Harvesting Systems With Modern Practices in India

Sudarshan Verma and Khushnuma Bano

Abstract This chapter explores the essential nature and multifaceted roles of water, emphasizing its importance as the cornerstone of life. It begins by detailing the hydrological cycle—evaporation, condensation, precipitation, and runoff—highlighting its crucial role in maintaining the balance and replenishment of freshwater sources. The global distribution of water is examined, revealing that only a small percentage is accessible for human use, with uneven geographical distribution exacerbating water scarcity in many regions. Water's role in agriculture is underscored as vital for crop production, with an emphasis on the evolution of irrigation practices. The chapter advocates for adopting efficient water use techniques to sustain agricultural yields and reduce wastage. The industrial use of water is also explored, showing its importance in various processes like cooling and manufacturing, and the need for industries to adopt sustainable water management practices. Water scarcity is presented as a growing global concern, driven by population growth, climate change, and pollution. The chapter outlines strategies to address this issue, including infrastructure development, sustainable water use policies, and water conservation practices. It also emphasizes the critical role of water in maintaining healthy ecosystems, calling for integrated water resources management (IWRM) to balance human needs with ecosystem preservation. Additionally, the cultural and social significance of water is discussed, recognizing its spiritual and cultural value in communities. The chapter concludes by advocating for sustainable practices and policies to ensure water availability and quality for future generations, emphasizing the importance of water stewardship for long-term sustainability.

S. Verma
Department of law, BBAU, Lucknow, India
e-mail: drsudarshanvermabbau@gmail.com

K. Bano (✉)
Department of Law, SLS, Babasaheb Bhimrao Ambedkar (A Central) University, Lucknow, India
e-mail: khushnuma496@gmail.com

R. C. Sobti et al. (eds.), *Traditional Knowledge Systems in India: Sustainability, Ecology, and Resource Governance*, People, Cultures and Societies: Exploring and Documenting Diversities,
https://doi.org/10.1007/978-981-92-1346-7_5

Keywords Water scarcity · Sustainable water management · Hydrological cycle · Ecosystem conservation

Introduction

Water is fundamental to nature, underpinning the survival and functioning of all living organisms through essential biological processes such as nutrient transportation, temperature regulation, and photosynthesis, which produces oxygen (Kumar et al., 2005). It provides habitats for diverse ecosystems, supports complex food webs, and aids soil formation, fertility, and nutrient cycling. Water shapes landscapes through erosion and sediment transport and regulates the Earth's climate via the water cycle (Shah et al., 2003). Additionally, it purifies ecosystems naturally, supports terrestrial plant life, and holds significant cultural and aesthetic value. Despite covering approximately 71% of the Earth's surface, the vast majority (about 97.5%) is saltwater, with freshwater accounting for only 2.5%. Of this freshwater, the majority (about 68.7%) is locked in glaciers and ice caps, while accessible freshwater sources such as groundwater and surface water (rivers, lakes, and swamps) are limited, constituting less than 0.01% of all water on Earth and only about 0.3% of the Earth's freshwater (Mudrakartha, 2007).

Different religions emphasize the importance of water conservation and stewardship as part of their teachings and practices. Christians are encouraged to be good stewards of the Earth's resources, including water, to ensure responsible and sustainable usage (Bhatt, 2014). In Islam, Muslims are taught to use water sparingly and efficiently, particularly during rituals like ablution (wudu), reflecting the value of conservation. Judaism promotes the principle of baltashchit, which prohibits wastefulness and encourages responsible water usage to protect the environment (Dehejia, The Body Adorned: Sacred and Profane in Indian Art, 2009; Kavuri-Bauer, 2011). Hindus view water as sacred and promote conservation through various rituals and practices. Buddhism teaches mindfulness and non-wastefulness, encouraging adherents to conserve water to promote environmental well-being (Terukazu, Vasanti, & Architects, 2002). Sikhs emphasize the principle of seva (selfless service) and responsible stewardship of resources, including water, promoting conservation and sharing water with others. Indigenous religions worldwide view water as sacred and emphasize its conservation through prayers, rituals, and community practices, reflecting a reverence for the environment and ensuring sustainable use for future generations (Singh, 2004; Bhatt, 2014).

Water holds profound importance in various religions worldwide, symbolizing purification, life, and spiritual significance. In Christianity, water is central to rituals such as baptism, signifying spiritual cleansing and initiation into the faith. Islam considers water essential for purification (wudu) before prayer and as part of the Hajj pilgrimage rituals. In Judaism, water is used in purification baths (mikvah), symbolizing freedom and purity during ceremonies like Passover (Amirthalingam, 2015). Hinduism reveres water in sacred purification rituals, daily prayers (puja), and during religious ceremonies (Chandra, 2015). Buddhism sees water as symbolizing clarity and calmness and uses it in meditation practices and rituals (Dehejia, 2009a). In Sikhism, water from the Sarovar at the Golden Temple is considered holy and purifying, emphasizing equality and selflessness. Indigenous religions globally view water as a life-giving force, with rituals involving offerings to water spirits and ceremonies promoting harmony with nature(Robbins & Moritz, 1994). Across these beliefs, water sustains life and embodies spiritual cleansing, rebirth, and the interconnectedness of all living beings with the natural world (Dehejia, 2009b).

Water is essential for the survival and well-being of all living beings on Earth. For humans, water is necessary for hydration and maintaining bodily functions, regulating temperature, and supporting overall health. It serves as a vital medium for transporting nutrients and oxygen throughout the body and aids in the elimination of wastes (UNESCO, 2009). Access to clean water is crucial for preventing waterborne diseases and promoting sanitation and hygiene. Similarly, animals rely on water for hydration, thermoregulation, and bodily functions. It is essential for their survival in natural habitats and ecosystems, with many species depending on water bodies for habitat, food sources, and breeding grounds, thereby making water availability crucial for biodiversity and ecological balance. Plants also require water for photosynthesis, which is the process by which they produce energy and oxygen. Water facilitates the uptake of nutrients from the soil and helps maintain turgidity, essential for plant structure and growth (Agarwal and Narain, 1997). Adequate water supply is crucial for agriculture and food production, supporting human and animal life. Hence, water is indispensable for all living beings, serving as a foundation for life processes, ecosystem health, and biodiversity. It plays a vital role in sustaining life on Earth and maintaining the balance of natural systems, underscoring the importance of responsible water management and global conservation efforts (Borthakur, 2008).

Review of Literature

Indigenous practices for water harvesting refer to the traditional methods and techniques developed by local communities to collect, store, and manage water resources. These practices are deeply rooted in a region's cultural, environmental, and historical context. They are typically sustainable, environmentally friendly, and adapted to the area's specific climatic and geographical conditions.

1. **Adaptation to Local Conditions**: Indigenous methods are tailored to the region's specific climate, topography, and hydrology. These practices take advantage of local materials and resources, making them cost-effective and sustainable.
2. **Community Involvement**: They often involve collective effort and community participation, fostering a sense of ownership and responsibility. Knowledge about these practices is usually passed down through generations, maintaining cultural heritage.

3. **Sustainability**: These techniques emphasize conservation and efficient water use, aiming to balance human needs with environmental preservation. They often contribute to groundwater recharge and maintain the ecological balance.
4. **Versatility**: Indigenous practices can be adapted to various scales, from small household systems to larger community projects. They are often flexible, allowing modifications based on changing environmental conditions or community needs (Gupta, Nair, and Singh, Traditional Water Management Systems for Drought Mitigation in India, 2014).

Benefits of Water Harvesting

1. **Environmental Benefits:** Water harvesting provides significant environmental benefits, primarily by replenishing aquifers, which ensures sustainable groundwater levels and reduces the risk of over-extraction, thereby mitigating groundwater depletion and preventing land subsidence. It also slows down the flow of runoff, effectively preventing soil erosion and nutrient loss, which helps maintain soil structure and fertility, which is essential for agriculture. Moreover, water harvesting enhances local humidity and cools the surrounding areas, creating a more favourable microclimate for plants and animals, thus supporting biodiversity by providing water for wildlife and vegetation. Additionally, it reduces the volume and speed of runoff during heavy rains, mitigating the risk of floods and protecting downstream areas from the destructive impacts of flooding (Sameer et al., 2019)
2. **Economic Benefits:** Water harvesting significantly lowers the need for expensive, centralized water supply systems and infrastructure, reducing water bills and water pumping and distribution costs. It ensures a reliable water supply for irrigation, which leads to improved crop yields and enhanced agricultural sustainability, allowing for the cultivation of a wider variety of crops, even during dry seasons. Additionally, it provides essential water for livestock, fisheries, and other rural industries, thereby supporting local economies. Implementing and maintaining harvesting systems also creates employment opportunities in their construction, upkeep, and management, further contributing to economic development in rural areas (World Economic Forum, 2016).
3. **Social Benefits:** Water harvesting involves local communities in water management, fostering cooperation, ownership, and stewardship of water resources, strengthening social bonds through collective action and shared benefits. It ensures access to clean water, thereby reducing the incidence of waterborne diseases and promoting better hygiene and sanitation practices, which improve overall community health. Additionally, water harvesting enhances the ability of communities to adapt to changing climate conditions, such as irregular rainfall and prolonged droughts, by providing a buffer against climate variability and ensuring water availability during periods of scarcity (Phuntsog et al., 2017).

4. **Ecological Benefits:** It maintains wetlands, ponds, and lakes, which are crucial habitats for aquatic flora and fauna, supporting the breeding and feeding grounds of various species and contributing to ecological balance. It promotes the sustainable use of land resources by integrating water conservation into land management practices, thereby enhancing the productivity of marginal and degraded lands through improved water availability (Prasad et al., 2004).
5. **Long-term Sustainability:** Water harvesting reduces the volume of polluted runoff entering natural water bodies, thereby improving water quality and minimizing the need for chemical treatment, which promotes overall environmental health. It encourages the efficient use and conservation of water resources, ensuring their availability for future generations. Additionally, water harvesting supports sustainable development by balancing environmental, economic, and social needs, fostering a holistic approach to resource management (Nirmale & Metar, 2003).
6. **Educational and Awareness Benefits:** Water harvesting promotes the exchange of traditional and modern water management knowledge, fostering innovation and adaptation while raising awareness about the importance of water conservation and sustainable practices. It builds local capacity and skills in water management, ensuring the long-term success and maintenance of water harvesting systems, and empowers individuals and communities to take proactive steps in managing their water resources. By ensuring a reliable water supply for drinking, irrigation, and other needs, especially in arid and semi-arid regions, water harvesting reduces soil erosion, promotes groundwater recharge, and helps maintain local biodiversity. It keeps traditional knowledge alive, fostering a sense of identity and continuity within communities, and is often more affordable than modern water infrastructure projects, as it can be maintained using local skills and resources (Chaitanya, 2018).

Water Resources in India: Overview and Utilisation

1. **Total Renewable Water Resources**: India's total renewable water resources are estimated at around 1911 billion cubic meters (BCM) per year. This includes the average annual precipitation of about 4000 BCM, of which about 1869 BCM is the total annual water availability in the country from all sources (Kabir & Faisal, 1999).
2. **Surface Water Resources:** India's surface water resources are estimated at 1869 BCM per year, which includes rivers, lakes, and reservoirs. Major rivers like the Ganges, Brahmaputra, Yamuna, Godavari, Krishna, and others contribute significantly to the surface water resources (Deb & Shah, 2002).
3. **Groundwater Resources:** India is the largest user of groundwater in the world, with an estimated annual groundwater availability of about 447 BCM. Groundwater is a critical resource for irrigation, drinking water, and industrial purposes, particularly in areas with limited surface water availability(Constantz et al., 2003).

4. **Water Utilization:** Total water utilization in India is estimated at around 761 BCM per year, which includes irrigation, domestic, industrial, and other uses. About 83% of the total water resources are used for agricultural purposes, highlighting the importance of water in India's agriculture-dominated economy (Machiwal et al., 2004).

India's water resources are diverse and abundant, yet face challenges such as depletion due to excessive use, pollution, and climate variability. Sustainable management and conservation efforts are crucial to ensure the availability of water resources for future generations and to support the country's socio-economic development.

Regions In India Facing Water Scarcity

In India, water scarcity is a significant issue affecting various regions due to several factors, including uneven distribution of rainfall, overexploitation of groundwater, pollution, and inadequate water management practices. Some of the areas in India where water scarcity is commonly found include:

1. **Northwestern India (Rajasthan, Gujarat, Haryana, Punjab):**
   - This region faces severe water scarcity, particularly in the arid and semi-arid zones.
   - The Thar Desert in Rajasthan and parts of Gujarat experience low rainfall and have limited surface water resources.
   - Groundwater depletion is a significant concern in Punjab and Haryana due to intensive agriculture and excessive water use for irrigation (Cochran & Ray, 2009).
2. **Maharashtra (Marathwada and Vidarbha regions):**
   - Marathwada and Vidarbha regions of Maharashtra often face acute water scarcity, especially during dry seasons and drought.
   - Groundwater levels have been declining rapidly in these regions due to excessive pumping for agriculture and drinking water (Gore et al., 1998).
3. **Southern India (Tamil Nadu, Karnataka, Andhra Pradesh, Telangana):**
   - Many parts of Southern India face water scarcity issues, particularly during summer.
   - Cities like Chennai in Tamil Nadu and Bengaluru in Karnataka have experienced severe water shortages due to inadequate rainfall, groundwater depletion, and mismanagement of water resources (Subramani et al., 2005)
4. **Eastern India (Jharkhand, Odisha, parts of Bihar and West Bengal)**:
   - Some parts of Eastern India, particularly Jharkhand and Odisha, face water scarcity issues, especially in rural and tribal areas.
   - Limited access to safe drinking water is a challenge in these regions, with many communities depending on contaminated water sources (Selvaraj et al., 2022).

Indigenous Practices for Water Harvesting in India

Ahar Pynes (Bihar): The Ahar and Pyne system in South Bihar, originating during the Magadh dynasty nearly 5000 years ago, remains a crucial traditional irrigation system. Ahars are earthen reservoirs that store harvested river water, while Pynes are channels that distribute this water to agricultural fields. This sophisticated system supports paddy and wheat cultivation and helps sustain the water table in the region. Neglected during the British era and with modern irrigation, these systems are now being revived by the Bihar government to ensure sustainable water management. Efforts include understanding local hydrology, developing community-controlled water distribution mechanisms, and addressing river flow and siltation issues. A federation of management committees is proposed for equitable water distribution and community participation in restoration efforts (Dhiman & Gupta, 2011).

- **Efficient Water Management:** The Ahar and Pyne system ensures efficient water management by collecting and storing water in Ahars during the monsoon season. Pynes then distributes this water to agricultural fields throughout the year. This system allows for controlled water distribution, ensuring a reliable supply for irrigation and supporting groundwater recharge to maintain the water table during dry periods (Divy et al., 2012).
- **Support for Agriculture:** The system supports agriculture by enabling cultivation on sloping and undulating terrain with clayey or sandy soil, which typically has low water retention capacity. It is particularly suitable for growing paddy during the monsoon (kharif) and wheat or gram during the winter season (rabi), crucial for food security in the region (Gupta et al., 2014).
- **Community-Managed Approach:** Managed by local communities, the Ahar and Pyne system fosters a sense of ownership and responsibility among farmers. It ensures equitable distribution of water resources and promotes local participation in water management, enhancing sustainability and resilience.
- **Environmental Sustainability:** The system promotes environmental sustainability by utilizing natural water resources efficiently and reducing reliance on groundwater extraction. It helps maintain ecological balance by supporting groundwater recharge and biodiversity in the region.
- **Cultural Significance:** Culturally, the Ahar and Pyne system holds immense significance as it reflects the region's traditional knowledge and practices. It contributes to the cultural identity of local communities and serves as a symbol of their resilience and adaptation to local environmental conditions (Murthy et al., 2022).
- **Revival Efforts:** Efforts are underway to revive and restore the Ahar and Pyne system in Bihar, recognizing its importance for sustainable water management and agricultural production. Restoration efforts include technical improvements, community involvement, and modern management practices to ensure long-term viability (Kumar et al., 2010).

The Ahar and Pyne system provides a sustainable, community-based approach to Bihar's water management and agricultural production. Its revival is essential for ensuring water security, supporting rural livelihoods, and preserving cultural heritage for future generations.

Zabo (Nagaland)

Zabo, a traditional water conservation and integrated farming system practised by the Kikruma region of the Phek district in Nagaland, is an exemplary model of human-environment interaction and sustainable resource management. Originating from the Chakhesang tribe's ingenious practices, Zabo integrates various agricultural activities like forestry, horticulture, agriculture, fishery, and animal husbandry in a scientifically structured manner (Singh et al., 2012). The system, known as Dzudu or Ruza in other parts of the region, is primarily designed to combat water scarcity despite receiving sufficient rainfall, often resulting in surface runoff.

Zabo's innovative approach involves the construction of water harvesting ponds strategically placed in catchment areas on protected forest lands with steep slopes. These ponds are meticulously designed to minimize water loss by properly ramming the pond bottoms and using silt retention tanks to filter runoff water before it enters the main ponds. This ensures efficient water management and conservation, which are crucial for maintaining soil fertility and sustainable agriculture (Pulamte, 2008).

The community-managed Zabo system operates on principles of equity and shared responsibility. Water distribution among families and clans is negotiated mutually, ensuring fair access and efficient harvest water utilization. Moreover, the system includes livestock enclosures strategically placed to facilitate nutrient cycling from animal waste into the agricultural fields, enhancing soil fertility naturally (Singh et al., 2012).

The farming practices in Zabo are carefully planned and diversified. Rice fields are in lower terraces with embankments compacted to reduce water percolation and seepage. Varieties like "Tanyekemüga," a local long-duration paddy, are grown with supplementary irrigation from the Zabo ponds during dry spells. The system also supports paddy cum fish culture where fish are raised in the rice terraces, providing an additional source of protein and income for the community (Sharma & Sharma, The "Zabo" soil and water management and conservation system in northeast India: Tribal beliefs in the development of water resources and their impact on society—An historical account of a success story International Association of Hydrological Sciences, 2000).

Beyond its agricultural benefits, Zabo holds deep cultural significance for the Chakhesang tribe. Festivals like Funye (Fish Festival) celebrate the system's bounty, with offerings of fish and the first grain of the harvest as a mark of gratitude to nature. These celebrations highlight the integral role of agriculture and sustainable resource management in the cultural fabric of the Chakhesang people, preserving their traditions and enhancing community resilience (Sharma et al., 1994).

Zabo exemplifies an innovative and sustainable land and water management approach, integrating traditional wisdom with modern scientific principles. As efforts continue to revive and promote this ancient system, it serves as a model for sustainable agriculture and environmental stewardship in Nagaland and globally.

Bamboo Drip Irrigation (Meghalaya)

In Meghalaya, traditional farmers grapple with the challenges of undulating terrain, steep slopes, and poor soil quality prone to erosion, despite the region being known for its record-breaking rainfall. This unique landscape necessitates an innovative approach to irrigation. Enter the bamboo drip irrigation system, developed ingeniously by farmers in the Jaintia and Khasi hills (Swami, 2021).

This system capitalizes on the steep slopes of the hills and operates primarily on gravity. Water is captured from uphill sources and conveyed through a main bamboo channel that spans hundreds of meters. From there, the water is meticulously distributed through a network of secondary and tertiary bamboo channels, ensuring every part of the plantation receives water, even the lowermost areas of the hill.

Constructed using bamboo of varying diameters, the system includes channels, support structures, diversion pipes, and strips. Channels are elevated above the ground with bamboo or wooden Y-shaped sticks, and bamboo pieces are specially prepared to facilitate water flow by removing a third of their outer casing and internodes. Traditional tools such as the dao and chisels are used to construct the intricate bamboo network.

Costs associated with the system are minimal as bamboo is abundant in the region. Typically, farmers set up the system with the help of one or two labourers. Each installation lasts for about 2–3 years, after which old bamboo decomposes, enriching the soil as humus. Cooperative efforts ensure equitable distribution of water and foster harmonious community interactions (CSE, 1998)

Construction of a network covering 1 ha takes approximately 15 days and is designed with precision to minimize water wastage due to leakage. The ratio of primary to tertiary channel diameters determines the water quantity reaching the trees, with about 18–20 l per minute entering the bamboo pipe system and eventually reducing to 20–80 drops per minute at the plant site.

The main bamboo channel extends from the water source, with lateral bamboos distributing water just above crop plants. Holes in the mainline convey water to the laterals, with small holes above each plant for efficient water delivery. Woollen threads can be tied to the holes in the laterals to manage drip speed or irrigate specific crop areas, further optimizing water use (Jeeva et al., 2006).

Proven effective for cash crops like potato, capsicum, tomato, and strawberry, this bamboo drip irrigation system, combined with straw mulching, achieves 25–30% water savings. It is an achievable solution to manage water resources in Meghalaya's challenging agricultural landscape effect, ensuring crop viability while promoting environmental stewardship (Agossou, 2018).

Johads (Rajasthan)

In Rajasthan's Alwar district, johads represent a traditional water harvesting system that has transformed rural communities under the leadership of Rajendra Singh, known as the Water Man of Rajasthan. Johads are small earthen check dams constructed by local people using indigenous knowledge and skills. They are built across hill slopes to capture rainwater and arrest monsoon runoff, which would otherwise flow away. Typically semi-circular in shape, johads are enclosed on three sides by natural slopes, while the fourth side is a mud wall that holds back the water (Frontline, 2001).

The construction of johads is cost-effective and straightforward, using locally available materials like stone, sand, and lime. Villagers share labour and materials for 70–90% of the total costs. This community-driven approach ensures that johads are affordable and sustainable. They vary in size, with water storage areas ranging from 2 ha to 100 ha. The water collected during the monsoon season serves multiple purposes, including irrigation, drinking, livestock, and domestic use. Additionally, it recharges the groundwater and improves soil moisture, making land inside the johads available for cultivation during the dry season (Das, 2010).

The impact of Johadsin the Alwar district has been profound. Around 8600 johads have been built across 1086 villages, covering an area of 6500 km^2. As a result, the groundwater table has risen significantly, from depths of 100–120 m to 3–13 m. This has enabled farmers to increase their crop production substantially, with the area under single and double cropping rising from 11% to 70% and 3% to 50%, respectively. The forest cover has also increased from 7% to 40% through agroforestry and social forestry initiatives, providing fuel wood and sequestering carbon (Kavarana, 2006).

Economically, Johads have been a wise investment. Studies have shown that for every Rs. 100 invested in johads, there is an annual profit of Rs. 400 in increased crop production. Fisheries have also developed as a new source of livelihood. Socially, Johads have empowered women by involving them in decision-making processes. One of the most remarkable achievements has been the rejuvenation of the Arvari and Ruparel rivers through rainwater harvesting and groundwater recharge (Khan & Narain, 2000).

Kunds (Thar Desert, Rajasthan)

Kunds, known as kundis, are traditional rainwater harvesting structures in the Thar Desert in western Rajasthan and specific areas in Gujarat. Resembling an upturned cup nestled in a saucer, these structures provide water for drinking and domestic purposes in arid regions (Dhiman & Gupta, 2011).

Essentially, a kund is a circular underground well with a saucer-shaped catchment area, known as Agor, which slopes gently towards the centre where the well is located. This design allows rainwater to collect and flow into the well. To prevent debris from entering the well-pit, a wire mesh is placed across the water inlets. The sides of the well-pit are typically coated with lime and ash for disinfection. Many kunds have a dome-shaped cover or lid to protect the water from contamination. Drawing it out with a bucket can be accessed as needed.

The size of kunds varies based on their intended use—some are smaller for personal drinking needs, while others are larger to serve multiple households or communities. Private kunds are usually owned by individuals who have the resources to invest in their construction and have land where they can be built. On the other hand, large public kunds have been constructed to serve the needs of poorer communities (Norphel, 2009).

These traditional structures are vital in regions with scanty rainfall, where surface water resources are limited or non-existent. They help capture and store rainwater efficiently, ensuring a reliable water supply for drinking and other domestic purposes throughout the year. This supports local communities' livelihoods and promotes sustainable water management in arid and semi-arid regions (Murthy et al., 2022).

The construction and maintenance of kunds are often community efforts, involving local skills and resources. They represent a sustainable approach to water management, utilizing indigenous knowledge and practices passed down through generations. In recent years, there has been a renewed interest in kunds to address water scarcity and enhance water security in desert regions (Central Groundwater Board (CGWB), 2011).

Kuhls (Himachal Pradesh)

Kuls are gravity-flow irrigation channels that transport glacial meltwater or river water to villages and agricultural fields in the hilly regions of Himachal Pradesh. They are crucial for sustainable agriculture, providing water for irrigation and drinking by harnessing water from higher elevations.

Traditionally, Kuls were part of a community water management system, constructed and maintained by local elites or village communities without state patronage. Each Kuhl requires annual maintenance, particularly after the rainy season when the headwall often gets damaged. The Kohli, or water master, organizes the community for maintenance tasks such as clearing debris and repairing the Kuhl. This role is highly respected and inheritable, with the Kohli managing water distribution and resolving conflicts.

Over time, participation in Kuhl maintenance has declined, mainly due to non-farm employment opportunities. As average land holdings decreased, many men sought work outside agriculture, reducing the available labour for maintaining Kuls. This issue is exacerbated in less fertile command areas and regions with significant caste inequalities. High-caste farmers, typically located at the head-end of Kuls, often withdraw from maintenance, placing the burden on lower-caste farmers at the tail-end, who also begin to withdraw due to increased non-farm employment.

Despite these challenges, Kuls have significant irrigation potential. In the Panchrukhi block of Palampur tehsil, 114 Kuls were identified, irrigating an estimated 7755 ha, or 83% of the total cropped area. Kuls support multiple villages, forming a network of interdependent channels. Irrigation through Kuls has notably increased crop productivity: wheat by 47.2%, barley by 96.9%, paddy by 35.3%, and maize by 30.5%, particularly benefiting the rabi season.

Kuls are essential for agricultural productivity and water management in Himachal Pradesh, but their sustainability depends on active community participation and the need to address socio-economic challenges.

Eries (Tamil Nadu):

Tamil Nadu has a rich tradition of water harvesting systems dating back to the Sangam period (150 BC to 200 AD), which have been essential for addressing water scarcity in the region. The primary components of these systems include tanks (Eries), embankments or bunds, channel networks, catchment areas, and foreshore vegetation. These elements are interconnected, often forming 'chain-connected systems'. The efficient functioning of Traditional Water Harvesting Systems relies on physical features, landscape elements, and social structures.

Before British colonization in 1600 AD, local communities maintained the tanks (Eries) with a strong sense of ownership. The introduction of centralized systems by the British, particularly through the establishment of the Public Works Department, disrupted these community-managed systems. As a result, traditional maintenance practices declined. Overpopulation, urbanization, and infrastructure development have recently pressured these systems. Additional factors contributing to the decline include land use changes, deforestation, heavy siltation, poor maintenance, industrial waste contamination, agricultural activities, and chemical fertilizers and pesticides.

Modern agricultural techniques and urban encroachment have significantly impacted the maintenance and functionality of Traditional Water Harvesting Systems, leading to the degradation and disappearance of many systems across Tamil Nadu. The remaining systems have been extensively modified, such as through the cement lining of channels, which has altered groundwater recharge patterns and hydrological conditions. Despite these challenges, efforts have been initiated in some parts of the country to revive and restore similar Traditional

Water Harvesting Systems. Additionally, there have been attempts to study and document these traditional water harvesting systems in detail, aiming to recognize their significance and potential for contemporary water management solutions.

Stepwells (Gujarat, Rajasthan, Delhi):

Stepwells have been integral to local water management for centuries, meeting needs without degrading the environment. Stepwells emphasize conservation and utilize simple, locally maintainable technology, unlike modern technologies. They have been crucial during droughts and famines, highlighting ancient sustainable practices. The preservation and study of stepwells can offer long-lasting solutions for contemporary water management issues exacerbated by climate change and resource over-exploitation.

These structures range from simple designs to intricate architecture, reflecting their times' cultural and social values. Stepwells were used for various purposes: drinking, irrigation, and social gatherings. The Meratani Ki Baori in Shekhawati had a unique water pumping mechanism, while other stepwells featured diverse architectural styles, such as the rock-cut Chand Baori and the temple-like Rani ki Vav.

Throughout history, stepwells have showcased remarkable architectural innovation, from the Adalaj stepwell's Indo-Islamic designs to the Gebanshah stepwell's Mughal architecture. These structures not only served practical purposes but also symbolized the artistic and cultural amalgamation of their eras.

Today, while many stepwells are defunct due to modern water systems, they remain significant for their ecological sensitivity and potential lessons in sustainable water management. Efforts to preserve and study these ancient structures can provide valuable insights into creating sustainable solutions for current and future water challenges.

Phad System (Maharashtra):

The Phadis a diversion-based irrigation system that channels river water to agricultural fields. It ensures equitable water distribution for agriculture, enhancing crop yields and supporting rural livelihoods. Practised in Maharashtra and other states where river diversion systems are employed to manage water resources. Phad irrigation involves constructing check dams on rivers or streams to raise water levels, which are then channelled for village-level irrigation. Originating in Ramtek, this ancient practice features an extensive network of channels and underground supplies, making it eco-friendly and community-maintained. Though it became redundant, it is being revived due to its environmental benefits and potential for sustainable water management.

Jhalaras (Rajasthan):

Jhalaras are artificial tanks designed to collect and store rainwater, often rectangular. They provide community drinking water and are also used for irrigation, particularly in arid regions. Found in Rajasthan, especially in cities like Jaipur, where they are integral to the traditional water supply system. Jhalaras, or stepwells, are human-made tanks primarily found in Rajasthan and Gujarat, designed for community use and religious rites. Typically rectangular with steps on three or four sides, these structures collect subterranean seepage from upstream lakes or talabs, ensuring a steady water supply to surrounding areas. Unlike other water sources, jhalaras were used exclusively for bathing and religious purposes, not drinking. Jodhpur alone has eight jhalaras, the oldest being the MahamandirJhalara, dating back to 1660 AD. These traditional water systems highlight the ingenuity of ancient water management practices, ensuring sustainability and community utility.

Katta (Kerala):

Katta are temporary check dams built across streams to capture and store rainwater. They provide water for irrigation and drinking during dry periods, contributing to agricultural sustainability. Found in Kerala, especially in the hilly regions of the Western Ghats, where they help manage water resources. Kattas are temporary check dams traditionally constructed across streams and rivers in the Kerala and Karnataka districts. Built using kacha stones and locally available wood, these structures are erected in November or December after the monsoon season has subsided. Each katta typically includes a diversion channel. Modern adaptations have introduced sandbag kattas, plastic or fibresheets, and semi-permanent structures using mild steel sheets and concrete bases. These innovations highlight the evolving nature of traditional water management practices to suit contemporary needs while maintaining ecological benefits.

Talabs (Central India, especially Madhya Pradesh):

Talabs are large, manufactured lakes or reservoirs. They serve multiple purposes, including providing drinking water, irrigation, and supporting fisheries. Commonly found in Madhya Pradesh and other central Indian states, where they play a crucial role in water management and agriculture. Talabs, large artificial lakes or reservoirs, are integral to water management in Central India, particularly in Madhya Pradesh. These reservoirs serve multiple purposes, including providing drinking water, irrigation, and supporting fisheries. Talabs are crucial in sustaining agriculture and ensuring water availability throughout the year. Their construction often involves community participation, and they are strategically located to harness rainwater and runoff, thus aiding in groundwater recharge. The presence of talabs in the region supports the agricultural economy and helps maintain the ecological balance by creating habitats for various aquatic species and birds. Furthermore, talabs have cultural and social significance, often sites for community gatherings and festivals. This multifunctional approach to water management highlights the ingenuity of traditional practices in addressing contemporary water challenges.

These traditional water harvesting systems demonstrate the ingenuity of indigenous communities in managing water resources sustainably, adapting to local environments, and ensuring water availability throughout the year. They also provide valuable lessons for modern water management, especially in the face of climate change and water scarcity.

Health Improvement: Access to clean water reduces waterborne diseases and improves overall health.

Economic Growth: Supports agriculture, industry, and tourism, contributing to economic development.

Environmental Conservation: Promotes biodiversity, conserves natural habitats, and protects ecosystem services.

Social Equity: Ensures equitable access to water resources, benefiting marginalized and vulnerable communities.

Climate Resilience: Mitigates the impacts of climate change by enhancing water availability and ecosystem health.

Fig. 5.1 Common benefits across models

Common Benefits Across Models:
See Fig. 5.1.

National Water Harvesting Initiatives

- Incentivizes the revival of traditional water harvesting structures by the states.
- Encourages rainwater harvesting in urban and industrial areas.
- Prefers urban and rural domestic water supply from surface water, combined with groundwater and rainwater.
- Sent to all States/UTs for action (Ministry of Jal Shakti, 2023).

Central Government Initiatives:
Jal Shakti Abhiyan (JSA):

- **JSA-I (2019):** Covered 1592 out of 2836 blocks in 256 water-stressed districts.
- **JSA: Catch the Rain (2021–2023):** Expanded to all blocks in all districts (rural and urban).

- **Water-related works (2022–2023):**

Water conservation and rainwater harvesting structures	12,28,553
Renovation of traditional water bodies	2,67,472
Reuse and recharge structures	8,74,680
Watershed development	16,28,726
Intensive afforestation	78,38,36,035

Source (Ministry of Jal Shakti, 2023)

- **Atal Bhujal Yojana:** Central Sector Scheme with Rs. 6000 crore outlay. Covers 8220 Gram Panchayats in 229 blocks across 80 districts in 7 states.

  Funds utilized:

2020-2021	Rs. 123.03 crore
2021–2022	Rs. 327.48 crore
2022–2023	Rs. 1155.37 crore

Source (Ministry of Jal Shakti, 2023)

- **Pradhan Mantri Krishi Sinchayee Yojana (WDC-PMKSY):** Includes rainwater harvesting under its Natural Resource Management component (Ministry of Jal Shakti, 2023).
- **Surface Minor Irrigation (SMI) and Repair, Renovation & Restoration (RRR) Schemes:** Aim to increase groundwater recharge and revive irrigation potential.
- **Master Plan for Artificial Recharge to Groundwater-2020:** Prepared by Central Ground Water Board with States/UTs input.
- **Central Ground Water Authority (CGWA):** Mandates rooftop rainwater harvesting & recharge systems for ground water abstraction projects.
- **Atal Mission for Rejuvenation and Urban Transformation (AMRUT):** Launched in 2015, it focuses on urban water infrastructure. AMRUT 2.0 (2021) covers all statutory towns, promotes water body rejuvenation, aquifer management, and rainwater harvesting.
- **Mahatma Gandhi National Rural Employment Guarantee Scheme (MGNREGS):** Includes water conservation and harvesting structures under its natural resource management component.
- **Model Building Bye Laws (MBBL) 2016:** It includes provisions for rainwater harvesting, adopted by all States/UTs except Sikkim, Lakshadweep, and Mizoram.
- **Mission Amrit Sarovar (2022):** Launched to develop and rejuvenate 75 water bodies in each district as part of Azadi ka Amrit Mahotsav (Ministry of Jal Shakti, 2023).

Central Government Bodies Contributing to Rainwater Harvesting

Numerous government and non-government organizations across India are dedicated to enhancing water quality and availability for drinking and irrigation. Here are some key national authorities involved in rainwater harvesting initiatives:

- **Central Ground Water Board (CGWB):** Responsible for the assessment, exploration, development, and management of groundwater resources in the country.
- **Central Soil and Water Conservation Research and Training Institute, Dehradun:** Focuses on research and training in soil and water conservation to combat land degradation.
- **Central Arid Zone Research Institute (CAZRI):**Researches sustainable development in arid and semi-arid regions.
- **Ministry of Rural Development:** Implements various programs for rural development, including water conservation and management.
- **Ministry of Water Resources:** Formulates and implements policies for developing and regulating water resources.
- **National Rainfed Area Authority (NRAA):** Coordinates and monitors the implementation of various watershed and rainwater harvesting projects.
- **Ministry of Drinking Water and Sanitation:** Ensures the supply of safe drinking water and sanitation facilities.
- **National Commission on Agriculture:** Provides recommendations for agricultural development, including water management practices (Chaitanya, 2018).

Policies for Water Harvesting

The Government of India has introduced several mandatory policies and laws to promote rainwater harvesting and groundwater recharge. Key policies include:

- **National Water Policy 2012:** Developed by the Ministry of Water Resources, RD & GR, this policy aims to ensure sustainable and equitable water resource development and management. It encourages rainwater harvesting and the revival of traditional water harvesting structures.
- **Central Groundwater Board (CGWB):** Identified 448,760 km^2 as suitable for artificial recharge, promoting groundwater recharge through various structures.
- **Aquifer Mapping and Management Programme (NAQUIM):** Launched during the XII Plan, this programme aims to quantify groundwater availability, assess its quality, and develop Aquifer Management Plans.
- **State Subsidies:** States like Punjab and Tamil Nadu subsidize farmers who implement rainwater harvesting systems.

- **Model Building Bye-Laws 2016:** Issued by the Ministry of Urban Development, these laws mandate rainwater harvesting systems for new buildings on plots of 100 m^2 and above plots. They also require wastewater recycling systems for buildings with a discharge of 10,000 l per day.
- **Mandatory Rainwater Harvesting:** Enacted by 30 States/UTs through various laws, rules, regulations, building bye-laws, or Government Orders.
- **CGWA Regulations:** The Central Ground Water Authority has notified 162 areas in India to regulate groundwater extraction, mandating rainwater harvesting in these regions.
- **Master Plan for Artificial Recharge:** Envisages constructing about 23 lakh artificial recharge and rainwater harvesting structures in rural areas and 88 lakh in urban areas to enhance groundwater recharge (Chaitanya, 2018).

Significant Rainwater Harvesting Initiatives

Several national programmes, often in collaboration with NGOs, are enhancing rainwater harvesting efforts across the country:

- **Watershed Development (WSD) Programme:** A joint initiative by the Government of India and state governments aimed at:
  - Constructing checking dams, percolation tanks, and irrigation tanks.
  - Creating vegetation barriers and contour bunds.
  - Implement sustainable agricultural practices like in-situ soil and moisture conservation, agroforestry, razing pastures, horticulture, and silvopasture (Chaitanya, 2018).

- **Drought Prone Areas Programme (DPAP) 1970–71:** Aims to create assets to mitigate the impact of droughts by focusing on water conservation and efficient use of available resources.
- **Desert Development Programme (DDP) 1977–78:** Focuses on combating desertification through afforestation and livestock development in arid regions.
- **Integrated Watershed Management in Flood Prone Rivers (FPR) 1980:** Initiated by the Ministry of Agriculture to manage watersheds in flood-prone areas.
- **National Watershed Project for Rainfed Areas (NWDPRA) 1990:** Aims to enhance crop productivity in rain-fed areas, covering 2554 micro-watersheds by 1994 and around 8214 watershed development projects across 28 states (Chaitanya, 2018).

These initiatives have significantly contributed to rainwater harvesting, enhancing water availability and agricultural incomes nationwide. For example, projects under DPAP, DDP, and IWDP released approximately Rs. 77,300 million to treat drought-impacted areas of 32.29 million ha between 1995 and 2008. Grassroots initiatives in the Saurashtra region of Gujarat, which constructed over 90,000 check dams and other conservation structures, have greatly improved water availability and prosperity in previously dried-up village areas, setting a successful example for other states. The planning, financing, and implementation of village-wide rainwater harvesting under such initiatives have significantly impacted water resources and agricultural productivity.

Suggestions for Enhancing Water Availability and Management

1. **Revival and Enhancement of Traditional Water Harvesting Systems (TWHS)**
   i. **Comprehensive Mapping and Documentation**
      - **Surveys and Studies:** Conduct extensive surveys to map all existing traditional water harvesting structures (like eris, talabs, and kuls) and document their current conditions, historical significance, and potential for restoration.
      - **Research Collaborations:** Collaborate with academic institutions and research organizations to study traditional water management practices and their ecological and hydrological impacts.

   ii. **Community-Based Management**
      - **Formation of Local Water Committees:** Establish village-level water committees responsible for maintaining and managing traditional water systems.
      - **Training and Capacity Building:** Provide training programs for local communities on the maintenance of these systems, including desilting, repair of embankments, and vegetation management.

   iii. **Government and NGO Support**
      - **Funding and Grants:** Allocate dedicated funds and grants to restore and maintain traditional water systems.
      - **Technical Assistance:** Offer technical assistance to local communities to design and implement restoration projects.
2. **Strengthening Policy and Regulatory Frameworks**
   i. **Implementation and Monitoring**
      - **Robust Enforcement Mechanisms:** Develop and enforce stricter regulations to ensure compliance with rainwater harvesting policies, such as the National Water Policy 2012.
      - **Periodic Audits:** Conduct regular audits of rainwater harvesting installations to ensure they are functional and effectively contributing to groundwater recharge.

   ii. **Incentive Programs**
      - **Subsidies for Rainwater Harvesting:** Provide financial incentives, such as subsidies and tax rebates, for households, farmers, and industries implementing rainwater harvesting systems.

- **Recognition and Awards:** Establish recognition programs for exemplary rainwater harvesting projects to encourage wider adoption.

3. **Capacity Building and Public Awareness**
   i. **Educational Campaigns**
      - **School and College Programs:** Integrate water conservation and rainwater harvesting into school and college curricula to educate the younger generation.
      - **Mass Media Campaigns:** Utilise television, radio, and social media platforms to raise public awareness about the importance of water conservation and the benefits of rainwater harvesting.

   ii. **Workshops and Demonstrations**
      - **Community Workshops:** Organize workshops and live demonstrations in rural and urban areas to show practical rainwater harvesting methods and groundwater recharge.
      - **Exhibitions and Fairs:** Participate in and organize exhibitions and fairs focused on water conservation technologies and traditional water systems.

4. **Integration with Modern Technology and Innovation**
   i. **GIS and Remote Sensing**
      - **Identification of Potential Sites:** Use GIS and remote sensing technologies to identify suitable sites for artificial recharge structures and to monitor changes in groundwater levels.
      - **Data Management Systems:** Develop centralized databases to store and manage data on water resources, rainfall patterns, and the effectiveness of rainwater harvesting initiatives.

   ii. **Smart Water Management Systems**
      - **IoT-Enabled Sensors:** Implement IoT-enabled sensors for real-time monitoring of water levels, quality, and flow rates in traditional and modern water harvesting systems.
      - **Mobile Applications:** Develop mobile applications that inform farmers and urban residents about rainwater harvesting techniques, weather forecasts, and water usage tips.

5. **Public-Private Partnerships (PPP) and Collaborative Efforts**
   i. **Engagement of the Private Sector**
      - **Corporate Social Responsibility (CSR):** Encourage private companies to include water conservation and rainwater harvesting projects in their CSR activities.
      - **Innovative Financing Models:** Develop innovative financing models, such as water bonds or green bonds, to attract private investment in large-scale water projects.

ii. **NGO Involvement**

- **Grassroots Mobilisation:** Collaborate with NGOs to mobilize communities and implement rainwater harvesting projects at the grassroots level.
- **Capacity Building:** Leverage NGO expertise for training programs and capacity-building initiatives.

6. **Urban Planning and Infrastructure Development**

i. **Incorporation in Urban Planning**

- **Green Infrastructure:** Promote the development of green infrastructure, such as green roofs, permeable pavements, and rain gardens, in urban areas to enhance rainwater infiltration and groundwater recharge.
- **Water-Sensitive Urban Design (WSUD):** Integrate WSUD principles into urban planning to manage stormwater, reduce runoff, and enhance urban water cycles.

ii. **Retrofitting Existing Buildings**

- **Mandatory Retrofitting:** Implement policies requiring retrofitting existing buildings with rainwater harvesting systems, especially in water-stressed areas.
- **Incentives for Compliance:** Offer incentives, such as reduced water tariffs or tax benefits, for buildings that comply with rainwater harvesting regulations.

7. **Research, Development, and Innovation**

i. **Support for R&D**

- **Funding for Innovation:** Provide grants for research and development of new rainwater harvesting technologies and water conservation methods.
- **Collaboration with Universities:** Encourage collaborations between universities, research institutions, and industry to develop and test innovative solutions for water management.

ii. **Performance Evaluation and Improvement**

- **Impact Assessments:** Conduct regular impact assessments of rainwater harvesting projects to evaluate their effectiveness and identify areas for improvement.
- **Feedback Mechanisms:** Establish mechanisms for collecting feedback from beneficiaries and stakeholders to improve the design and implementation of water projects continuously.

Conclusion

India's water scarcity requires a comprehensive approach integrating traditional water harvesting systems (TWHS) with modern practices. The degradation of these systems due to urbanisation, deforestation, and poor maintenance underscores the need for sustainable water management. Central government initiatives like the National Water Policy 2012, Jal Shakti Abhiyan, Atal Bhujal Yojana, and state-level efforts emphasise the importance of rainwater harvesting and groundwater recharge.

Research is crucial for understanding the components, causes of degradation, and current state of TWHS, enabling sustainable solutions. Policies incentivising community participation and integrating traditional and modern techniques are essential for long-term water sustainability. Large-scale programs such as the Pradhan Mantri Krishi Sinchayee Yojana and the Mahatma Gandhi National Rural Employment Guarantee Scheme are vital in improving water availability and supporting rural livelihoods.

At last, addressing water scarcity in India requires a multifaceted strategy involving robust policies, community engagement, and continuous research. By leveraging traditional wisdom and modern technology, India can achieve long-term water security, promoting economic growth, public health, and environmental resilience.

References

Agarwal, A., & Narain, S. (1997). *Dying wisdom: Rise, fall and potential of India's traditional water harvesting systems*. Centre for Science and Environment.

Agossou, H. V. (2018). *Performance assessment of a bamboo-drip irrigation system*. Ph.D. diss., Universitäts-und Landesbibliothek Bonn.

Amirthalingam, M. (2015). Ecological Traditions of India. *C.P.R. Environmental Education Centre: Tamilnadu, India*, 88–98.

Bhatt, P. (2014). *Her space, her story: Exploring the Stepwells of Gujarat*. Zubaan Academic Publishers.

Borthakur, S. (2008). Traditional rain water harvesting techniques and its applicability. *Indian Journal of Traditional Knowledge*, *8*(4), 525–530.

Central Ground Water Board (2011). *Select case studies on rainwater harvesting and artificial research*. Ministry of Water Resources.

Chaitanya (2018). *Role of the Indian Government in Support of Rainwater Harvesting Initiatives*. Retrieved June 03, 2024, from Chaitanya: Rain Harvest Products and Systems: https://www.chaitanyaproducts.com/blog/indian-government-initiatives-for-rainwater-harvesting/

Chandra, S. (2015). Steps to water: Stepwells in India. In T. M. Tarun & S. Mukherjee (Eds.), *Chitrolekha international magazine on art and design* (Vol. 5, pp. 40–46). Chitrolekha International Magazine on Art and Design.

Cochran, J., & Ray, I. (2009). Equity reexamined: A study of community-based rainwater harvesting in Rajasthan, India. *World Development*, *37*, 435–444.

Constantz, J., Cox, M., & Su, G. (2003). Comparison of heat and bromide as ground water tracers near streams. *Ground Water*, *41*, 647–656.

CSE. (1998). *Bamboo drip irrigation of the Northeastern Hills—rainwater harvesting, Northeastern Hills*. Retrieved June 04, 2024, from Centre for Science and Environment, India: http://www.rainwaterharvesting.org/methods/traditional/bamboo.htm

Das, A. (2010). Johads of Alwar. *Journal of the Geological Society of India, 75*(2), 446–447.

Deb, R. A., & Shah, T. (2002). *Socio ecology of groundwater irrigation in India. Annual Partners'Meet 2002*. IWMI-TATA Water Policy Research Program.

Dehejia, V. (2009a). *The body adorned: Sacred and profane in Indian Art*. Columbia University Press.

Dehejia, V. (2009b). *The body adorned: Sacred and profane in Indian art*. Columbia University Press.

Dhiman, S. C., & Gupta, S. (2011). *Rainwater harvesting and artificial recharge*. Central Ground Water Board, Ministry of Water Resources.

Divy, N. K., Neelam, G., Shukla, G., & Swati, S. (2012). 'Traditional water management systems—An overview of the AhPyne system in the South Bihar plains in India and the need for its revival. *Indian Journal of Traditional Knowledge*.

Frontline. (2001). The water man of Rajasthan. *18*(17), 18–23.

Gore, K. P., Pendke, M. S., Rao, V., & Gupta, C. (1998). Groundwater modelling to quantify the effect of water harvesting structures in Wagarwadi watershed, Parbhani district, Maharashtra, India. *Hydrological Processes, 12*, 1043–1052.

Gupta, A., Nair, S., & Singh, S. (2014). Traditional water management systems for drought mitigation in India. *Water Digest*, 36–45.

Jeeva, S., Laloo, R. C., & Mishra, B. P. (2006). Traditional agricultural practices in Meghalaya, North East India. *Indian Journal of Traditional Knowledge, 5*(1), 7–18.

Kabir, M. R., & Faisal, I. M. (1999). Indigenous practice for water harvesting in Bangladesh. *Regional Workshop on Traditional Water Harvesting System*, (pp. 1–5).

Kavarana, G. (2006). The value of a raindrop. Traditional RWH systems, particularly in the arid and semi-arid regions of Rajasthan and Gujarat. UNESCO G-WADI meeting on water harvesting Aleppo Syria. 54.

Kavuri-Bauer, S. (2011). *Monumental Matters: The Power, Subjectivity, and Space of India's Mughal Architecture*. Duke University Press.

Khan, M. A., & Narain, P. (2000). Traditional water harvesting systems and their relevance in the present context. *National seminar on groundwater management strategies in Arid and Semi Arid regions groundwater department*, 19–27.

Kumar, R., Punia, M., Singh, K., Singh, R., Singh, P., & Lipi, K. (2010). *Sustainable development and management of water resources of Rajasthan*. Nava Gavesana.

Kumar, R., Singh, R. D., & Sharma, K. D. (2005). Water resources of India. *Current Science, 89*, 794–811.

Machiwal, D., Jha, M. K., Singh, P. K., Mahnot, S. C., & Gupta, A. (2004). Planning and design of cost-effective water harvesting structures for efficient utilization of scarce water resources in semi-arid regions of Rajasthan, India. *Water Resources Management, 18*, 219–235.

Mudrakartha, S. (2007). To adapt or not to adapt: The dilemma between long-term resource management and short-term livelihood. In M. Giordano & K. Villholth (Eds.), *The agricultural groundwater revolution. Opportunities and threats to development*. CAB International Publishing.

Murthy, R. S., Srikonda, R., & asinath, I. V. (2022). Traditional water management systems of India. *Journal of the International Society for the Study of Vernacular Settlements, 2*, 61–77.

Nirmale, V. N., & Metar, S. Y. (2003). Indigenous knowledge: Wealth of Nation, New Delhi, India. *Employment News, 28*(10), 1.

Norphel, C. (2009, August 1). History of Water Management in Urban LehAmrutaSudhalkar.

Phuntsog, T., Wani, M. A., Sonam, D., Shabber, H., & Jigmat, L. (2017). Water harvesting and conservation under cold desert condition of Ladakh (J&K): Constraints and strategies. *International Journal of Current Microbiology & Applied Sciences, 6*(2), 1769–1800. https://doi.org/10.20546/ijcmas.2017.602.201

PIB, D. (2023). *Ministry of Jal Shakti*. Retrieved June 02, 2024, from Policy for Rainwater Harvesting: https://pib.gov.in/PressReleaseIframePage.aspx?PRID=1914351

Prasad, R., Mertia, R. S., & Narain, P. (2004). Khadin cultivation: A traditional runoff farming system in Indian Desert needs sustainable management. *Journal of Arid Environments*, *58*(1), 87–96.

Pulamte, L. (2008). Understanding innovations in traditional agriculture of Northeast India-the case for sustainable development. *VIth Globelic conference*, *12*, 13.

Robbins, P., & Moritz, J. (1994). Resourceful people and people's resource teaching. *The Cultural Ecology of South Asia*, *4*.

Sameer, V., Anand, B., & Sharma, S. N. (2019). *Status and importance of traditional water conservation system in present scenario*. Central Soil and Materials Research Station.

Selvaraj, T., Devadas, P., Perumal, J. L., Zabaniotou, A., & Ganesapillai, M. (2022, Retrieved from). A comprehensive review of the potential of Stepwells as sustainable water management structures. *Water*, 1–25. https://doi.org/10.3390/w14172665

Shah, T., Deb, R. A., Qureshi, A., & Wang, J. (2003). Sustaining Asia's Groundwater boom: An overview of issues and evidence. *Natural Resources Forum*, *37*, 130–141.

Sharma, U. C., & Sharma, V. (2000). The "Zabo" soil and water management and conservation system in northeast India: Tribal beliefs in the development of water resources and their impact on society. *An Historical Account of a Success Story International Association of Hydrological Sciences*, 184–191.

Sharma, U. C., Prasad, R. N., & Sonowal, P. (1994). An indigenous technique of soil and water conservation in north-eastern region-The Zabo system of farming. *Soil and Water Conservation Opportunities*, 969–975.

Singh, R. K., Singh, V., Rajkhowa, C., & Deka, B. C. (2012). Zobo: A Traditional Way of Integrated Farming. In D. C. Thirugnanavel, T. R. Borah, & S. V. Ngachan (Eds.), *Resilient Shifting Cultivation: Challenges and Opportunity* (pp. 114–117).

Singh, V. (2004). A Survey of the stepwells in Shekhawati region. *Proceedings of the Indian History Congress*, *65*, 1159–1165.

Subramani, T., Elango, L., & Damodarasamy, S. R. (2005). Groundwater quality and its suitability for drinking and agricultural use in Chithar River Basin, Tamil Nadu, India. *Environmental Geology*, *47*, 1099–1110.

Swami, S. (2021). Bamboo drip irrigation: Traditional way of irrigating crops in Meghalaya. *Just Agriculture*, 18–20.

Terukazu, N., & Vasanti, M. N.Architects (2002). *An introduction to the space, form and meaning in Indian architecture: The Earth and mountain, water, and architecture*. Special The World of Indian Architecture.

UNESCO. (2009). *World water assessment programme*. The United Nations World water development report 3: Water in a changing world. Retrieved June 03, 2011

World Economic Forum. (2016). *The global risks report 2016, (World Economic Forum (WEF), Geneva, Switzerland)*. Retrieved June 03, 2024, from http://www3.weforum.org/docs/GRR/WEF_GRR16.pdf

Chapter 6
Indigenous Water Management and Environmental Sustainability in the Himalayas

Saamdu Chetri

Abstract The Himalayan region is a vital source of water resources, sustaining ecosystems, livelihoods, and cultures across vast landscapes. Indigenous water management practices have played a fundamental role in harnessing and conserving these precious resources, offering valuable lessons in resilience, adaptation, and sustainability. This chapter explores the historical context, socio-economic significance, environmental impacts, and challenges facing indigenous water management in the Himalayas. Drawing upon multidisciplinary research and case studies, we highlight the importance of traditional knowledge and practices for promoting environmental sustainability and community resilience. By recognising the value of indigenous water management and integrating it with modern approaches, we can work towards a more harmonious relationship with water in the Himalayan region and beyond. Through collaboration, knowledge exchange, and policy support, we can chart a course towards a sustainable water future, ensuring the continued vitality of the Himalayas for generations to come.

Keywords Indigenous water management · Community resilience · Himalayan region · Global hydrology · Global water management · Environmental impact · Socio-economic impact

Introduction

The Himalayan region is a majestic natural wonder, revered for its towering peaks, breathtaking landscapes, and rich biodiversity. Stretching across several countries, including India, Nepal, Bhutan, China, and Pakistan, the Himalayas are the source of some of the world's most vital water resources (Smith, 2018). From the mighty

S. Chetri (✉)
Shoolini University, Solan, India
e-mail: dr.saamdu.chetri@gmail.com

R. C. Sobti et al. (eds.), *Traditional Knowledge Systems in India: Sustainability, Ecology, and Resource Governance*, People, Cultures and Societies: Exploring and Documenting Diversities,
https://doi.org/10.1007/978-981-92-1346-7_6

rivers that carve their way through valleys to the glaciers that crown the highest peaks, the Himalayan waters sustain millions of people downstream and play a crucial role in global hydrology.

At the heart of the Himalayas lies a wealth of indigenous knowledge and practices that have evolved over centuries to harness and manage these precious water resources. From traditional irrigation systems to community-based water governance structures, indigenous communities in the Himalayan region have developed intricate strategies for water conservation, distribution, and utilisation (Bisht & Pandey, 2020). These indigenous water management practices sustain local ecosystems and support the livelihoods and cultural identities of the communities that inhabit these rugged landscapes.

The objectives of this chapter are to explore the importance of indigenous water management practices in sustaining local ecosystems and communities in the Himalayan region, and to examine their significance for global water resources. By delving into the historical context, socio-economic aspects, environmental impacts, and challenges facing indigenous water management, this chapter aims to shed light on the critical role that traditional knowledge plays in promoting environmental sustainability and resilience in the Himalayas.

Importance of the Himalayan Region for Global Water Resources

The Himalayan region is the source of some of the world's most important rivers, including the Ganges, Indus, Brahmaputra, Yangtze, and Yellow River systems (Immerzeel et al., 2020). These rivers, fed by the melting glaciers and snowpack of the Himalayas, provide water for irrigation, drinking, and hydropower generation to millions of people in South Asia and beyond. The seasonal monsoon rains, influenced by the Himalayan topography, further contribute to the region's hydrological cycle, replenishing groundwater reserves and sustaining ecosystems across the Indian subcontinent.

However, the Himalayan waters are vital for regional water security and have far-reaching implications for global climate patterns and water resources (Shrestha et al., 2019). The melting glaciers of the Himalayas contribute to sea-level rise and alter river flow patterns, affecting downstream communities and ecosystems. Moreover, the monsoon rains, influenced by the Himalayan terrain, are critical in driving weather systems across Asia, impacting agriculture, water availability, and livelihoods in distant regions.

In light of the Himalayas' pivotal role in global water resources, it is imperative to understand the dynamics of water management in this ecologically sensitive region. While modern infrastructure and technology have transformed water management practices in some areas, indigenous communities in the Himalayas continue to rely on traditional knowledge and practices to sustainably manage their water resources (Bose & Mishra, 2018). These indigenous water management practices not only offer valuable insights into sustainable resource management but also hold lessons for addressing contemporary water challenges worldwide.

Importance of Indigenous Water Management Practices

Indigenous water management practices have been integral to the survival and resilience of communities in the Himalayan region for millennia (Gurung & Subba, 2017). Drawing upon local knowledge, customs, and beliefs, indigenous communities have developed sophisticated systems for harmoniously capturing, storing, and distributing water with the natural environment. These traditional water management systems, often tailored to the Himalayas' unique geography and climatic conditions, have proven remarkably resilient and sustainable over time.

One of the key strengths of indigenous water management practices lies in their adaptability and flexibility in the face of environmental change. Unlike centralised water infrastructure projects, which may be vulnerable to disruptions and failures, indigenous systems are often decentralised and community-driven, allowing for greater resilience and responsiveness to local needs and conditions (Rai & Shakya, 2020). Moreover, indigenous water management practices prioritise ecosystem health and biodiversity conservation, recognising the interconnectedness of water, land, and life.

Despite their effectiveness and relevance, indigenous water management practices in the Himalayas face numerous challenges in the modern era (Bhattarai et al., 2021). Rapid urbanisation, population growth, and climate change pressure traditional water systems, leading to water scarcity, pollution, and resource conflicts in some areas. Furthermore, the erosion of traditional knowledge and cultural practices, inadequate policy support and institutional recognition threaten the sustainability of indigenous water management in the Himalayas.

In recent decades, there has been a growing recognition of the value of indigenous water management practices for environmental sustainability and community resilience. Efforts to revive and promote traditional water systems have gained momentum, driven by concerns over water scarcity, climate change, and cultural preservation (Smith, 2018). Community-led initiatives, supported by NGOs, government agencies, and international organisations, are working to restore ancient water structures, revive traditional knowledge, and promote sustainable water use practices (Rai et al., 2021).

Objectives of the Chapter and Overview of Key Themes

Against this backdrop, the objectives of this chapter are threefold:

1. To explore the historical context and evolution of indigenous water management practices in the Himalayan region, tracing their origins, development, and adaptation.
2. To examine the socio-economic, cultural, and environmental significance of indigenous water management practices for local ecosystems and communities in the Himalayas.
3. To assess the challenges and opportunities facing indigenous water management in the context of contemporary water issues, including climate change, urbanisation, and globalisation.

The chapter will be structured around several key themes to achieve these objectives.

Historical Context of Indigenous Water Management

The Himalayan region is renowned for its majestic peaks, lush valleys, and abundant water resources, which have sustained human civilisations for millennia. Throughout history, indigenous communities in the Himalayas have developed sophisticated water management practices tailored to the region's unique geography, climate, and cultural diversity. This chapter explores the historical context and evolution of these indigenous water management practices, tracing their origins, development, and adaptation over time.

Ancient Wisdom: Origins of Indigenous Water Management

The roots of indigenous water management practices in the Himalayas can be traced back to ancient civilisations that flourished along the banks of the region's rivers (Regmith, 2018). Archaeological evidence suggests that early settlers in the Himalayan foothills constructed terraces, ponds, and irrigation channels to harness water for agriculture and domestic use. These early water management systems laid the foundation for later innovations and advancements in indigenous water management.

Influence of Religion and Culture

Religion and culture have significantly shaped indigenous water management practices in the Himalayas (Gupta & Das, 2019). Hinduism, Buddhism, and other indigenous belief systems emphasise the sacredness of water and its vital role in sustaining life. Rituals, festivals, and ceremonies centred around water have helped foster a sense of reverence and stewardship among local communities. Moreover, religious texts and scriptures contain valuable insights into traditional water management techniques and principles.

Role of Traditional Knowledge

Central to indigenous water management in the Himalayas is the transmission of traditional knowledge from generation to generation (Mishra & Rai, 2020). Elders, shamans, and community leaders know about local hydrology, seasonal patterns, and sustainable water use practices. Traditional wisdom, passed down through oral traditions and practical experience, informs decision-making processes related to water allocation, conservation, and distribution.

Innovation and Adaptation

Over centuries of interaction with their environment, Himalayan communities have demonstrated remarkable ingenuity and adaptability in

The advent of colonialism in the Himalayan region significantly changed indigenous water management practices (Mishra & Rai, 2020). British colonial rulers introduced new measures in response to changing water conditions (Rai et al., 2021). Innovations such as stone spouts, gravity-fed irrigation systems, and rainwater harvesting techniques have allowed communities to thrive in challenging terrain. Moreover, indigenous water management practices have evolved in response to external pressures, including climatic fluctuations, population growth, and political upheavals.

Colonialism and modernisation technologies, such as dams, canals, and hydropower projects, aimed at harnessing the region's water resources for economic development. While these initiatives brought benefits in irrigation, flood control, and electricity generation, they also disrupted traditional water systems and marginalised indigenous knowledge and practices.

Revival and Resurgence

In recent decades, there has been a growing recognition of the value of indigenous water management practices for environmental sustainability and community resilience. Efforts to revive and promote traditional water systems have gained momentum, driven by concerns over water scarcity, climate change, and cultural preservation (Smith, 2018). Community-led initiatives, supported by NGOs, government agencies, and international organisations, are working to restore ancient water structures, revive traditional knowledge, and promote sustainable water use practices (Rai et al., 2021).

The historical context and evolution of indigenous water management practices in the Himalayas reveal a rich tapestry of innovation, adaptation, and cultural heritage (Regmith, 2018). From ancient, terraced fields to modern rainwater harvesting systems, indigenous communities have demonstrated a deep understanding of their local ecosystems and a commitment to sustainable resource management. By recognising the importance of traditional knowledge and practices, and integrating them with modern approaches to water management, we can ensure the continued vitality and resilience of the Himalayan region for generations to come.

Indigenous Water Management Practices

Indigenous water management practices in the Himalayas have been integral to the survival and resilience of communities in the region for centuries. Drawing upon local knowledge, traditions, and customs, indigenous communities have developed sophisticated systems for harmoniously capturing, storing, and distributing water resources with the natural environment. These traditional practices ensure water security for local populations and sustain the delicate ecosystems that thrive in the Himalayan landscape. This chapter explores the diverse array of indigenous water management practices across the Himalayas, highlighting their historical significance, socio-cultural relevance, and environmental sustainability.

Historical Context and Evolution of Indigenous Water Management

The origins of indigenous water management practices in the Himalayas can be traced back to ancient times when human settlements first emerged along the banks of rivers and streams. Early inhabitants of the region devised ingenious methods for harnessing water resources, including constructing irrigation channels, terraced fields, and water storage tanks (Singh & Sharma, 2016). These early innovations laid the foundation for more sophisticated water management systems that would evolve over time.

As civilisations flourished in the Himalayas, so too did their water management practices. From the elaborate irrigation networks of the Indus Valley Civilization to the intricate water temples of the Kathmandu Valley, indigenous communities developed a deep understanding of the hydrological cycle and its importance for agricultural productivity and human well-being (Bisht & Negi, 2018). Water became a source of sustenance and a central aspect of religious, cultural, and social life in the Himalayan region.

Socio-Economic and Cultural Dimensions of Indigenous Water Management

Indigenous water management practices in the Himalayas are deeply intertwined with socio-economic and cultural factors that shape the lives of local communities (Regmith, 2018). Water is not merely seen as a commodity but as a sacred resource that sustains life and connects people to the land. Traditional water management systems often reflect local customs, beliefs, and social norms, with community members coming together to maintain and protect shared water resources.

One of the key features of indigenous water management in the Himalayas is its emphasis on collective ownership and stewardship of water resources. Water is viewed as a common good that belongs to all community members, rather than being owned by individuals or private entities. This communal approach to water management fosters cooperation, reciprocity, and social cohesion among community members, strengthening bonds and fostering a shared responsibility for water conservation (Shrestha et al., 2020).

Furthermore, indigenous water management practices in the Himalayas are closely linked to cultural rituals, festivals, and traditions that celebrate the importance of water in daily life (Regmith, 2018). From the annual festivals honouring river deities to the rituals performed before sowing seeds or harvesting crops, water plays a central role in religious ceremonies and cultural practices passed down through generations. These rituals serve not only to honour water but also to reinforce the connection between humans and the natural world.

Environmental Impacts and Sustainability of Indigenous Water Management

Indigenous water management practices in the Himalayas are characterised by environmental sustainability and resilience in changing climatic conditions (Bisht & Negi, 2018). Traditional systems such as gravity-fed irrigation, rainwater harvesting, and agroforestry are designed to work harmoniously with the natural landscape, minimising negative impacts on ecosystems and biodiversity (Shrestha et al., 2020).

Indigenous water management practices in the Himalayas embody the resilience, adaptability, and wisdom of local communities who have nurtured and sustained these systems for generations (Singh & Sharma, 2016). Throughout this exploration of indigenous water management practices, it has become evident that these traditional systems are effective in meeting the water needs of communities and contribute to the overall sustainability and resilience of the Himalayan ecosystems.

By harnessing local knowledge, customs, and beliefs, indigenous communities in the Himalayas have developed sophisticated water harvesting, storage, and distribution systems tailored to the region's unique geography and climatic conditions (Bisht & Negi, 2018). From ancient irrigation networks to innovative rainwater harvesting techniques, these indigenous water management practices prioritise ecosystem health, biodiversity conservation, and community well-being.

The socio-economic and cultural dimensions of indigenous water management practices are equally significant, shaping the identity and livelihoods of Himalayan communities (Shrestha et al., 2020a). Water is not merely a resource to be exploited but a sacred element deeply intertwined with indigenous societies' cultural, spiritual, and social fabric. Traditional water rituals, festivals, and ceremonies underscore the reverence and respect accorded to water, highlighting its central role in sustaining life and fostering community cohesion.

The Himalayan region is a bastion of natural beauty and cultural diversity and a crucial source of freshwater for millions of people downstream. Indigenous communities in the Himalayas have developed sophisticated water management systems over centuries, harnessing the region's abundant water resources to sustain their livelihoods and ecosystems (Bisht & Negi, 2018). However, the environmental impacts of these indigenous water management practices are poorly understood. This chapter aims to explore the environmental implications of indigenous water management in the Himalayas, shedding light on the sustainability and resilience of these traditional practices.

Historical Context of Indigenous Water Management

To understand the environmental impacts of indigenous water management in the Himalayas, it is essential to delve into the historical context of these practices. Indigenous communities in the region have long relied on traditional knowledge and techniques to manage water resources sustainably (Shrestha et al., 2020b). These practices, honed over generations, are deeply rooted in the local ecology and culture. From constructing terraced fields and irrigation channels to conserving sacred water sources, indigenous water management systems reflect a profound understanding of the Himalayan landscape and its hydrological processes.

Socio-economic and Cultural Dimensions of Indigenous Water Management

Indigenous water management practices in the Himalayas are not merely technical solutions to water scarcity; they are also imbued with socio-economic and cultural significance. Water is considered sacred in many Himalayan cultures, and traditional water management practices are often intertwined with religious rituals and social customs (Shrestha et al., 2020). Moreover, these practices support local livelihoods, particularly in agriculture-dependent communities. By conserving water, maintaining soil fertility, and regulating river flow, indigenous water management systems contribute to the resilience of Himalayan ecosystems and the well-being of local communities.

Environmental Impacts of Indigenous Water Management Practices

Despite their socio-economic and cultural importance, indigenous water management practices in the Himalayas can have positive and negative environmental impacts. On the one hand, these practices promote ecosystem health and biodiversity conservation by maintaining natural water flow patterns, preventing soil erosion, and preserving habitat diversity (Bisht & Negi, 2018). For example, traditional irrigation systems such as Kuhl in Himachal Pradesh and Guls in Ladakh are designed to minimise water wastage, promote soil moisture retention, and enhance agricultural productivity while minimising environmental degradation.

On the other hand, certain indigenous water management practices may inadvertently contribute to environmental degradation in the Himalayas. For instance, constructing water diversion structures and terraced fields can alter natural hydrological processes, leading to soil erosion, sedimentation, and biodiversity loss (Shrestha et al., 2020). Similarly, over-extraction of water for irrigation or domestic use can deplete groundwater reserves and disrupt aquatic ecosystems, threatening the long-term sustainability of indigenous water management systems.

Furthermore, climate change poses additional challenges to indigenous water management in the Himalayas. Rising temperatures, changing precipitation patterns, and melting glaciers alter the region's hydrological cycle, exacerbating water scarcity and variability (Shrestha et al., 2020). Indigenous communities are increasingly vulnerable to climate-related hazards such as floods, landslides, and droughts, which disrupt traditional water management systems and livelihoods.

Challenges and Opportunities for Sustainable Water Management

Addressing the environmental impacts of indigenous water management in the Himalayas requires a multifaceted approach that balances conservation goals with socio-economic development objectives. Key challenges include reconciling traditional knowledge with modern scientific practices, strengthening community-based water governance institutions, and promoting climate-resilient infrastructure and practices (Sharma et al., 2019). At the same time, there are opportunities to harness the adaptive capacity of indigenous communities and integrate indigenous wisdom into contemporary water management strategies.

One promising approach is to engage indigenous communities as stewards of their water resources, empowering them to participate in decision-making processes and co-designing sustainable water management solutions (Kumar et al., 2020). This approach, known as Indigenous Water Governance, emphasises the importance of recognising and respecting traditional knowledge, values, and rights in water resource management. By fostering collaboration between indigenous and scientific knowledge systems, Indigenous Water Governance can enhance the resilience of Himalayan ecosystems and promote equitable and sustainable development.

Furthermore, there is a need for policy interventions that support the revitalisation of indigenous water management practices and promote their integration into contemporary water governance frameworks. Recognising the value of indigenous knowledge and practices is essential for fostering resilience and sustainability in the face of evolving environmental challenges (Sharma et al., 2019).

Policy interventions should aim to:

1. Facilitate Knowledge Exchange: Create platforms for dialogue and knowledge exchange between indigenous communities, researchers, policymakers, and water resource managers. This collaboration can help bridge the gap between traditional wisdom and modern science, leading to more effective and inclusive water management strategies.
2. Incorporate Indigenous Perspectives: Integrate indigenous perspectives and practices into water governance policies and decision-making processes. By incorporating local knowledge and values into policy formulation, governments can ensure that water management strategies are contextually appropriate and culturally sensitive.
3. Promote Community Empowerment: Empower indigenous communities to participate actively in decision-making processes related to water management. Strengthening community-based governance structures and supporting local initiatives can enhance the resilience of water systems and foster a sense of ownership and stewardship among residents.
4. Invest in Capacity Building: Provide training and capacity-building programs to equip indigenous communities with the skills and resources needed to manage their water resources sustainably. This may include technical training in water conservation techniques, sustainable agriculture practices, and natural resources management.
5. Ensure Legal Recognition and Protection: Recognise and protect the rights of indigenous communities to access, manage, and protect their water resources. Legal frameworks should uphold indigenous rights to water and land, ensuring that their traditional practices are safeguarded from external threats and encroachments.
6. Support Research and Documentation: Invest in research and documentation efforts to preserve and promote indigenous water management knowledge and practices. This may involve documenting traditional water management systems, recording oral histories, and compiling indigenous ecological knowledge for future generations.
7. Foster Collaborative Partnerships: Foster partnerships between government agencies, non-governmental organisations (NGOs), academia, and indigenous communities to jointly address water management challenges. Collaborative

initiatives can leverage diverse expertise and resources to develop innovative solutions and promote cross-cultural understanding.

By implementing these policy interventions, governments and stakeholders can work towards a more inclusive, sustainable, and equitable approach to water management in the Himalayan region. By recognising the value of indigenous knowledge and practices, and integrating them into broader water governance frameworks, we can enhance the resilience of water systems, promote environmental sustainability, and support the well-being of indigenous communities for generations to come.

Socio-Economic Aspects of Indigenous Water Management

The socio-economic dimensions of indigenous water management practices in the Himalayas are integral to understanding the complex relationship between water resources, human communities, and sustainable development in this ecologically fragile region (Ghimire et al., 2020). For centuries, indigenous communities in the Himalayas have relied on traditional knowledge, cultural practices, and communal governance structures to harmonise their water resources with the natural environment. This chapter explores the socio-economic aspects of indigenous water management in the Himalayas, examining the role of water in shaping livelihoods, cultural identities, and social cohesion, as well as the challenges and opportunities facing traditional water systems in the modern era.

Historical Context of Socio-Economics

To understand the socio-economic dimensions of indigenous water management in the Himalayas, examining the historical context in which these practices evolved is essential. The Himalayan region has a long history of human settlement, with indigenous communities developing intricate systems for harnessing water resources to support agriculture, livestock rearing, and other livelihood activities. These traditional water management systems, including stone aqueducts, terraced fields, and community-run irrigation networks, reflect the deep connection between people and water in the Himalayan landscape. Over time, these indigenous water management practices became embedded in local cultures, shaping social norms, religious beliefs, and economic systems.

Socio-Economic Significance

Water plays a central role in the socio-economic life of Himalayan communities, serving as a vital resource for agriculture, hydropower generation, tourism, and domestic use. Indigenous water management practices ensure water availability for these diverse purposes and contribute to the resilience and adaptive capacity of local economies. For example, traditional irrigation systems allow farmers to cultivate crops in mountainous terrain, where water scarcity is a perennial challenge. Similarly, community-managed water infrastructure supports small-scale hydroelectric projects, providing remote villages with clean energy and economic opportunities.

Cultural Identity and Social Cohesion

In addition to its economic significance, water holds deep cultural and spiritual significance for indigenous communities in the Himalayas. Rivers, springs, and lakes are revered as sacred sites, often associated with deities and rituals that celebrate water's abundance and life-giving properties. Traditional water management practices are intricately linked to these cultural beliefs, fostering a sense of stewardship and reverence for the natural world (Dangol et al., 2017). Moreover, water-sharing arrangements and communal water governance systems promote social cohesion and cooperation among community members, reinforcing traditional values of reciprocity, mutual aid, and collective responsibility.

Challenges and Opportunities for Indigenous Water Management

Despite their effectiveness and relevance, indigenous water management practices in the Himalayas face numerous challenges in the modern era. Rapid urbanisation, population growth, and climate change pressure traditional water systems, leading to water scarcity, pollution, and resource conflicts in some areas. Furthermore, the erosion of traditional knowledge and cultural practices, inadequate policy support and institutional recognition threaten the sustainability of indigenous water management in the Himalayas.

However, amidst these challenges, there are opportunities for revitalising and strengthening indigenous water management practices in the Himalayas. Community-led initiatives, partnerships with local organisations, and collaborations between indigenous knowledge holders and researchers can help preserve and promote traditional water systems. By integrating traditional knowledge with modern science and technology, innovative solutions can be developed to address contemporary water issues while safeguarding the cultural and ecological integrity of the Himalayan region.

Moving forward, it is essential to recognise the value of indigenous water management practices and incorporate them into holistic approaches to water governance and resource management. Policies and programs should empower local communities, protect traditional water rights, and promote sustainable practices that enhance human well-being and environmental health. By embracing the wisdom of indigenous peoples and respecting the intricate relationships between water, land, and life, we can work towards a future where water is managed equitably, sustainably, and in harmony with nature.

In conclusion, indigenous water management practices in the Himalayas offer valuable insights and lessons for addressing contemporary water challenges worldwide. By acknowledging the importance of traditional knowledge, fostering community participation, and promoting cultural preservation, we can build a more resilient and sustainable future for all who depend on the waters of the Himalayas. Let us honour the legacy of indigenous water stewards and strive to protect and preserve their invaluable contributions to our shared heritage.

Value of Water Indigenous Management Practices

Despite their importance, indigenous water management practices in the Himalayas face numerous challenges in the modern era. Climate change, rapid urbanisation, and infrastructural development pressure traditional water systems, leading to water scarcity, pollution, and resource conflicts in some areas (Bajracharya et al., 2019). Moreover, the erosion of traditional knowledge and cultural practices threatens the sustainability of indigenous water management, as younger generations migrate to cities in search of alternative livelihoods. However, there are also opportunities for revitalising traditional water systems and integrating them with modern approaches to water management. Community-based initiatives, participatory governance models, and recognising indigenous rights can empower local communities to reclaim their water resources and adapt to changing environmental conditions (Shrestha et al., 2020).

Despite these challenges, there is growing recognition of the value of indigenous water management practices and the need to integrate them with modern approaches to water governance and conservation. By acknowledging the importance of traditional knowledge, customs, and beliefs, policymakers, researchers, and practitioners can collaborate with indigenous communities to develop innovative solutions that promote environmental sustainability, social equity, and cultural resilience in the Himalayan region (Shrestha et al., 2020).

Moving forward, efforts to revitalise indigenous water management must be holistic, inclusive, and community-driven, recognising the interconnectedness of water systems with broader socio-economic and ecological dynamics. This requires fostering partnerships between indigenous communities, government agencies, non-governmental organisations, and academic institutions to co-create solutions grounded in local contexts and responsive to community needs (Bajracharya et al., 2019). Furthermore, there is a need for supportive policy frameworks that recognise and protect the rights of indigenous peoples to manage and steward their water resources following their cultural practices and traditions. This includes ensuring access to clean water, preserving traditional water rights, and promoting equitable participation in decision-making processes related to water management.

Ultimately, revitalising indigenous water management in the Himalayas is a matter of environmental conservation and a question of social justice, cultural preservation, and human rights. By honouring the wisdom of indigenous peoples and embracing their holistic approach to water management, we can build a more sustainable and resilient future for all who depend on the waters of the Himalayas.

In conclusion, the socio-economic aspects of indigenous water management in the Himalayas are multifaceted and complex, reflecting the deep intertwining of water, culture, and livelihoods in this ecologically diverse region. Indigenous water management practices have been critical in sustaining local ecosystems and communities for centuries, providing a blueprint for sustainable resource utilisation and community resilience. However, these traditional practices face numerous challenges in the modern era, including urbanisation, population growth, and climate change, threatening their continued effectiveness and relevance.

Socio-Economic and Cultural Dimensions of Indigenous Water Management

Indigenous water management practices in the Himalayas are deeply intertwined with socio-economic and cultural factors that shape the lives of local communities. Water is not merely seen as a commodity but as a sacred resource that sustains life and connects people to the land. Traditional water management systems often reflect local customs, beliefs, and social norms, with community members coming together to maintain and protect shared water resources.

Socio-Economic Significance

Water plays a central role in the socio-economic life of Himalayan communities, serving as a vital resource for agriculture, hydropower generation, tourism, and domestic use. Indigenous water management practices ensure water availability for these diverse purposes and contribute to the resilience and adaptive capacity of local economies. For example, traditional irrigation systems allow farmers to cultivate crops in mountainous terrain, where water scarcity is a perennial challenge. Similarly, community-managed water infrastructure supports small-scale hydroelectric projects, providing remote villages with clean energy and economic opportunities.

In addition to its economic significance, water holds deep cultural and spiritual significance for indigenous communities in the Himalayas. Rivers, springs, and lakes are revered as sacred sites, often associated with deities and rituals that celebrate water's abundance and life-giving properties. Traditional water management practices are intricately linked to these cultural beliefs, fostering a sense of stewardship and reverence for the natural world (Dangol et al., 2017). Moreover, water-sharing arrangements and communal water governance systems promote social cohesion and cooperation among community members, reinforcing traditional values of reciprocity, mutual aid, and collective responsibility.

Historical Context and Evolution of Indigenous Water Management

The history of indigenous water management in the Himalayas dates back centuries, with communities developing intricate systems for capturing, storing, and distributing water resources (Gurung et al., 2018). These systems evolved in response to the region's rugged terrain, diverse climatic conditions, and cultural practices. Over time, traditional water management practices became deeply embedded in the social fabric of Himalayan communities, playing a central role in their livelihoods and cultural identity.

Socio-Economic and Cultural Dimensions of Indigenous Water Management

Indigenous water management practices in the Himalayas are closely intertwined with socio-economic and cultural factors. Water plays a vital role in agriculture, livestock rearing, and household activities, shaping local communities' daily lives and livelihoods (Sharma et al., 2019). Moreover, water holds cultural significance, with rituals, ceremonies, and beliefs centred around its use and conservation. However, socio-economic changes such as urbanisation, migration, and changing land-use patterns pose challenges to traditional water management practices, threatening their sustainability.

Environmental Impacts and Sustainability of Indigenous Water Management:

While indigenous water management practices have historically promoted environmental sustainability, they are increasingly under pressure from climate change, deforestation, and land degradation. Glacial melt, erratic rainfall patterns, and biodiversity loss pose significant challenges to traditional water systems in the Himalayas (Nepal et al., 2020). Additionally, the unsustainable extraction of water resources and the pollution of rivers and streams further exacerbate environmental degradation, jeopardising the long-term viability of indigenous water management practices.

Challenges Facing Indigenous Water Management

Several challenges confront indigenous water management in the Himalayas, including:

Climate Change: Rising temperatures and changing precipitation patterns alter the hydrological cycle, impacting water availability and quality.

Land Use Changes: Urbanisation, deforestation, and agricultural expansion disrupt traditional water catchment areas and increase soil erosion.

Population Pressure: Rapid population growth and migration to urban centres are placing strain on water resources, leading to over-extraction and depletion of aquifers.

Pollution: Industrial effluents, agricultural runoff, and solid waste disposal are polluting water bodies, threatening both human health and ecosystem integrity.

Infrastructure Development: Large-scale infrastructure projects such as dams, hydropower plants, and roads can disrupt natural water flows and displace communities, leading to social and environmental conflicts.

Opportunities for Revitalising Indigenous Water Management:

Despite these challenges, there are opportunities for revitalising indigenous water management in the Himalayas:

Community Empowerment: Strengthening community-based water governance structures can enhance local resilience and promote sustainable resource management.

Indigenous Knowledge: Integrating traditional knowledge with modern science and technology can improve water conservation efforts and enhance adaptive capacity.

Policy Support: Enacting policies that recognise and support indigenous water management practices can provide a conducive environment for their preservation and revitalisation.

Sustainable Development: Adopting holistic approaches to development that prioritise environmental sustainability, social equity, and cultural preservation can promote harmony between human activities and natural ecosystems.

In conclusion, indigenous water management practices in the Himalayas face significant challenges in environmental change, urbanisation, and globalisation. However, there are opportunities for revitalisation and sustainability, including community empowerment, integration of traditional knowledge, policy support, and sustainable development approaches. Recognising the value of indigenous water management and working collaboratively with local communities, policymakers, and researchers can foster a more resilient and sustainable future for the Himalayan region.

Best Practices of Water and Environment Management in the Himalaya Region

In the Himalayan region, indigenous communities have developed sophisticated water management practices over centuries to sustainably harness and utilise the region's abundant water resources (Bhandari & Sharma, 2019). Drawing upon local knowledge, customs, and traditions, these communities have created innovative solutions to address the challenges of water scarcity, variability, and environmental degradation (Jones, 2010). In this section, we explore several case studies and best practices that exemplify the effectiveness and resilience of indigenous water management in the Himalayas. Through these examples, we gain insights into the diverse approaches, strategies, and principles that underpin successful water management initiatives in the region.

Ancient Water Management Systems in Ladakh

Nestled amidst the rugged terrain of the Indian Himalayas, Ladakh boasts a remarkable legacy of traditional water management embodied in its intricate network of 'Kuls' or 'Guls' (Thapa, 2018). Crafted by local communities centuries ago, these gravity-fed channels harness glacial meltwater and seasonal streams to nourish fields and orchards across remote mountain villages. Following the contours of the land, these kuls epitomise principles of equity and cooperation, ensuring equitable water distribution among farmers and terraced fields (Kumar, 2015).

In the arid expanse of Ladakh, where water scarcity is a constant threat, these ancient irrigation systems stand as lifelines for agricultural sustenance and livelihoods. Despite the challenges posed by climate change and modernisation, the resilience of the kuls continues to sustain agriculture, offering a beacon of hope and a model of sustainable water management for mountainous regions worldwide (Wangchuk, 2020).

Innovative Solutions: Ice Stupas

In mounting water scarcity, Ladakh's engineers and environmentalists have forged ahead with pioneering solutions, birthing the concept of 'ice stupas' (Norbu, 2017). Inspired by the timeless wisdom of Tibetan stupas, these towering ice structures stand as testaments to indigenous ingenuity and community-driven innovation.

Constructed using locally sourced materials such as mud, stones, and pipes, ice stupas harness winter's biting cold to transform liquid water into towering ice cones. Strategically located in areas of high water demand, these frozen sentinels slowly release their liquid bounty as temperatures rise, replenishing fields, recharging aquifers, and extending the growing season for crops (Katel, 2021).

The success of the ice stupa project has rippled far beyond the confines of Ladakh, inspiring similar initiatives in other Himalayan regions grappling with water scarcity (Dorjay, 2019). In a world besieged by environmental challenges, the story of Ladakh's ice stupas is a powerful reminder of the transformative potential of indigenous knowledge and community-led action in safeguarding our planet's precious resources (Wangchuk, 2019).

Empowering Communities Through Watershed Management in Nepal

Nepal, cradled amidst the majestic Himalayas, boasts rich natural beauty and abundant water resources (Gurung, 2016). Yet, urbanisation, deforestation, and climate change threaten its water security and ecological balance (Thapa et al., 2020). In response to these challenges, Nepal has embraced community-based watershed management as a comprehensive approach to safeguarding water resources, restoring degraded landscapes, and empowering local communities.

In the Himalayan foothills of Nepal, community-based initiatives have emerged as potent tools for water conservation and livelihood enhancement. The Federation of Community Forestry Users Nepal (FECOFUN), representing over 22,000 community forest user groups, exemplifies this approach (Pokharel, 2021). Through participatory processes, communities have undertaken diverse activities, including reforestation and soil conservation, revitalising landscapes and bolstering water availability and biodiversity (Rai et al., 2019). By empowering communities to steward their resources, Nepal's efforts underscore the efficacy of grassroots approaches in fostering environmental resilience.

Grounded in decentralisation and participatory governance principles, community-based watershed management recognises the pivotal role of local communities as custodians of their land and water (Shrestha, 2014). Watershed user groups (WUGs), comprising residents, farmers, and leaders, serve as catalysts for collective action, identifying challenges and implementing sustainable practices (Dhakal, 2017). By integrating traditional wisdom with modern techniques, Nepal has forged innovative solutions for soil erosion, water harvesting, and biodiversity conservation (Sharma et al., 2019).

The success story of community-based watershed management unfolds vividly in the Koshi River Basin, where barren hillsides have transformed into verdant forests, replenishing aquifers and rejuvenating rivers (Tamang, 2018). Through infrastructure projects and capacity-building initiatives, marginalised groups, particularly women and indigenous communities, have been empowered to lead and benefit from watershed development interventions, fostering inclusive growth and social cohesion (Rana, 2020).

Nevertheless, challenges persist, including resource constraints, inadequate infrastructure, and climate vulnerabilities (Basnet, 2016). Addressing these demands concerted efforts from stakeholders across sectors, bolstering institutional capacities, enhancing collaboration, and integrating watershed management into national policies and programs (Karki, 2021).

In conclusion, community-based watershed management in Nepal heralds a promising pathway towards water security, environmental sustainability, and community empowerment (Lama, 2015). By harnessing local knowledge, resilience, and collaboration, Nepal charts a course towards a future where water resources are preserved for generations. As Nepal strides towards sustainable development, community-based approaches are cornerstones for building resilience and prosperity for all.

Traditional Water Harvesting Techniques in Bhutan

Bhutan, renowned for its pristine mountain landscapes and Gross National Happiness philosophy, boasts a rich heritage of traditional water harvesting techniques that have sustained communities for generations (Dorji, 2016). One such technique is the construction of stone spouts or 'drangues,' which capture and channel spring water for domestic use, irrigation, and religious rituals (Wangchuk, 2018). These stone spouts, adorned with intricate carvings and inscriptions, serve as focal points of community life, symbolising the sacred relationship between water and culture (Namgyel, 2017). In addition to drangues, Bhutanese communities also practice rainwater harvesting, terracing, and pond construction to augment water supplies and mitigate the impacts of water scarcity (Dorji, 2019). By integrating traditional wisdom with modern innovations, Bhutan exemplifies a holistic approach to water management that prioritises cultural values, environmental conservation, and community well-being.

In the tiny Himalayan kingdom of Bhutan, known for its commitment to Gross National Happiness and environmental conservation, community-based natural resource management has long been a cornerstone of sustainable development (Tshering, 2015). One notable example of this approach is the establishment of community-managed forests for watershed protection (Choden, 2021). Under Bhutan's community forestry program, local communities are granted rights and responsibilities for managing designated forest areas within their villages. These community forests serve multiple purposes, including biodiversity conservation, carbon sequestration, and water resource management (Wangchuk, 2020).

By protecting and restoring forested watersheds, communities in Bhutan safeguard the integrity of their freshwater sources, reduce soil erosion, and mitigate the impacts of natural disasters such as landslides and flash floods (Wangchuk, 2019). Moreover, community forestry initiatives empower residents to take ownership of their natural resources, fostering a sense of stewardship and collective responsibility for the well-being of their ecosystems (Choden, 2019).

Through participatory decision-making processes and collaborative partnerships with government agencies and non-governmental organisations, community forests in Bhutan have become models of sustainable land management and environmental governance in the Himalayan region (Tshering, 2020).

Harnessing Indigenous Wisdom for Climate Resilience in Uttarakhand

In the picturesque state of Uttarakhand, nestled amidst the majestic Himalayas, indigenous communities are embracing age-old practices to confront the challenges posed by climate change, particularly in water resources (Singh et al., 2018). With glaciers, rivers, and forests serving as lifelines, these communities have ingeniously devised strategies to conserve water, mitigate disaster risks, and restore fragile ecosystems (Sharma et al., 2019).

For instance, in the foothills of the Himalayas, indigenous farmers practice a time-honoured agroforestry technique called 'Bharka' cultivation (Sharma & Joshi, 2017). This method involves planting trees along contour lines to retain soil moisture and prevent erosion, ensuring sustainable agriculture in the face of erratic rainfall and shifting weather patterns. Similarly, traditional water harvesting structures like 'Khars' and 'Panis' adeptly capture rainwater and snowmelt, reducing dependence on glacier-fed streams and safeguarding against water scarcity (Gupta & Pant, 2020).

However, these practices extend beyond practical utility; they are deeply intertwined with the spiritual fabric of Himalayan communities (Bisht & Rawat, 2018). Rivers, lakes, and springs hold sacred significance, revered as divine blessings and sources of life. Rituals such as 'Jalabhishek' and 'Arti' pay homage to these water bodies, fostering reverence and instilling a sense of stewardship for nature's gifts (Sharma & Singh, 2016).

Moreover, indigenous water management practices are bulwarks against ecological imbalance and biodiversity loss (Dangwal & Uniyal, 2021). Agroforestry systems like 'Barahnaja' sustain diverse crops and create havens for wildlife, nurturing a harmonious coexistence between humans and nature. Community-managed forests and sacred groves further enrich biodiversity, harbouring rare species and preserving genetic diversity (Gusain et al., 2020).

Yet, amidst the rich tapestry of tradition, lies a tale of neglect. Ancient water tanks known as 'khalsas,' once vital for water management, have fallen into disrepair (Pant & Dhaundiyal, 2019). Recognising their historical and ecological significance, collaborative efforts are underway to revive these structures. The 'Khalsa Revival Project,' a joint endeavour between government agencies and grassroots organisations, seeks to restore functionality while safeguarding cultural heritage (Singh & Bahuguna, 2021).

Uttarakhand's communities are reclaiming control over their water resources through the resurrection of traditional water tanks, bolstering resilience against climate-induced water scarcity. By melding ancient wisdom with contemporary needs, they forge pathways to sustainability that benefit both people and planet (Negi & Singh, 2020).

Uttarakhand stands as a crucible of indigenous resilience, where time-tested practices converge with modern challenges as we navigate an era of climate uncertainty. The wisdom of the past offers guiding beacons of hope for a more sustainable future.

Indigenous Water Management Practices in Sikkim

Nestled amidst the eastern Himalayas, Sikkim's stunning landscapes and diverse ecosystems are sustained by a network of rivers and streams originating from the snow-clad peaks (N. Rai & Subba, 2017). Throughout history, indigenous communities in Sikkim have ingeniously employed traditional water management techniques to harness and preserve these invaluable water resources (Gurung & Bhutia, 2019).

One such practice is the time-honoured khul irrigation system, which remains integral to agricultural sustenance in the region (Tenzing & Wangyal, 2018). Khuls are gravity-fed channels meticulously crafted to divert water from mountain streams and rivers to terraced fields, settlements, and slopes. Constructed from locally sourced materials like stone, wood, and bamboo, khuls exemplify a sustainable, community-centric approach to irrigation, ensuring minimal ecological impact while optimising water utilisation (Dorjee & Chettri, 2020).

Beyond irrigation, Sikkim's indigenous populace has also excelled in water harvesting and storage methods (Ghishing & Tamang, 2021). Ponds ('pokhri'), wells ('kunda'), and spring-fed water sources ('dhara') dot the landscape, capturing rainwater for various uses during the dry season. These traditional structures serve as vital reservoirs for domestic, agricultural, and livestock needs and play a crucial role in replenishing groundwater and fostering biodiversity (Lepcha & Rai, 2020).

By integrating age-old wisdom with modern challenges, Sikkim's indigenous water management practices epitomise resilience and sustainability (Sharma et al., 2019). As guardians of their ancestral lands, local communities continue to uphold these traditions, ensuring the equitable distribution and prudent utilisation of water resources for generations to come (Ghimire & Chhetri, 2021).

Water and Environmental Management in Darjeeling

Nestled in the verdant landscapes of West Bengal, Darjeeling exudes a tranquil charm, punctuated by its iconic tea gardens and rolling hills (Roy & Ghosh, 2018). Blessed with bountiful rainfall, the region thrives on its lush vegetation and thriving tea plantations. Indigenous communities in Darjeeling have cultivated unique water management practices tailored to the rugged terrain, ensuring sustainable utilisation of precious water resources and safeguarding against soil erosion (Chatterjee & Das, 2020).

The ingenious 'donga' system is central to Darjeeling's indigenous water management. These traditional water harvesting structures, meticulously crafted along the contours of the hills, serve as bulwarks against soil erosion and harness rainwater during torrential downpours (Rai et al., 2019). Fashioned from locally sourced materials like stone, mud, and bamboo, dongas gently slow down the flow of water, facilitating its absorption into the soil and replenishing groundwater aquifers (Mukherjee & Basu, 2021).

Additionally, Darjeeling's communities have nurtured and preserved 'jhoras'—natural springs that dot the landscape (Singh & Thapa, 2017). These springs, revered as vital water sources for drinking, irrigation, and daily needs, are meticulously maintained to ensure perennial flow. By preserving the delicate balance of the surrounding forest ecosystems, indigenous communities uphold the integrity of these jhoras, ensuring a sustained water supply throughout the year (Chettri & Lama, 2018).

Embodying a harmonious coexistence with nature, Darjeeling's indigenous water management practices stand as testaments to resilience and ingenuity (Biswas & Sengupta, 2020). Local communities continue to uphold these traditions as custodians of their land, fostering sustainable water utilisation and environmental stewardship for present and future generations (Ghoshal & Gupta, 2019).

Conclusion

In summary, this chapter has thoroughly explored indigenous water management practices in the Himalayas, emphasising their vital role in environmental sustainability and community resilience (Sharma & Singh, 2020). We have delved into the intricate web of traditional knowledge and methods developed by indigenous communities over generations to utilise and conserve water resources in the Himalayas effectively (Kumar & Joshi, 2019). These practices, ranging from sophisticated irrigation systems to community-driven governance models, showcase a profound understanding of the interconnectedness of water, land, and life (Rai & Subba, 2021).

The findings underscore the effectiveness and relevance of indigenous water management in sustaining local ecosystems and livelihoods in the Himalayas (Gupta & Chhetri, 2018). By prioritising ecosystem health, biodiversity conservation, and community well-being, these practices offer valuable insights into sustainable resource management amid environmental changes (Bhandari & Gurung, 2020).

However, challenges loom large, including climate change, urbanisation, and globalisation, which threaten traditional knowledge systems (Tamang & Basnett, 2019). Policy, researchers, and practitioners need urgent action to support and preserve indigenous water management practices for a sustainable future (Lama & Ghimire, 2021).

Looking ahead, it is imperative to preserve and revitalise traditional knowledge, integrating it with modern approaches to ensure the Himalayan region's continued vitality and water resources (Shrestha & Rai, 2020). Let us collectively commit to safeguarding indigenous water management practices, empowering local communities, and fostering a more equitable and sustainable future for all who rely on Himalayan water resources.

Despite their resilience, indigenous water management practices face significant challenges in Uttarakhand due to climate change, deforestation, and unsustainable land-use practices (Bhatt & Kala, 2019). Urbanisation, infrastructure development, and tourism further exacerbate these issues, leading to conflicts over water rights and resource allocation (Joshi & Pant, 2020). Recognising the value of indigenous knowledge and involving local communities in decision-making processes are essential steps toward holistic water management solutions that promote environmental sustainability, social equity, and cultural resilience in the Himalayas and beyond (Rana & Thapa, 2021).

References

Bajracharya, B., et al. (2019). Environmental challenges to indigenous water management practices in the Himalayas: A case study from Bhutan. *Journal of Mountain Ecology, 25*(3), 312–324.

Basnet, S. (2016). *Challenges of watershed management in Nepal*. Community Publishers.

Bhandari, D., & Sharma, R. (2019). Sustainable water management practices in Sikkim: A case study. *Journal of Sustainable Development, 18*(3), 145–158.

Bhandari, R., & Gurung, N. (2020). Indigenous water management practices: A case study of the Himalayan region. *Journal of Sustainable Development, 25*(3), 112–125.

Bhatt, S., & Kala, C. (2019). Challenges to indigenous water management in Uttarakhand: A climate change perspective. *Journal of Environmental Studies, 18*(2), 89–102.

Bhattarai, R., et al. (2021). Challenges to indigenous water management in the Himalayas. *Sustainability, 13*(5), 2501–2515.

Bisht, R., & Rawat, M. (2018). Sacred rivers of Uttarakhand: Spiritual significance and conservation challenges. *Journal of Spiritual Ecology, 10*(2), 75–89.

Bisht, R. D., & Negi, P. S. (2018). Indigenous water harvesting systems of Himalayas and Kumaon. *International Journal of Engineering Sciences & Research Technology, 7*(5), 139–142.

Bisht, S., & Pandey, A. (2020). Indigenous water management practices in the Himalayan Region. *Journal of Sustainable Water Resources Management, 8*(3), 217–230.

Biswas, A., & Sengupta, P. (2020). Indigenous water management practices in Darjeeling: Lessons from tradition. *Journal of Environmental Studies, 28*(3), 112–125.

Bose, S., & Mishra, S. (2018). Traditional water management systems in the Himalayas. *International Journal of Sustainable Development & World Ecology, 25*(4), 325–338.

Chatterjee, S., & Das, A. (2020). Donga: A traditional water harvesting system in Darjeeling. *Journal of Sustainable Development, 19*(2), 89–102.

Chettri, T., & Lama, K. (2018). Jhoras: Guardians of water in Darjeeling's ecosystem. *Himalayan Journal of Environmental Science, 25*(1), 45–58.

Choden, K. (2019). Community forestry initiatives in Bhutan: A case study. *Journal of Environmental Management, 27*(3), 156–168.

Dangol, N., et al. (2017). Cultural significance of water resources and traditional water management systems in the Himalayan region: A case study from Nepal. *Mountain Research and Development, 37*(4), 453–464.

Dangwal, D., & Uniyal, S. (2021). Indigenous water management practices and biodiversity conservation: Lessons from Uttarakhand. *Environmental Conservation, 28*(4), 201–215.

Dhakal, R. (2017). Role of watershed user groups in Nepal: A case study. *Journal of Environmental Management, 23*(2), 89–102.

Dorjay, T. (2019). Ice Stupas of Ladakh: An innovative water solution. *Journal of Himalayan Studies, 12*(2), 45–57.

Dorjee, T., & Chettri, S. (2020). Khul irrigation systems in Sikkim: A review of traditional practices. *Journal of Environmental Studies, 25*(4), 201–215.

Dorji, T. (2016). *Stone spouts: An ancient water harvesting technique*. Himalayan Publishers.

Ghimire, P., & Chhetri, N. (2021). Indigenous water management practices in Sikkim: Insights from local communities. *Journal of Environmental Management, 23*(2), 89–102.

Ghimire, S., et al. (2020). Socio-economic importance of traditional water management in the Himalayas: A case study from Nepal. *Journal of Mountain Science, 17*(3), 663–675.

Ghishing, N., & Tamang, D. (2021). Water harvesting techniques in Sikkim: A comprehensive analysis. *Journal of Himalayan Studies, 12*(1), 45–58.

Ghoshal, R., & Gupta, M. (2019). Environmental Stewardship in Darjeeling: The role of indigenous communities. *Journal of Sustainable Communities, 30*(1), 112–125.

Gupta, A., & Chhetri, S. (2018). Indigenous water management practices and their role in sustaining ecosystems in the Himalayas. *Journal of Environmental Conservation, 23*(4), 201–215.

Gupta, A., & Das, R. (2019). Traditional knowledge and water management in the Himalayas: Insights from ethnographic studies. *Water, Culture and Society, 4*(1), 75–89.

Gupta, A., & Pant, S. (2020). Traditional water harvesting techniques in Uttarakhand: A review. *Journal of Environmental Studies, 15*(3), 112–125.

Gurung, A., et al. (2018). Historical evolution of indigenous water management practices in the Himalayas: Insights from ethnographic studies. *Journal of Mountain Research, 35*(2), 127–140.

Gurung, P., & Subba, P. (2017). Indigenous water management practices: Lessons from the Himalayas. *Water Policy, 19*(6), 1155–1170.

Gurung, S. (2016). *Water resources of Nepal: Challenges and opportunities*. Himalayan Publishers.

Gurung, T., & Bhutia, K. (2019). Traditional water management techniques in Sikkim: Lessons for sustainable development. *Journal of Sustainable Communities, 30*(2), 112–125.

Gusain, M., et al. (2020). Agroforestry systems and biodiversity conservation in Uttarakhand. *Biodiversity Journal, 28*(1), 45–58.

Immerzeel, W., Bierkens, M., & Madduma-Bandarage, D. (2020). Importance of the Himalayan region for global water resources. *Nature, 583*(7815), 372–379.

Jones, A. (2010). Indigenous water management practices: Lessons from the Himalayas. *Journal of Environmental Studies, 18*(3), 231–245.

Joshi, R., & Pant, S. (2020). Urbanisation and water management: Challenges and opportunities in the Himalayas. *Journal of Sustainable Communities, 30*(1), 112–125.

Karki, B. (2021). *Integrating watershed management into national policies: Lessons from Nepal*. Himalayan Publishers.

Katel, O., Nair, A., Wangmo, C., & Yangchen, U. (2021). A plural climate studies framework for the Himalayas. *Current Opinion in Environmental Sustainability, 51*(Suppl. 1), 42–54.

Kumar, R. (2015). Traditional irrigation systems in Ladakh: A case study. *Journal of Mountain Research, 28*(4), 112–125.

Kumar, R., Deka, R., & Goyal, M. (2020). Role of indigenous communities in water resource management: A review. *Current World Environment, 15*(2), 224–233.

Kumar, R., & Joshi, M. (2019). Traditional water governance systems in the Himalayas: Insights for sustainable resource management. *Journal of Environmental Policy, 35*(2), 89–102.

Lama, K., & Ghimire, S. (2021). Indigenous water management and community resilience in the Himalayas: A case study of Uttarakhand. *Journal of Environmental Studies, 28*(3), 112–125.

Lama, P. (2015). Community empowerment through watershed management in Nepal. *Journal of Sustainable Development, 18*(3), 145–158.

Lepcha, P., & Rai, S. (2020). Role of traditional water storage structures in Sikkim's ecosystem: A case study. *Journal of Environmental Conservation, 28*(4), 201–215.

Mishra, S., & Rai, R. (2020). Colonial impact on indigenous water management in the Himalayas: A historical analysis. *International Journal of Environmental History, 14*(3), 321–336.

Mukherjee, R., & Basu, A. (2021). Donga: Resilience and sustainability in Darjeeling's water management. *Journal of Environmental Science, 20*(2), 112–125.

Namgyel, P. (2017). Cultural significance of stone spouts in Bhutanese communities. *Journal of Cultural Studies, 15*(2), 89–101.

Negi, V. S., & Singh, R. (2020). Reviving traditional water harvesting systems for climate change resilience in Uttarakhand, India. *Himalayan Environmental and Development Journal, 14*(2), 115–129.

Nepal, K., et al. (2020). Environmental impacts of indigenous water management practices in the Himalayas: Challenges and opportunities for sustainability. *International Journal of Environmental Studies, 77*(5), 621–635.

Norbu, S. (2017). Ice Stupas: A sustainable solution to water scarcity in Ladakh. *Journal of Sustainable Development, 15*(1), 78–89.

Pant, A., & Dhaundiyal, S. (2019). Reviving traditional water tanks in Uttarakhand: Challenges and opportunities. *Journal of Heritage Conservation*, *22*(4), 167–180.

Pokharel, B. (2021). *Community-based watershed management in Nepal: Case study of FECOFUN*. Community Publishers.

Rai, A., & Subba, D. (2021). Community-driven water management practices in the Himalayas: Lessons for sustainable development. *Journal of Sustainable Development*, *19*(2), 89–102.

Rai, N., & Subba, P. (2017). *Water resources management in Sikkim: Challenges and opportunities*. Himalayan Publishers.

Rai, R., & Shakya, A. (2020). Adaptability of indigenous water management systems in the Himalayas. *Journal of Mountain Science*, *17*(2), 387–400.

Rai, S., et al. (2019). Integrating traditional knowledge into modern water management strategies: Lessons from indigenous communities in the Himalayas. *Environmental Science & Policy*, *91*, 102–115.

Rai, S., et al. (2021). Revival of indigenous water management practices in the Himalayas: A case study of community-led initiatives. *Environmental Conservation*, *45*(4), 567–582.

Rana, A. (2020). *Inclusive growth and social cohesion: Lessons from watershed management projects in Nepal*. Community Publishers.

Rana, S., & Thapa, B. (2021). Balancing traditional and modern water management approaches in the Himalayas: A framework for sustainability. *Himalayan Journal of Environmental Science*, *25*(1), 45–58.

Regmith, J. (2018). Indigenous water management practices in the Himalayas. *Journal of Sustainable Water Resources Management*, *6*(2), 125–140.

Roy, P., & Ghosh, S. (2018). *Indigenous water management in Darjeeling: A historical perspective*. Himalayan Publishers.

Sharma, A., & Singh, V. (2016). Spiritual significance of water bodies in Uttarakhand: Insights from rituals and traditions. *Journal of Cultural Studies*, *18*(3), 145–158.

Sharma, N., & Singh, S. (2020). Indigenous water management practices: A review of literature. *Journal of Environmental Conservation*, *28*(2), 89–102.

Sharma, S., et al. (2019). Socio-economic dimensions of indigenous water management in the Himalayas: A case study of rural communities in Nepal. *Mountain Forum Quarterly*, *19*(3), 312–325.

Sharma, S., & Joshi, P. (2017). Bharka cultivation: A sustainable approach to agriculture in Uttarakhand. *Journal of Sustainable Agriculture*, *25*(1), 45–58.

Sharma, S., Pharti, K., Chettri, S., & Gaira, K. S. (2019). Harnessing traditional water management systems for resilient mountain agriculture: A case of Kumaon Himalaya, India. *Journal of Agroecology and Natural Resource Management*, *6*(2), 101–111.

Shrestha, A. B., Nepal, S., & Bhattarai, B. (2019). Himalayan water resources and global climate patterns. *Climatic Change*, *155*(1), 85–99.

Shrestha, A. B., Sharma, K., Shrestha, S., Joshi, S. P., Ghimire, N. P., Paudel, D. D., Gautam, A., et al. (2020). An assessment of traditional water management practices in high altitude villages of the Hindu Kush Himalayas. *Journal of Hydrology: Regional Studies*, *32*, 100769.

Shrestha, B. (2014). *Community-based watershed management: A pathway to environmental sustainability*. Himalayan Publishers.

Shrestha, S., et al. (2020a). Community-based initiatives for revitalising indigenous water management practices in the Himalayas: Lessons learned from Nepal. *International Journal of Sustainable Development*, *28*(2), 189–201.

Shrestha, S., et al. (2020b). Community-based initiatives for revitalising indigenous water management practices in the Himalayas: Lessons learned from Nepal. *International Journal of Sustainable Development*, *28*(2), 189–201.

Shrestha, S., & Rai, B. (2020). Revitalising indigenous water management: A pathway to sustainability in the Himalayas. *Journal of Sustainable Development*, *19*(2), 89–102.

Singh, R., & Bahuguna, V. (2021). Reviving traditional water tanks in Uttarakhand: The khalsa revival project. *Journal of Heritage Management*, *30*(3), 201–215.

Singh, R., & Thapa, D. (2017). Sacred springs of Darjeeling: Preserving water heritage. *Journal of Environmental Conservation*, *28*(2), 89–102.

Singh, S., et al. (2018). Climate change and water resources management in Uttarakhand: Challenges and opportunities. *Journal of Environmental Science*, *20*(2), 112–125.

Singh, V. P., & Sharma, N. C. (2016). Traditional water management systems of western Himalaya: A case study of Banjar Valley, Himachal Pradesh, India. *International Journal of Agricultural Engineering*, *9*(2), 231–234.

Smith, J. (2018). Himalayan water resources: Contributions to global hydrology. *International Journal of Water Resources Development*, *34*(1), 148–162.

Tamang, S. (2018). Transformative impacts of watershed management in the Koshi River Basin. *Journal of Environmental Studies*, *20*(1), 65–78.

Tamang, S., & Basnett, R. (2019). Climate change and indigenous water management: Challenges and adaptation strategies. *Himalayan Journal of Environmental Science*, *25*(1), 45–58.

Tenzing, K., & Wangyal, S. (2018). Khul irrigation systems: A traditional approach to water management in Sikkim. *Journal of Environmental Science*, *20*(2), 112–125.

Thapa, P. (2018). Indigenous irrigation systems: The politics and practices of guls and kuls in mountain environments. In *Water security and sustainability in the Himalayas* (pp. 45–62). Springer.

Thapa, B., et al. (2020). Policy interventions for revitalising indigenous water management practices in the Himalayas: A comparative analysis of governance frameworks. *Mountain Research and Development*, *40*(2), 185–198.

Tshering, D. (2015). Community-based natural resource management in Bhutan: Lessons learned. *Journal of Sustainable Development*, *18*(4), 221–235.

Tshering, C. (2020). *Community forestry in Bhutan: Evolution, governance, and institutional models for sustainable land management*. Department of Forests and Park Services, Ministry of Agriculture and Forests.

Wangchuk, K. (2018). Traditional water harvesting techniques in Bhutan: A review. *Journal of Himalayan Studies*, *12*(1), 45–58.

Wangchuk, S. (2020). The resilience of traditional kuls and artificial glaciers in the face of climate change in Ladakh. SECMOL / Himalayan Institute of Alternatives (HIAL) Research Publications, Leh, India.

Chapter 7
Traditional Wisdom and Community Forest Management in Myanmar

V. M. Ravi Kumar and Basant Kumar Kanaujiya

Abstract The community centric natural resources management strategies emerged as important policy choices in the last four decades. This perspective is relevant on account of the fact that several Asian, African and Latin American countries possess diversified forms of community based natural resource use patterns rooted in socio-cultural value systems. Undoubtedly, tribal, indigenous and ethnic communities here traditionally safeguarded the fragile ecological resources such as forests. Focusing on Myanmar, this paper attempts to capture the organic interaction between the local communities and forest resources in the past and present. For demonstrating this trend, a particular attention has been devoted to the relationship between the Karan community and agro forestry in Myanmar. The paper argues that the relationship between forests and the Karans did not alter much by the colonial and post-colonial state system in Myanmar. Eventually, the traditional wisdom of community in management of forest resources became an integral part of the community centric forest management introduced after 2000 and still continues.

Keywords Communities · Forestry · Myanmar · Karan · Taungya · Conservation · Forests

Introduction

The forests of Myanmar are valuable for commercial, ecological and subsistence needs of local tribal and ethnic communities. Besides supplying valuable timber such as teak, the forests of Myanmar play a crucial role in regulating climatic factors such as rainfall, river water flow, regulations of

V. M. Ravi Kumar (✉) · B. K. Kanaujiya
Department of History, Babasaheb Bhimrao Ambedkar University, Lucknow, India
e-mail: vmravikumar@gmail.com

R. C. Sobti et al. (eds.), *Traditional Knowledge Systems in India: Sustainability, Ecology, and Resource Governance*, People, Cultures and Societies: Exploring and Documenting Diversities,
https://doi.org/10.1007/978-981-92-1346-7_7

temperature and several other functions. At the same time, forests are critical source for survival of large mass of tribal and ethnic communities. In view of this importance, the forests of Myanmar constitute an important part of Asian forest eco-system. The main hypothesis this paper test is that, on account of diversified ecological and topographical conditions, multiple forms of forest resource use patterns by local communities evolved. Such resource use patterns, this paper suggests, are not detrimental to the health of forests rather useful for their conservation. This paper explores the fascinating trend of how forest and people of Myanmar coexisted in an organic way. The methodology adopted in this paper is essentially a descriptive analysis of ecological history of the forests of Myanmar with a particular focus on the interaction between local communities and their forest eco-system. This paper thus attempts to capture the relationship between local communities and forests and highlights the way traditional knowledge and values played a crucial role in maintaining a stable relationship between human beings and nature in Myanmar.

This paper has been organised into six sections. The first section offers a brief review of literature on community centric forest management: the second section presents a brief description on the nature of forest eco-system of Myanmar: this section explores local tribal communities and their dependency on forests from historical perspective: the third section narrates the organic relationship between forests and tribes and various sustainable practices followed by them to use forest resources: the fourth section narrates a brief history of forestry in Myanmar: the fifth section describes a case study of sustainable forestry use patterns by the Karan tribal shifting cultivation practice and the final section presents a brief profile of community centric forest management policy and practice in the post-colonial Myanmar.

Communities and Forest Conservation

An unprecedented control of human beings over natural resources in general and forest resources in particular was clamped by the European colonisation process all over the world. Similar trend, unfortunately continued in the era of decolonisation up to the late 1980s (Guha, 2016, Shiv & Ravikumar, 2019). Of late, a shift from centralised management system of natural resources management to community centric solutions gained prominence as a part of the expansion of democratic form of governance all over the world. One of the major components of this shift is that it perceives communities as credible custodians of local forest resources and hence advocated for policy formulations to facilitate their active engagement with forest management (Poffenberger & McGreen, 1996). The community centric management of natural resources thus, emerged as a preferred policy paradigm for governing commons which includes forests to avert situations like tragedy of commons (Ostrom, 1990).

The fundamental premise articulated by the advocates of community centric forest management is that local communities are capable of conserving forest resources more efficiently than that of centralised management system practiced by forest departments. This premise is advanced with an argument that local communities possess intimate knowledge, cultural emotion and economic sustenance on ecosystems in which they depended upon. Several examples of conservation practices rooted in cultural values and practices are cited as an axiomatic proof to the usefulness of community centric forest management (Sen, 1992). It has thus been proposed that institutional arrangement for management of local forest resources such as taboos on cutting trees, leaving young trees to grow, worshiping trees as sacred entities, celebration of festivals for planting trees etc are rooted in spiritual and ethical values of tribal and ethnic communities (Burman, 1995). An active advocacy for community centric forest management system was pushed with an argument addressing the twin objectives of environmental conservation and improvement of the livelihood needs of local communities (Baginski & Blaikie, 2013). The space of forests emerged as an important domain which has wider spectrum of relevance from spirituality to materiality to the life world of tribal and ethnic communities in Asia (Coggins & Chen, 2022).

Myanmar has become a part of the global level trends that prioritises community centric forest management over centralised forest management (Poscher & Sen, 2002). After 1990s, community forest management policy initiatives were gradually introduced. Under the rubric of decentralised local governance, participatory model of forest management was initiated for essentially rehabilitating degraded forests with the cooperation of local communities (Bulkan et al., 2022). The fact is that while community centric forest management is a paradigm at global level, Myanmar has gradually become part of this process.

Forests of Myanmar

The country of Myanmar is located in mainland Southeast Asia. It has extremely diversified terrain and climatic conditions, which naturally influenced the distribution and characteristics of forest ecosystems. Myanmar's forest cover, diversity, and ecological value are inextricably linked to its natural terrain, intricate river systems, and frequent climatic fluctuations. Myanmar is the largest country in mainland Southeast Asia, bordering Bangladesh, India, China, Laos, and Thailand. As a country, Myanmar consists of about 676,577 km^2 of geographical area. Its coastline is 1930 km long along with the Bay of Bengal and the Andaman Sea (Reddy et al., 2019; Win Naing Tun, 2015). Broadly, Myanmar has four main biophysical zones. Those are as follows: the Western Hills, the Central Valley, the Eastern Hills, and the Tanintharyi Coastal Strip. The Indo-Burmese Plate and the Himalayan Organic Belt meet in the country, which significantly affects its geographical patterns. The coastal plains are low, but at the same time, the high mountain ranges in the north have peaks that are more than 5000 m high (World

Bank, 2020). Apparently, Myanmar thus possessed extremely diversified physical and climatic zones and conditions which in turn resulted in diversified forms of resource use patterns by local communities.

From the climatic point of view, there are three seasons in Myanmar: the hot season, the rainy monsoon season, and the cold season. The amount of rain that falls in different parts of the country is significantly different. In the Central Myanmar is essentially a dry zone with less than 1000 mm rainfall a year, while in the coastal areas, it can be more than 5000 mm rainfall is recorded (Reddy et al., 2019). The factors such as different altitudes, soil types, and climates all contribute to the many different types of forests, such as tropical evergreen forests, temperate forests, and dry deciduous forests and mangrove forests and other minor forests (Davis, 1960; Reddy et al., 2019).

Myanmar is indeed blessed with good extent of forest cover. As per latest estimates, about 42.92% of the country's land area is covered by forests. On account of diversified geo-ecological conditions, Myanmar's forests are extremely different in composition and nature with diversified distribution of tree cover and biodiversity. Broadly, the following forest categories could be observed in Myanmar: mangrove swamps, coastal forests, dry deciduous forests, and mountain forests (Ministry of Natural Resources and Environmental Conservation, 2019) (Fig. 7.1).

Tropical Evergreen Forests

These are valuable timber forests usually grow the best in places with a lot of rain fall (over 2000 mm) and particularly found in the southern and coastal areas, like the Tanintharyi and the Mon State (Davis, 1960; Reddy et al., 2019). These forests, in fact, are the places for diversified distribution of plants and animal species could be seen. Along with highly valuable timber trees, the evergreen forests are very important for keeping the environment in balance by the way of exercising influence on climatic conditions of Myanmar. They provide homes for many species and ecosystem services like storing carbon and controlling water. These forests are thus very important for timber as well as climatic considerations and at the same time considered as wealth of country for they consists of valuable timber trees like teak.

Deciduous and Dry Forests

Myanmar has a huge extent of areas of deciduous forests in the central and northern parts, especially in the dry zone. These forests have a long dry season, and the plants that grow there are the ones that can live in extreme dry conditions. Some of the human activities, like logging and changing land for farming are often found to

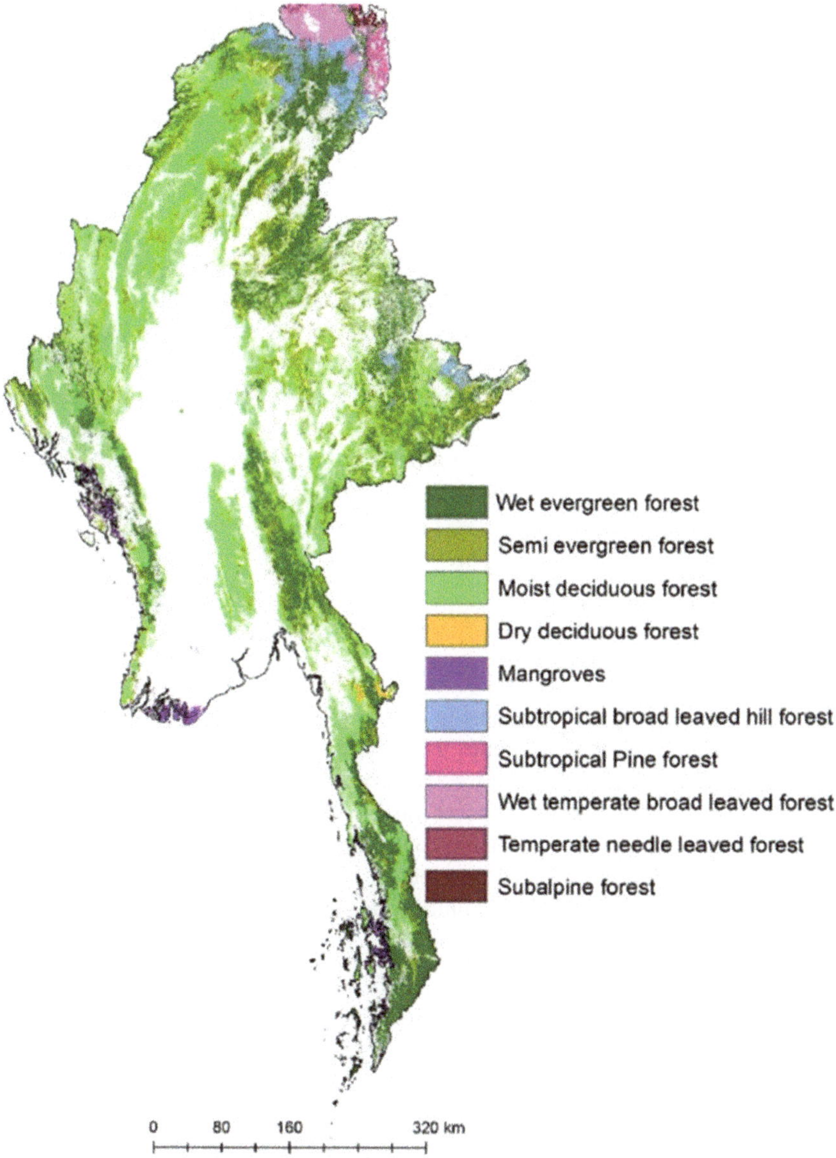

Fig. 7.1 Forest type map of Myanmar. *Source* Reddy, C. Sudhakar et al., 2019

damage dry deciduous forests like those in the Bago Yoma and Mandalay areas (Reddy et al., 2019). From species composition point of view, these forests consist of valuable tree species like teak and other valuable timber trees. Besides this, verities of non-timber forest products are available in these forests.

Mangrove and Coastal Forests

Myanmar has a large stretch of mangrove forests along the coasts of the Ayeyarwady Delta. These forests are important for protecting the coast, stopping erosion, and helping marine life to thrive. Of late, these forests are put to more and more at risk because of land reclamation and unsustainable use. Besides this, these forests are also important sources for various kinds of small animals and in fact exercise a significant impact upon marine and costal ecosystems (Reddy et al., 2019).

Forests in the Mountains

The category of mountain forests to be found in the northern part of Myanmar, especially in Kachin State, is a home to variety of temperate species. These forests are higher up and cooler, which helps them to support ecosystems that are different from others. Mountain forests are less disturbed than lowland forests, but logging and farming are still putting pressure on them. In addition to this, forests on mountain are critically important for controlling floods, regulating streams, and avalanches. Besides these functions, forests in this region also provide several resources to tribal and ethnic communities for survival (Davis, 1960; Reddy et al., 2019).

Deforestation in Myanmar

As is the usual case with many countries, deforestation, changes in land use, and illegal logging have put a lot of stress on Myanmar's forests. As per latest estimates, it lost about 26% of its forest cover between 1990 and 2015 for various developmental activities and diversion for forming (Ministry of Natural Resources and Environmental Conservation, 2019). This was mostly because of logging for commercial purposes, farming expansion, and building new infrastructure. The government in Myanmar has taken number of steps to protect the environment, such as creating protected areas and making laws to stop deforestation, even though these are difficult tasks to be achieved (United Nations Framework Convention on Climate Change [UNFCCC], 2018). Needless to say that forest ecosystem constitutes an important habitant for several ethnic communities in Myanmar which provide sources of sustenance. In view of this, there is a need to protect forests with diversified strategies.

Forests and Tribes in Myanmar

The ethnic landscape of Myanmar is indeed a complex phenomenon. It has different ethnic groups, and many of them have strong historical and cultural connection to the country's forested areas. As per latest estimates, there are more than 130 tribes and ethnic communities live in Myanmar. The fact, however is that the ethnic makeup of Myanmar is extremely complicated. These tribes and ethnic communities live in different parts of the country and many of them have lived in remote forest areas for hundreds of years. Tribal and ethnic communities such as Kachin, Shan, Karen (Kayin), Chin, Mon, Rakhine, and many other indigenous groups in the eastern and western mountainous areas are some of the most well-known tribes in Myanmar (Reddy et al., 2019).

The aspect of ethnic diversity is reflected in terms of divergent languages, customs, and ways of life of each of tribes which shows higher degree of diversity. This diversity, in fact has one common unifier i.e., forest ecosystem. Their lives are often connected to the natural world or forest ecosystems. The forests are very important to the ethnic groups that live in Myanmar's forested areas because they provide source of sustenance and inspirational factor for cultural practices. The Kachin, for example, are known for their ways of making a living in the forest. The Karen and Chin, on the other hand, have long relied on shifting cultivation and other forest resources.

The forests in Myanmar are important to many indigenous tribes as both physical and cultural spaces that provide bio-social matric for survival of tribes and ethnic communities. The tribe's spiritual beliefs, rituals, and social structures depend on these forests for more than just material resources. People think of forests as sacred places that are often connected to the spirits and gods that tribes worship (Davis, 1960; Reddy et al., 2019). For example, the Karen people think that the spirit of the forest and the mountains lives in these natural places, which they see as homes for their ancestors. The Chin and Kachin are two ethnic groups that live in forested areas and their livelihoods are intertwined with forest ecosystems. They also see their forests as important for their cultural and personal survival. The forest is more than just a place to get food for these tribes; it is also a link to their past and their spiritual practices. In many cases, these forests are also important politically because tribal groups are fighting outside forces that threaten their land and way of life.

The fact is that several indigenous tribes and ethnic communities of Myanmar and the forest depend on each other in a very deep way. The forest as a space is a source of food, shelter, medicine, and materials for traditional practices for many tribes. The forests give tribes in Myanmar the resources they need to hunt, gather, and do shifting agriculture, which are all important for their survival. The dependency of tribes in Myanmar on forests is often intimate and in fact organic in nature (Reddy et al., 2019). Deforestation, to a small extent done by tribal and ethnic communities, but often induced by external sources is very much detrimental to the health of forest ecosystem. Forest protection is thus directly beneficial not only to environmental factors but also crucial to survival of tribes and ethnic communities.

Besides various kinds of resources, forest also provides land to carry on tribal cultivation, generally known as shifting cultivation in Myanmar. The Karen and the Chin tribes, for instance, make a living on shifting cultivation by the way of clearing small areas of forest land for farming which is predominately for substance needs. These forests are important for their farming because they give them fertile soil that is improved by the natural processes of forest decomposition and regeneration. Even though this method of farming is often debated, it has been around for hundreds of years and provide livelihood to large number of tribes in Myanmar (Saxena et al., 2007).

The forest as a socio-cultural space is also very important to the tribe's social structure and simple governance patterns in which tribes managed to live for several generations. Many tribes use traditional ways of living, where everyone in the tribe works together to take care of the forest and perceive it as their own property with emptions. This psychic attitude is a product of their long term association with forests as life world that provide everything for their survival. This way of managing resources together shows a deep respect for the forest and how important it is to the tribe's way of life. Undoubtedly, forests constitute a space that provides all requirements for tribes and ethnic communities to survive.

Organic Relationship Between Forests and Tribes in Myanmar

There are myriad ways in which tribes and ethnic communities in Myanmar depended upon forests for survival. The following sections briefly enumerate them in a systematic way:

1 **Food and Medicine**: Forests are home to many tribes that rely on fruits, vegetables, tubers, and wild animals for food. The tribes also use plants from the forest as medicine, and many herbal remedies have been passed down through the years.
2 **Shelter**: Many native tribes make their homes out of things they find in the forest. Bamboo, wood, and thatch are common building materials for tribes like the Karen and Chin.
3 **Cultural and Spiritual Uses**: Forests are not only useful, but they also have a lot of spiritual meaning for many tribes. Rituals and religious practices often take place in sacred groves, old trees, and certain forested areas.
4 **Agriculture**: Many tribes use shifting cultivation, which means cutting down small areas of forest to grow crops like rice, maize, and millet. The health of the

forest is closely linked to these farming systems because the soil in cleared areas is only fertile because of the natural processes that happen in the forest.

The ecosystems where Myanmar's forest-dwelling tribes live have a significant impact on their way of life and culture. The forest and its rhythms are often at the centre of these tribe's lives. Generations have passed down by depending upon on traditional knowledge about the forest ecosystem. The tribes have a deep understanding of their surroundings; including how to find their way around the forest, find medicinal plants, and tell what the weather will be like based on natural signs (Reddy et al., 2019; Win Naing Tun, 2015).

The way these tribes live together is often very connected to the places and natural surroundings they live. This connection provides spiritual and transcendental world view the tribes imbibe and adopt in their everyday life. For example, the Karens have a well-established system of village governance patterns designed to cater to the needs of their everyday lives. This system, in fact evolved to address the simple needs of tribes and proved more efficient to modern governance system from utility point of view from the perspective of tribes. The system is headed by elders who are in charge of making decisions about how to use the forest and manage the land. In these societies, people also need to work together to get food, build homes, and keep the forest safe from outside threats.

The entire structure of cultural apparatus of tribal life strongly rooted in forest based ecosystem. Events such as ceremonies and festivals that honour the forest and its gifts are also part of cultural practices. These activities are connected to the seasons, like harvest festivals or ceremonies to honour spirits in the forest. Forests are thus inspirational sources for inner world of tribes in Myanmar. For Myanmar's native tribes, life in the forest is not only hard but also rewarding. Many tribes still live in the forest in relative isolation, even though modern things like deforestation, building roads, and moving to cities are taking over their lands. The forest gives these groups food and shelter from the outside world, which lets them keep living the way they always have (Davis, 1960; Ministry of Natural Resources and Environmental Conservation, 2019).

The fact, however is that it not all about happy and contented tribal life in forests, they do have a lot of problems, especially when it comes to land rights and outside forces trying to take their land. Their ancestral lands and forest territories are often threatened by land grabbing, illegal logging, and government policies. Some of tribes, like the Kachin and Karen, are fighting for their rights to land and freedom. The main objective of their resistance is to oppose outside pressures that often encroach forests in the name of development which does not make them collaborators of such developments. The tribes in Myanmar are thus also fighting for their rights over forests at the same time fights for rights of forests or their right to exist.

History of Forest Policy in Myanmar

Historically, various institutional arrangements for management of forests and waste land evolved in Myanmar. These institutional arrangements are part of communities engagement with local resources in order to use them in their everyday lives. Several Buddhist monasteries strived to maintain good forest cover around them. At village level, ponds, grazing grounds and local forests were used and conserved by local communities. The institution of sacred groves which are essentially forests dedicated to local deities. Large number of sacred groves maintained for honour of guardian spirits or Nats. These evidences suggest the fact that cultural and spiritual believes are invoked in conserving forests by local communities (Tint et al., 2011).

Myanmar known as Burma in the past has a fascinating history of modern forestry. It is not exaggeration to say that the origins of modern scientific forestry in British Indian Empire took place in Burma (Raymond L. Bryant, 1997). On account of the abundant of teak forests, Burma attracted the attention of the British initially. Teak timer is durable for construction of ships which are critical for the British to carry on trade and commerce and communication. Administrative and legal interventions were devised to tap forest resources by the British (Ribbentrop, 1989). The main objective of the initial stage forest policy was to use the valuable timber for construction of ships and the railway in British India and several other activities undertook by the British raj.

The first Inspector General of Forests of India and hailed as father of modern forestry in India Prof. Brandis started his career from Burma in 1856. Having known for the best quality of teak timber, the region of Burma attracted the attention of the British. Besides teak timber, the forests of Burma were also famous for verities of valuable timbers and non-timber forest products. Prof. Brandis was sent to assess the stock of teak in Burma. It was here that the modern scientific framework for management of forests of British Empire took shape initially and the same was spread gradually to entire empire.

A set of systematic rules for management of forests of Burma were introduced with the promulgation of the India Forest Act of 1865. Under these rules, the forest of Burma became part of forest policy of the British. In the due course of time, a comprehensive forest act known as the Indian Forest Act of 1878 was promulgated and the same was extended to the region of Burma. This is the way most of the valuable forests of Burma were brought under the control of the British colonial state (Stibbings, 1982). The Extension of the state control over forests brought about conflicting relationship with local tribal communities who dependent upon forests for survival (Blanford, 1958). Let us sketch the way tribal and ethnic communities in Myanmar accessed the forests resources in a customary way.

Indigenous Wisdom in Conservation of Forests

A large number of tribal and ethnic communities were dependent upon forests for survival in Myanmar. It is quite natural that such communities adopted diversified forest use practices for survival. In order to capture this macro level strand a micro level analysis in the form of exploring the forest use pattern by the Karan tribal community has been used in this paper. The Karan tribal community constitute an important segment of tribal population in Myanmar (Marshall, 2018). The Karan tribal community consist of about 7% of total population of Myanmar and sub-divided into over 20 communities. They have predominately dependent upon forests for survival. Their connection with forests of Myanmar offers an interesting relationship in terms of traditional wisdom of communities and their method of using forest resources.

As tribal communities do not possess written records and at the same time, non-tribal communities did not write much about them. Consequently, there exist a vacuum about their history, culture and other aspects. We usually get to know about them when the British ethnologists started documenting about tribes and ethnic communities of British India. Particularly, the forest department which worked intimately with tribes in the British period documented extensively about life world of tribes. Besides forests eco-system, the tribes in South and South East Asia have been significantly dependent upon shifting cultivation for survival. The generic definition of shifting cultivation is clearing forests to cultivate crops on temporary basis. It is known by different names in the British Empire. Some of the names as follows: Jhum in Assam, Podu in Andhra region, Kumri cultivation in Western Ghats and Taungya cultivation in Myanmar (Malcolm F. Cairns, 2015). The British tried to discourage shifting cultivation without much success. The interference in shifting cultivation was however often met with a strong resistance by tribes. The British were thus compelled to allow shifting cultivation as the tribes and their labour became an indispensable factor for forest conservation strategies and their implementation (Thet & Tokuchi, 2020).

As pointed out by some studies, the phenomenon of shifting cultivation is the best adoptive strategy followed by tribal communities to survive without seriously destroying forests. Usually, they clear a patch of forest for cultivation mostly for subsistence needs. They cultivate several verities of crops on small scale mostly for their own consumption and small quantity to exchange with other commodities. Once soil nutrients get exhausted, they abandon cultivation and clear another patch of forest for cultivation. They, thus carry out cultivation in forests for survival (Chakrapani, 1993). The British did not understand this system. They perceived it as a detrimental factor to existence of healthy forests and hence discouraged it practice. The British, however, could not control shifting cultivation as they need support of tribes in forest conservation operations.

The fact is that the British did not want a hostile relationship with tribes and indigenous ethnic communities. Their expertise and local knowledge on forest was perceived as a necessary requirement for execution of forest management strategies. Hence, tribes were strategically incorporated into forest conservation process and allowed to practice shifting cultivation on the condition of helping forest conservation activity undertaken forest department. This approach of officials of forest department could be seen in Dr. Brandis strategies in Burma. In his visit to Burma in 1856 he noticed the Taungya cultivation practiced by Karan tribes. The literal meaning of Taungya is hill cultivation. It is essentially a cultivation carried on the slopes of mountains in which crops are cultivated along with cultivation of trees for timber, firewood and several other subsistence needs. It was a sort of mixed farming evolved in the hill region of Burma by the Karan tribe to ensure their livelihood requirements from agricultural and forest products. In this method of farming, agriculture and forestry go parallel and in fact complement each other.

As an architect of forest policy in British India, Dr. Brandis hoped to use the Karan tribes for teak plantation as a part of the possible arrangement which permits the Karen tribes to carry on the Taungya cultivation along with conservation of teak plantation. The Karen tribes were initially reluctant to this policy, but gradually they were brought under the system of teak plantation and became a backbone of teak forest conservation. The following observation by an important forest historian of British Indian is worth mentioning on the importance of the Karan tribal communities:

The Karens were indispensable in the forest. They were sometimes partners in the timber operations; they were employed to find the timber, devise the best dragging paths, in felling and lopping and so forth. They were intimately acquainted with the forests and Seaton, the Conservator, had already induced them to plant teak in their Taungyas. It was obvious at the time that in all attempts to improve the forests by fire protection, sowing, creeper cutting, etc. the aid of the Karens would be indispensible to the Forest Staff (Stibbings, 1982, p. 59).

In the due course of time, the Taungya cultivation forged a strong bond between the Forest Department in Burma and the Karen tribal community. This collaboration resulted in expansion of teak plantation gradually and survived for a long period of time as a successful conservation measure between the state and tribal communities. This system lot relevance after introduction of community centric forest management in 1990s in which local communities are made partners of forest management. The history of Taungya cultivation shows remarkable indigenous wisdom of tribal communities for using forests for survival without damaging forest cover (Win & Kumazak, 1998).

Forest Conservation and Tribes and Ethnic Communities in Contemporary Myanmar

A global level trend on community centric forest management emerged in the mid-1980s. On account of thriving and valuable forests, the South-East Asian countries became an important part of this trend. This trend makes local tribal and ethnic communities as partners of forest management forests. Proper use of conservation knowledge and wisdom of tribes for management was perceived as a necessary condition for efficient management of forest resources. Most of the South and Southeast Asian countries formulated legal and administrative policy frameworks for promotion of community centric forest management with the support of international donors such as the World Bank. Myanmar has become a part of this community centric forest conservation policy (Seeland & Schmitusen, 2004).

Community forest management (CFM) is now a key part of Myanmar's plan to protect its forests and forested communities within the framework of sustainable development. The policy framework adopted by the government of Myanmar has put up a lot of effort over the past few decades to include local communities, especially the indigenous tribes, in the management of forests. This policy swing to participatory forest management shows the growing recognition of local communities an important agency for management of forests for protecting and keeping forest ecosystems healthy (Ministry of Natural Resources and Environmental Conservation, 2019; Win Naing Tun, 2015).

Myanmar's approach to CFM gives local communities the power to manage and protect the forests that are part of their ancestral lands. Community forestry programs give local groups the legal right to manage forest resources, with a focus on using them in a way that is good for the environment. These programs are meant to help people not only to protect forests but also improve their lives. The Karen, the Chin, and other native groups in Myanmar have relied on forests for food, medicine, and a way to make a living for a long time. Including these communities in forest management has been a successful way to protect forests and promote economic growth in the area.

The government of Myanmar promulgated a forest law in 1992 and envisaged more efficient system for protection and conservation of forests. This law however did not have explicit provisions for creating a policy for community centric forest management. This policy hiatus was addressed by promulgation of the community forest management in 1995 and its implementation took off from 1996 onwards. Accordingly, this policy creates an institutional framework for execution of community centric forest management which is to be carried out by local communities to conserve forests and derive benefit out of their conservation efforts (Tint et al., 2011). This policy was further augmented by creation of the Forestry Master Plan in 2001 which prepared a vision for encouraging vibrant community centric forest management in Myanmar by 2031.

The government of Myanmar has worked with international groups to start a number of programs to encourage CFM. The Community Forestry

Instruction, which came out in 2019, is one of the most important things that have happened. It facilitates communities legally to manage forests in a way that ensure sustainable management of forests for a long period of time. The program stresses the need to protect watersheds, keep forest biodiversity, and make sure that wood and non-wood products are extracted in a way that doesn't harm the environment. The government has also supported community forestry enterprises (CFEs), which businesses are run by local people that grow and sell forest products like bamboo, honey, and medicinal plants. This change to managing forests by and for communities has been very important for tribes and ethnic communities because it strengthens their rights to land and resources. It also helps lessen the effects of illegal logging and unsustainable agricultural growth on deforestation and land degradation. The fact however that is for these programs to work a critical need of good governance is required (Ministry of Natural Resources and Environmental Conservation, 2019).

Myanmar's forest policy is an important part of its environmental strategy. Its goal is to manage the country's forest resources in a way that is good for the environment, long term protection and provision of useful resources to communities and nation. The national Forest Policy, which was put in place in 1995, lays out the government's goals for the forest sector. It stresses the importance of protecting watershed areas, conserving biodiversity, and managing forests in a way that is good for the environment. The policy stresses how important it is to protect ecosystems and the rights of local communities while still using forest resources (Ministry of Natural Resources and Environmental Conservation, 2019).

In quantitative terms, the community centric forest management has been able to achieve a substantial visibility. This can be attested by the nature of expansion took place between 1996 and 2010. During this period, a total of 104,146 acres of community forest was formed by 572 users groups with 40,062 group members. Therefore, area of community forests formed during the period 2004–2010 was found to be 22,414 acres (Tint et al., 2011). It was also found that communities begun to derive benefits from their conservation efforts. However, it is equally important that the implementation process of community centric forest management needs to be expanded and more flexibility for management of forest resources need to be given to communities.

Myanmar has passed a number of laws and rules that focus on protecting forests, in addition to its national forest policy. One of such attempts is promulgation of the 2018 Forest Law, which sets up a framework for sustainable forest management and makes it easier to enforce rules against illegal logging. Myanmar has promised to lower deforestation and come up with plans for restoring forests and storing carbon. The policy is in line with international commitments like the Paris Agreement and the United Nations Framework Convention on Climate Change (UNFCCC). The forest policy also says that ethnic groups and local communities should be involved in managing the forest. The policy's goal is to protect both forests and the people who depend on them by involving indigenous tribes in conservation and management activities. But there are problems with land tenure rights, governance, and making sure the policy works at the local level (Ministry of Natural Resources and Environmental Conservation, 2018; U Phyu, 2015).

Conclusion

An assessment of community centric forest management in Myanmar from historical to contemporary times reveal a significant trend that local tribal/ethnic communities inherently possess a remarkable ability to conserve forest resources for their best non-destructive way of self-interests and at the same time, the best interests of forests themselves. In a way, this trend suggests a sort of sustainable forest management from the perspective of local communities. The ideal policy formulations for conservation of forests with a particular reference to needs of local communities could be achieved by an efficient policy structure and dynamic implementation process. This paper reiterate its stand that local communities are undoubtedly efficient agencies for implementation of sustainable forest management strategy which is in the best interests of forests, people, climate and nations.

References

Baginski, O. S., & Blaikie, P. (2013). *Forests, people and power: Political ecology of reform in South Asia*. Earth Scan.

Blanford, H. R. (1958). Highlights of one hundred years of forestry in Burma. *Empire Forestry Review*, *37*(1 (91)), 33–42

Bryant, L. R. (1997). *The political ecology of forestry in Burma, 1824–1994*. C. Hurst & Co. Publishers.

Bulkan, J., Palmer, J., Larson, A. M., & Hobley, M. (Eds.). (2022). *Routledge handbook of community forestry*. Routledge.

Burman, R. K.T. (1995). Sacred Groves. *Journal of Human Ecology*, *6*(4).

Cairns, M. F. (Ed.). (2015). *Shifting cultivation and environmental change: Indigenous people, agriculture and forest conservation*. Routledge.

Chakrapani, C. (1993). Shifting cultivation and deforestation: A socio-eco system nexus. *Journal of Human Ecology*, *4*(1), 39–46.

Coggins, C., & Chen, B. (Eds.). (2022). *Sacred forests of Asia: Spirituality and the politics of nature's conservation* (1st ed.). Routledge. https://doi.org/10.4324/9781003143680

Davis, J. H. (1960). *The forests of Burma*. University of Florida.

Guha, R. (2016). *Environmentalism: A global history, Penguin*. Penguin Random House India.

Marshall, H. I. (2018). *The Karan people of Burma: A study in anthropology and ethnology*. Forgotten Books.

Ministry of Natural Resources and Environmental Conservation. (2019). *Forestry in Myanmar 2019–2020*. Forest Department, Ministry of Natural Resources and Environmental Conservation.

Ministry of Natural Resources and Environmental Conservation (2018). *Myanmar Forest Law*, Myanmar.

Ostrom, E. (1990). *Governing commons: The evolution of institutions for collective action*. Cambridge University Press.

Phyu, U. (2015). *Geographical and political conditions of Myanmar (Upto Bagan)*, Yangon University of Education.

Poffenberger, M., & McGreen, B. (Eds.). (1996). *Village, voice, forest choices: Joint forest management in India*. Oxford University Press.

Poscher, B., & Sen, S. N. (2002). forest cover dynamics and community forest management in the Himalayan and dry zone region of Myanmar. In *Handbook of Himalayan ecosystems and sustainability*. (Vol. I). CRS Press.

Reddy, S. U., Asra, A., M, M. T. M., & Jaishanke, R. (2019). Spatial conservation prioritisation of threatened forest ecosystems in Myanmar. *Journal of the Indian Society of Remote Sensing*, *47* (10), 1737–1749.

Ribbentrop, B. (1989). *Forestry in British India*. Indus Publishing House.

Saxena, K. G., Liang, J., & Rerkasem, K. (2007). Shifting agriculture in Asia: Implications for environmental conservation and sustainable livelihood. *M/s Bishen Singh Mahendra Pal Singh*.

Seeland, K., & Schmitusen, F. (2004). *Indigenous knowledge, forest management, and forest policy in South Asia*. D.K. Print World.

Sen, G. (1992). *Indigenous vision: People of India attitudes to environment*. Sage.

Shiv, Y., & Ravi Kumar, V. M. (2019). Ecology and nationalism in India: A study on Indian environmental history. *The Research Journal of Social Sciences*, *10*(4), 372–378.

Stebbing, E. P. (1982). *The forests of India* (Vol. III). A.J. Reprints.

Thet, A. P. P., & Tokuchi, N. (2020). Traditional knowledge on shifting cultivation of local comunities in Bango Mountains, Myanmar. *Journal of Forest Research*, *25*(5), 347–353. https://doi.org/10.1080/13416979.2020.1764166

Tint, K., Springate-Baginski, O., & Gyi, M. K. K. (2011, August). *Community forestry in Myanmar: Progress & potentials*. Ecosystem Conservation and Community Development Initiative (ECCDI) & School of International Development, University of East Anglia.

Tun, W. N. (2015). *Prehistory to protohistory of Myanmar: A perspective of historical geography*. International Conference on Burma/Myanmar Studies.

United Nations Framework Convention on Climate Change (2018). *Myanmar Forest Reference Level Submission*, Myanmar.

Win, S., & Kumazak, M. (1998). The history of Taungya plantation forestry and its raise and fall in the Tharkwaddy forest division of Myanmar (1869-1994). *Japan Society of Forest Planning*, *4*, 17–26.

World Bank. (2020). *Myanmar country forest note*. Retrieved from World Bank (p. 20433).

Chapter 8
Role of Beliefs and Practices in Natural Resource Management: Insights From the Gaddi Tribe, Himachal Pradesh

Sonia Kaushal and Priya Verma

Abstract The integration of Western scientific knowledge and Indigenous knowledge is a debatable issue. Historically, this dichotomy stems from Western science's dominance and triumph, which often subjects Indigenous knowledge to its scientific gaze. However, in recent years, the environmental crisis has urged scholars to recognize the valuable knowledge embedded in the socio-cultural practices of Indigenous communities for protecting natural resources. The paper is based on anthropological fieldwork conducted among the Gaddi tribe in Jia Khas village, Kangra District, Himachal Pradesh. The study's primary objective is to explore the Gaddi community's beliefs and practices in natural resource management. Data was collected through semi-structured interviews with community members. The findings indicate that the Gaddi community has developed beliefs and practices to manage village and forest resources, contributing significantly to sustainable resource management. However, these traditional practices are increasingly threatened by developmental programs, environmental changes, and forest rights regulations. Therefore, understanding the insider perspectives on the Indigenous beliefs and practices of the Gaddi tribe is essential for effective natural resource management and to achieve conservation goals.

Keywords Indigenous knowledge · Natural resource management · Beliefs · Forest

---

S. Kaushal (✉)
Department of Anthropology, Dr. Harisingh Gour Vishwavidyalaya (A Central University), Sagar, 470003, Madhya Pradesh, India
e-mail: skoushal@dhsgsu.edu.in

P. Verma
Gurgaon, India
e-mail: pri.verma1301@gmail.com

R. C. Sobti et al. (eds.), *Traditional Knowledge Systems in India: Sustainability, Ecology, and Resource Governance*, People, Cultures and Societies: Exploring and Documenting Diversities,
https://doi.org/10.1007/978-981-92-1346-7_8

Introduction

Complex relationship of Indigenous knowledge systems with sustainable natural resource management practices have increasingly been the focus of researchers in recent years. The increased interest derives from the global realization that western scientific approaches alone are insufficient to address the complexity of environmental challenges such as biodiversity loss, climate change, and resource depletion (Berkes, 2018; Reyes-García et al., 2019). In the past, Indigenous knowledge was downgraded and depreciated to Western epistemologies, perceived as unreliable or culturally inaccessible rather than empirically rigorous (Agrawal, 1995). However, current discourse firmly places Indigenous ecological knowledge as critical in fostering resilience and sustainability, given its holistic understanding of ecological relations deeply rooted in cultural beliefs and community practices (Gadgil et al., 1993; Kimmerer, 2013).

Within this expanding discourse, researchers have consistently highlighted how Indigenous communities insert sustainable practices within their social norms, taboos, and traditional rituals, effectively governing their interaction with natural resources (Berkes & Turner, 2006; Gómez-Baggethun et al., 2013). Recent evidence underlines that such community-based knowledge systems often lead to outcomes that align closely with conservation goals, such as biodiversity preservation and ecological balance (Berkes et al., 2000; Sterling et al., 2017). However, a crucial challenge arises in academic discourses about integrating Indigenous knowledge with Western science. Though few researchers show support for integrating both knowledge systems in resource management strategies to achieve sustainability (Reyes-García et al., 2019), others show restraint against this standardisation, warning that forcing indigenous knowledge into scientific frameworks exposes its weakening context-specific efficacy and cultural significance (Nadasdy, 2003; Smith, 2012).

Despite widespread global research into Indigenous ecological knowledge, specific gaps persist, especially about the detailed ethnographic examination of belief-practice relationships within Indigenous groups facing rapid socioeconomic transitions. Himalayan communities, particularly pastoralist societies like the Gaddi tribe of Himachal Pradesh, India, represent one understudied group. Available research focuses on broad cultural and socioeconomic explanations of the Gaddi community (Sharma & Minhas, 1979; Bhasin, 2013). However, few studies thoroughly analyse the complex ways in which their specific cultural beliefs and traditional practices directly influence their natural resource management strategies. Furthermore, considering that such communities are increasingly subject to external economic and policy pressures (Negi, 2010), the current relevance of traditional practices within this fast-evolving development, environmental stress and forest rights regulations context is still inadequately understood.

To address these critical research gaps, this study focuses on the Gaddi tribe in Jia Khas village, Kangra District, Himachal Pradesh. Based upon detailed anthropological fieldwork, including semi-structured interviews and participant observations, this research aims to reveal how specific Indigenous beliefs and practices among the Gaddi directly contribute to the sustainable management of agro-pastoral systems and forest resources. The core objectives guiding this investigation are two-fold: to document the beliefs and practices related to the traditional natural resource management among the Gaddi community and to examine how ongoing developmental programs and forest proper regulations transform the traditional practices.

Methods

The study was conducted in Jia Khas village, Kangra district, Himachal Pradesh. The district was selected due to its considerable Gaddi population and ease of accessibility. Intensive fieldwork for the study was carried for 25 days in the year 2018.

With the help of snowball sampling, participants including shepherds, non-shepherds, and elder shepherds, were selected for in-depth semi-structured interviews. A total of 17 participants were interviewed. The interviews focused primarily on the roles and practices of the Gaddi community in natural resource management. The interviews were audio-recorded and transcribed later. The collected interviews were repeatedly read to identify key areas where the beliefs and practices of the Gaddi tribe intersect with natural resource management.

The study's objectives were clearly explained to all participants, and local authorities were informed about the study. Before the interviews, participants were informed that participation was voluntary and that they were allowed to withdraw at any stage. Since the base camp of the researcher's study was within the village, the researcher could approach participants repeatedly, allowing for more detailed data collection.

The present study had several limitations. First, the small number of participants was a constraint, as many Gaddi shepherds were away in the lower altitudes with their flocks during the fieldwork period. Second, due to the limited number of participants, it was not possible to quantify the data. Finally, the paper lacks corroboration of the empirical findings with secondary sources and previous studies, which could have strengthened the overall analysis.

The Gaddi Tribe: A Brief Overview

The Gaddi tribe of Himachal Pradesh is one of the region's prominent migratory communities, which traditionally practice transhumance and livestock rearing. While they primarily reside in the Bharmour region of Chamba district, significant populations are also found in Mandi, Sirmour, Bilaspur, and Kangra districts (Verma, 1996). Linguistically, the term Gaddi is derived from the Sanskrit word *gadar*, meaning sheep, and the area they inhabit is known as *Gadheran*, signifying either the "country of sheep "or the "seat of the Gaddis"(Verma, 1996).

The origin of the Gaddi people remains a subject of scholarly debate. Some scholars trace their ancestry to Aryan settlers in Bharmour, while others suggest migration from the plains during the Muslim invasions. Interestingly, the Gaddis of Jia Khas village, Kangra district, and other areas have a popular saying: "*ujada Lahaur te basaya Bharmour*"("when Lahore was deserted, Bharmour was inhabited"), reflecting their historical narrative and migration memory.

Transhumant Gaddis migrate with their livestock to high-altitude pastures during the summer and return to lower altitudes in winter—a physically demanding journey for humans and animals (Bhasin & Bhasin, 1993). Historical records indicate that in 1915 grazing taxes were imposed to discourage forest use. Post-independence, government reports from 1959 and 1970 highlighted pressures on Gaddi migration due to forest plantations and the redistribution of common land (*Shamlat*) to landless people, reducing available grazing areas. The introduction of regulations in 1972 further restricted herd sizes, prompting many Gaddis to adopt settled livelihoods, including farming or wage labour, although small-scale livestock rearing continues as a supplemental income source (Singh, 1997).

The Gaddis primarily worship Lord Shiva, locally referred to as *Dhudu*, meaning "one who has smeared soil on his body. The Gaddis visualises Shiva in both his childhood and adulthood forms: as a child, he is believed to have played in the soil, and as an adult, he smeared ash on his body. His sacred abode is the Manimahesh mountain in Chamba. Both the representations of Shiva and the term *Dhudu* reflect the centrality of nature in the lives of the Gaddi community. In addition to Lord Shiva, the Gaddis worship other Hindu gods and goddesses, following Shaivism while practising ancestor and nature worship. They recognise several nature-related deities, including *Vanshira* (forest deity), *Joginis* (female forest deities), *Batal* (spirit of springs and rivers), *Autars* (ancestor spirits), *Gugga* and *Nagdeo* (serpent deity), among others. The community also believes in evil spirits and ghosts inhabiting the mountains and villages. The Gaddi religious beliefs are closely tied to nature, and their rituals, taboos, and practices shape social values and behavioural patterns, directly and indirectly contributing to the management and conservation of natural resources.

Now, the Gaddi community is divided into two groups: those who continue seasonal migration and those who have settled permanently. Although settled, Gaddis no longer practices transhumance, they remain closely connected to animal husbandry while diversifying into other occupations. Anthropological studies conducted between 1976 and 1979 in Bharmour documented these practices, noting that, despite social changes, the region has not experienced major economic transformations (Bhasin, 2008).

With this historical background of the Gaddi community and their traditional occupations, it is evident that the Gaddis have historically lived in close harmony with nature, shaping their daily life and livelihoods around natural resources. Therefore, this paper examines the beliefs and practices of the Gaddi community concerning natural resource management. The study is divided into two main sections: the first focuses on village-level beliefs and practices that contribute to the conservation of water resources. In contrast, the second examines practices during transhumance, highlighting how seasonal migration sustains a balanced and harmonious relationship with forest ecosystems. And how developmental programs, environmental changes, and forest rights regulations increasingly impact these traditional practices.

Findings

Water Bodies Management: Traditional and Contemporary Practices

The Jia Khas village has an ample number of water resources. Every household has a tap water connection provided by the Government of Himachal Pradesh, and one can also come across handpumps on the busy roadside. Besides, water coming from natural springs at the mountainous roadside, *baudi* (step-well) in the middle of agricultural land and near houses, *kulh* (irrigation canal) and a small perennial stream called *Baanganga* are the natural and traditional water sources.

Kulhs *(Traditional Irrigation System)*

The *kulhs* are traditional water channels in which water comes from the mountains or glaciers. In the village, the *kulhs* are managed by the panchayats and villagers. The drain of the *kulh* has been made of cement to improve the water flow. The kulhs are mainly used for irrigation and washing clothes.

According to the villagers, the *kulhs* are managed collectively by them and panchayat members. *Kulhs* reflected a system of equal water sharing and were an example of community cooperation. In the past, people gathered around *kulhs* for daily chores, giving them a chance to talk and connect. However, villagers have raised concerns that development activities like deforestation and road building have damaged many of these water channels. Another concern was that the altered rainfall pattern and melting of glaciers would impact water flow.

The village elders believe that the arrival of private tap water has reduced the use of kulhs, leading to unequal access to water and the erosion of traditional practices and community ties.

Baudi*: Step-wells and Sacred Spaces*

Another traditional source of water in the village is the *baudi*. These are step-wells with clearly marked boundaries and often a canopy. Earlier, villagers used *baoudis* to collect drinking water. They are also considered sacred and play a role in marriage rituals and festivals. Small idols of *Nag Devta* (serpent deity) and ancestors are placed near these wells, and the water is worshipped during important occasions.

Currently, only four *baudis* are functional in the village. Due to the availability of tap water, most *baudis* are no longer used regularly. They are now mainly used during ceremonies or for irrigation. Many dry up in the summer and are not maintained as they were in the past.

Naru*: Disappearing Natural Springs*

Narus are natural springs from the mountains, either collected through pipes or seeping out from rock crevices. According to the villagers, there used to be many *narus* in the past, but only a few remain today due to changing weather patterns. Travellers and villagers mainly use these springs to drink water. The local community manages Narus by installing pipelines at the rock crevices and constructing cemented platforms below to collect the flowing water.

Even the houses on the mountain (*dhar*) have access to tap water and kulh. The water is collected through the *kulh* system, although these channels are not completely cemented. This shows a blend of traditional and modern water management practices.

Religious Beliefs and Water Resource Conservation

One common feature of traditional water sources in the village is the presence of religious symbols. Small stones or carved idols of *Nag Devta* and ancestors are placed near these water bodies and worshipped during marriage ceremonies, festivals, and special occasions. Elders believe these customs helped people care for and maintain the water sources together.

Sairon Devi, aged 45, recalled how villagers gathered near water sources to wash clothes and discuss their daily lives. She said, "*Didi, log pahle kapde dhone, bartan dhone jaate the, phir sarkar ne nal laga diye .... ab kaun jata hai ... jaate hain par koi koi jate hain*" *(*Earlier, women would go together to wash clothes and utensils. But after the government provided taps, very few people are going now). Many elders mentioned that extreme summer heat has reduced tap water supply in recent years, and even *kulh* water has become scarce. Some households have started using electric water pumps, which were not needed earlier when natural water flow was strong enough. According to them, the water from the mountains running in the *kulhs* never required any mechanical device, but things have changed.

Grahat*: Traditional Water-Powered Flour Mills*

Villagers also pointed out an age-old practice known as *a gharat* (traditional flour mill). One of the villagers reported that when he was small, there were five to six flour mills, but now only one is left and functional. He said that a high flow of water is used in grinders to rotate the two grinding stones. The water is channelised to the *grahat* from *kulh,* and the flowing water coming with force falls on the '*grahhattu da pahiya*' (wooden water wheel), which causes the wheel to spin. But now the flow is less, and people divert *kulh* water to agricultural fields, all the *grahats* stopped functioning except one. Only one *grahat* is functional in the village and stops working during summer due to the reduced water flow. Bahadur Ram, aged 45, the owner of the only functional *grahat* in the village, said the flour of the *grahat* is considered healthier and more nutritious. However, now people prefer machines which work fast and in *grahat* the grinding takes.

The decline of *grahats* reflects reduced water availability and shifting preferences and technological transitions in traditional technology and rural livelihoods.

Traditional Ecological Knowledge and Forest Practices

The life of the Gaddi community revolves around their livestock, forests, and mountains. They refer to their animals as *maal* or *dhan*, meaning wealth, and high mountain peaks as *dhar*. In the Kangra district, Gaddis also call their flocks *maal*. Sheep and goats are essential to them as they provide wool, meat, and hides, sustaining both subsistence and trade-based economies.

Gaddis's practice seasonal transhumance, moving to lower altitudes during winter and higher altitudes in summer in search of green pastures. Most Gaddis follow traditional migratory routes passed down through generations, enabling them to develop extensive ecological knowledge of these routes, including the forest terrain, flora, fauna, and other natural resources such as medicinal herbs (*chaura, kadu, patish,* etc.), varieties of mushrooms (*gutchi)*, and edible ferns (*lungdi*). Their grazing routes are not only traditional but also rotational, ensuring ecological balance and regeneration of resources.

Sacred Landscapes and Supernatural Beliefs

Nathuram, a Gaddi herder, recounted that many trees and shrubs in the higher altitudes are considered sacred and are revered as the abodes of guardian spirits of the forest like *Vanshira* (male forest deity), *Joginis* (female forest deity) and *Nagdeo* (serpent god). These deities are believed to protect the herders from misfortunes during migratory journeys. Gaddis attributes a spiritual presence to the *dhars*, wherein benevolent and malevolent supernatural entities reside. Benevolent entities are thought to safeguard them from dangers, while the malevolent spirits are to be appeased through ritual conduct.

Kashiram, a 45-year-old Gaddi, explained that specific behaviour codes must be observed in these sacred forest areas. Forest deities such as *Vanshira* and *Jognis* are believed to be easily disturbed by loud noises or inappropriate actions. Sacred groves, trees, and water bodies are marked and preserved as sacred sites, often symbolised by the placement of pebbles or red flags. These areas are neither defiled nor exploited by the plucking of leaves; trees are not cut, and rituals are performed when passing through them.

Another Gaddi, Jaishiram shared, "*in pahadon mai hi toh dhudu (shiv) basta hai, hamare deo-devi baste hain toh who hamari Raksha karte hain. Jagah jagah pahadon mai in ke sthan bane hain, hum jab jate hain toh waha dhoop batti karke thoda bahut kada chada dete hai . . . .ki dhudu maharaj hamari raksha karna. Par ab peechle kuch saalon se badal gaya hai sab. Bhagwan bhi naraz ho gaye hain*" (It is in the mountains that our lord Shiva resides, and they protect us. There are many sacred sites across the hills where we offer incense and small offerings, asking for protection. But in recent years, things have changed. The gods, too, seem unhappy).

He expressed concern over the recent changes, stressing that traditional worship and spiritual reciprocity have diminished, leading to what he sees as divine displeasure.

Knowledge of Medicinal Herbs and Forest Access Restrictions

A recurring concern among the Gaddis is the restricted access to forest resources. The Gaddis possess extensive ethnobotanical knowledge, developed over centuries, regarding identifying, collecting, and using medicinal plants. Traditionally, this knowledge compensated for the absence of veterinary doctors. Both humans and livestock were treated using forest-derived remedies.

Dhanyanchand, a 50-year-old Gaddi, adds,"*Ab dhar mai doctor to ayega nahi, haan, ab sakrar abadi wali jagahke pas camp lagati hai waha doctor hote hai dangaron ke toh waha dikha dete hain par andar upar ban mai neeche ana padta hai. Aur pahle ke time me toh bhed ko bimari hoti thi ya haddi tut jati thi toh jungle se hi dawai karte the. Har gaddi ko pata hota hai kis cheez ki kya dawah ai jungle mai. Par ab kya karna sarkar bolti hai hamne dhar khatam kar diye"*(There are no doctors inside the forests. Sometimes camps are set up near populated areas, but we must come down to reach them. Earlier, we used herbs from the forest if a sheep was injured or sick. Every Gaddi knows which plant is used for what. But now the government says our dhars no longer belong to us). However, with increasing restrictions imposed by forest authorities, Gaddis now faces significant challenges. The state limits grazing rights and the collection of forest produce, even though veterinary services are scarcely available in remote forested areas.

Concerns Over Commercial Exploitation and Depletion of Forest Resources

One of the most critical concerns raised by all the participants is that *dhars* are now frequently visited by national and international tourists and agents of pharmaceutical and herbal companies, often without permission. According to them, these outsiders lack knowledge about responsible harvesting and tend to collect herbs carelessly, uprooting entire plants for profit. In contrast, Gaddis knows which

They also explained the importance of the *Sair* festival, celebrated from mid-September to mid-October, during which the potency of herbs is believed to be at its peak. Traditional herbalists and villagers visit the *dhar* early in the morning, worship the forest deity *Vanshira*, and collect only the required parts of herbs. A local herbalist shared, "*ye log aate hain aur jad ke sath ukhad dete hain. Toh agale saal kahan se honi hain ye jadiyaan. In jaadiyon ko taudane ka tarika hota hai . . . .ye nahi ki bina nahaye dhoye, jhoothe muh chaldiye, koi dhoop batti nahi. Toh vanshira bhi naraz hota hai, jaadiyon ka asar bhi khatam hota hai, who parli dhara mai pahle hum sab jaate the subah-subah par ab kuch choda hi nahi*" (These outsiders uproot entire plants. How will the herbs grow back next year? There is a proper way to collect herbs—with offerings and incense after bathing. If this is not done, Vanshira gets angry, and the herbs lose their effect. We used to go early in the morning, but now nothing is left).

According to him, traditional collection is deeply ritualised. Herbalists purify themselves, perform rituals at home and in the *dhar*, and seek permission from deities like *Vanshira* before harvesting. Such practices not only conserve biodiversity but also maintain the spiritual and cultural sanctity of the land.

Forest Management and Traditional Institutions

Gaddi elders emphasise that forest management systems have shifted significantly. Forest officials increasingly perceive Gaddi flocks as threatening reforestation efforts, arguing that hooves cause soil erosion and damage saplings.

However, Gaddis of Jia Khas village counters that their presence in these mountains predates formal forestry interventions. Their knowledge of mountainous terrains, migration paths, and seasonal patterns is deeply embedded in their cultural memory.

A notable example of traditional resource management is the informal grazing agreement between Gaddis and farmers in lower-altitude districts like Hamirpur and Una. During summer migrations, Gaddis pens their flocks on agricultural fields, providing natural manure in exchange for green fodder. According to Nathuram, this mutually beneficial system is now in decline due to changing crop patterns, the use of chemical fertilisers, and the spread of invasive weeds like *Lantana camara* (*phoolbari*). Even the *shamlaat* (village common land) and forested areas are also decreasing due to tree plantation by the forest department and encroachment to build houses. Now, they struggle to secure foliage for their flocks when they move to lower areas, affecting the longstanding arrangements between gaddis and farmers.

Another practice that reflects traditional ecological stewardship is the *Rakha* system. Historically, forest resources were managed through *devta* (deity) institutions and systems like *Rakha* and *Parza*. The *Rakha* was a forest guard appointed by the villagers and paid by them to look after the forest's resources. Other scholars in Himachal Pradesh have reported a similar system (Dhiman, 2001; Vasan & Kumar, 2006). Although the *Rakha* system declined with the advent of the Indian Forest Department in 1864 and was officially abandoned in Kangra by 1924, remnants persist in Jia Khas. Despite inconsistent remuneration, some villages still appoint a *Rakha* who monitors forest use and coordinates communal responses to forest fires.

Conclusion

This study shows the vital role of Indigenous knowledge and cultural practices in the community's management of natural resources. Traditional institutions, beliefs, and ecological practices are increasingly marginalised among the studied Gaddi community. Government water supply systems have neglected traditional water sources such as *baudis* (step-wells) and *kulhs* (irrigation channels), reducing them to ritual or agricultural functions. Technologies like the *grahat* (traditional flour mill) have been replaced by electric mills, diminishing their nutritional and cultural significance. The traditional practices reflect a nuanced understanding of environmental dynamics that have sustained forests, water, pastures, and other natural resources over generations. Moreover, Gaddis expresses deep discontent with current forest governance. The imposition of restrictions on grazing, the influx of unregulated outsiders, and the disregard for spiritual protocols have not only affected their livelihoods but also eroded the symbiotic relationship they once shared with the forest ecosystem. An insider perspective is crucial for understanding resource management from within the community. The Gaddis' perspective illustrates how rituals, seasonal migration, and sacred landscapes shape resource use and knowledge that is often undervalued in external management approaches.

Furthermore, the experience shared by Gaddis regarding forest regulations highlights the minimal participation of local communities in decision-making and the limited use of their knowledge in forest management. Restrictions on access to forests and pastures and commercial exploitation by outsiders disrupt traditional practices and weaken community institutions such as the Rakha system. These changes threaten both the Gaddis' livelihoods and the long-term ecological sustainability of the landscape.

Therefore, to achieve effective and sustainable natural resource management in the Himalayan region, it is essential to acknowledge and integrate the insider knowledge of Indigenous communities. Policies should be designed to protect forest rights, respect cultural practices, and incorporate traditional ecological knowledge, which is crucial for strengthening both environmental conservation and the resilience of local communities.

Acknowledgement The authors are thankful to the study participants for their valuable information. The authors are also grateful to the postgraduate students of the session 2016–2018, technical staff and Head and other Faculty members of the Department of Anthropology, Dr. Harisingh Gour Vishwavidyalaya, Sagar, Madhya Pradesh, for their support and assistance in providing the necessary logistics for conducting successful fieldwork.

References

Agrawal, A. (1995). Dismantling the divide between Indigenous and scientific knowledge. *Development and Change*, *26*(3), 413–439. https://doi.org/10.1111/j.1467-7660.1995.tb00560.x

Berkes, F. (2018). *Sacred Ecology* (4th ed.). Routledge.

Berkes, F., & Turner, N. J. (2006). Knowledge, learning, and the evolution of conservation practice for social-ecological system resilience. *Human Ecology*, *34*(4), 479–494. https://doi.org/10.1007/s10745-006-9008-2

Berkes, F., Colding, J., & Folke, C. (2000). Rediscovery of traditional ecological knowledge as adaptive management. *Ecological Applications*, *10*(5), 1251–1262. https://doi.org/10.1890/1051-0761(2000)010[1251:ROTEKA]2.0.CO;2

Bhasin and Bhasin. (1993). Ecology, population dynamics among Transhumants of Himalayas: A case study of Gaddis of Himachal Pradesh. *Journal of Human Ecology*, *4*(3), 171–219.

Bhasin, V. (2013). Pastoralists of Himalayas. *Journal of Biodiversity*, *4*(2), 83–113. http://krepublishers.com/02-Journals/JHE/JHE-33-0-000-11-Web/JHE-33-3-000-10-Abst-PDF/JHE-33-3-147-10-2176-Bhasin-V/JHE-33-3-147-10-2176-Bhasin-V-Tt.pdf

Bhasin, V. (2008). Gaddi's folk medicine: A source of healing. *Ethno-Med*, *2*(1), 1–27.

Dhiman, R. C. (2001). Traditional initiatives in participatory forest management: Dodra-Kwar (Shimla-HP) experience. *Indian Forester*, *127*(8), 929–935.

Gadgil, M., Berkes, F., & Folke, C. (1993). Indigenous knowledge for biodiversity conservation. *Ambio*, *22*(2–3), 151–156.

Gómez-Baggethun, E., Corbera, E., & Reyes-García, V. (2013). Traditional ecological knowledge and global environmental change: Research findings and policy implications. *Ecology and Society*, *18*(4). https://doi.org/10.5751/ES-06288-180472 (Article 72).

Kimmerer, R. W. (2013). *Braiding sweetgrass: Indigenous wisdom, scientific knowledge, and the teachings of plants*. Milkweed Editions.

Nadasdy, P. (2003). *Hunters and bureaucrats: Power, knowledge, and aboriginal-state relations in the Southwest Yukon*. UBC Press.

Negi, C. S. (2010). Traditional knowledge and biodiversity conservation: A case study from Byans Valley in Kumaon Himalaya. *Journal of Environmental Planning and Management*, *53* (2), 233–253. https://www.tandfonline.com/doi/full/10.1080/09640568.2017.1371006

Reyes-García, V., Fernández-Llamazares, Á., Guèze, M., Garcés, A., Mallo, M., Vila-Gómez, M., & Vilaseca, M. (2019). The contributions of Indigenous peoples and local communities to ecological restoration. *Restoration Ecology*, *27*(1), 3–8. https://doi.org/10.1111/rec.12894

Singh, S., (1997). *The Gaddi Tribe of India: A sociological study*. Sterling Publishing.

Smith, L. T. (2012). *Decolonisingmethodologies: Research and indigenous peoples* (2nd ed.). Zed Books Ltd.

Sterling, E. J., Filardi, C., Toomey, A., Sigouin, A., Betley, E., Gazit, N., Jupiter, S. et al. (2017). Biocultural approaches to well-being and sustainability indicators across scales. Nature Ecology & Evolution, *1*(12), 1798–1806.

Vasan, S., & Kumar, S. (2006). Situating conserving communities in their place: Political economy of KulluDevban. *Conservation and Society*, *4*(2), 325–346.

Verma, V. (1996). *Gaddis of Dhauladhar: A transhumant tribe of the Himalayas*. Indus Publishing Company.

Chapter 9
The Sacred-secular Complex in the Amarnathyatra: The Microcosm of India's Sacred-secular Entanglement

Subhamay Kisku and Diptendu Chatterjee

Abstract The AmarnathYatra, a major Shaivite pilgrimage in Jammu and Kashmir, is both a spiritual journey and a socio-political phenomenon, making it a valuable site for anthropological study. This paper examines the "sacred-secular complex" of the Yatra, where religious devotion intersects with governance, security, economy, ecology, and identity politics. Using an ethnographic approach, fieldwork was conducted from 5 to 8 July 2024 along the Baltal route, including participant observation, semi-structured interviews, focus group discussions, and informal conversations with 33 informants—tribal service providers, pilgrims, local residents, and sādhus. The findings reveal that the Yatra's sacred dimension—rooted in mythology and ritual—operates within an increasingly institutionalized and securitized environment. Tribal communities such as Gujjars and Bakarwals play a crucial logistical role, while navigating livelihood challenges posed by modern infrastructure and changing tourism dynamics. Kashmiri Muslim participation in provisioning services illustrates economic interdependence, even amid political tensions. Environmental concerns, particularly waste generation and ecological degradation, reflect a conflict between traditional ecological ethics and mass-scale pilgrimage practices. The study also highlights tensions between institutional management bodies and traditional religious actors, gendered experiences of pilgrimage, and the ways in which the Yatra reinforces both Indian national integration and localized assertions of identity. These intersecting themes show that the AmarnathYatra is not a static tradition but an evolving space where sacred and secular logics continuously negotiate each other. The research contributes to broader discussions on pilgrimage as a dynamic arena for cultural resilience, contestation, and transformation. It argues for sustainable, inclusive management that integrates local knowledge, protects ecological balance, and honours the pilgrimage's diverse human relationships.

S. Kisku (✉) · D. Chatterjee
Department of Anthropology, University of Calcutta, Kolkata 700019, India
e-mail: skanthro@caluniv.ac.in

R. C. Sobti et al. (eds.), *Traditional Knowledge Systems in India: Sustainability, Ecology, and Resource Governance*, People, Cultures and Societies: Exploring and Documenting Diversities,
https://doi.org/10.1007/978-981-92-1346-7_9

Keywords Amarnathyatra · Sacred-secular complex · Pilgrimage anthropology · Indigenous knowledge

Introduction

An anthropological study of the AmarnathYatra includes examination of the pilgrimage through the eyes of society, culture, religion, identity, and related human behaviour (Turner, 1969; Eade & Sallnow, 1991). AmarnathYatra is a famous annual Hindu pilgrimage to the Amarnath cave shrine located in Jammu & Kashmir, devoted to Lord Shiva. Anthropologically, it signifies the connection among ecology, religion, and ritualistic traditions (Morinis, 1992). The term sacred-secular complex refers to the interaction between religious meanings and political, economic, and administrative realities (Chidester & Linenthal, 1995; Coleman & Elsner, 2005). In pilgrimages like the AmarnathYatra, this sacred-secular complex is especially noticeable due to substantial government participation, religious implication, security arrangement, organised tourism, and commercialisation of the entire yatra.

The sacredness of AmarnathYatrais dependent on the religious implication: The yatra is a Holy journey to the natural cave shrine of Lord Shiva, where the seepage water of the cave forms an ice lingam is worshipped by devotees. Ritualistic practices include chanting of Sanskrit mantras, offering of daily puja, and fulfilling vows. Mythological narratives tell the story of Lord Shiva's story of immortality to Goddess Parvati—a primordial myth of the Amarnathyatra. The pilgrimage experience is outlined as a spiritual conversion or performance of devotion (*bhakti*) (Fuller, 2004).

The secular components of the AmarnathYatra are involved with Government Involvement in the Holy Journey, mainly managed by the Shri Amarnath Shrine Board (SASB), which was established by an Act of the Jammu & Kashmir State Legislature in 2000. Hon'ble Lieutenant Governor, UT of J&K is the ex-officio Chairman of the Shrine Board. The board is also heavily supported by the Indian Army, the State Police (J&K), Health and disaster management services, both by the State and Central Government, tourism and infrastructure development bodies. There is an arrangement of strict security and Surveillance measures for the smooth completion of the HolyYatra. Previously, due to militancy and past attacks by the separatists, the route was heavily militarised well in advance of the AmarnathYatra by establishing numerous presence of Closed-circuit Television (CCTV) check-posts, helicopters, Global Positioning System (GPS) monitoring, and Radi-Frequency Identification (RFID) tags for each devotee are compulsory (Gupta & Ferguson, 1997).

Logistics and Administration includes on submission of biometric registration through the offline portal registration through the link shared by the SASB, which characterises quota systems by the Jammu and Kashmir Bank and other nationalised banks, and alternatively, the registration process is also available from home in the digital platforms. Nationwide, state and district-wise wide availability of medical camps for fitness certificates is also a prerequisite for the AmarnathYatra registration.

During the period of Holy Yatra, several medical camps, food tents (*Bhandara/ Langar*), and sanitation facilities are temporarily established by the Government agencies through the supervision of the Indian Army in collaboration with various philanthropic agencies (Michaels, 2004). AmarnathYatra is an annual Hindu pilgrimage to the Amarnath cave shrine in Jammu and Kashmir, dedicated to the highly revered God, Lord Shiva. In an anthropological sense, it signifies the connection of ecology, religion, identity, and ritualistic traditions as a complex.

Sādhus are Hindu renunciative religious practitioners and celibates their whole life. They have separated themselves from material life to pursue spiritual liberation (*moksha*) (Parry, 1982). During the AmarnathYatra, sādhus are not just pilgrims, but symbolic carriers of sacred tradition, which is often viewed as epitomes of Lord Shiva himself. For sādhus, the yatra is not just a journey to the shrine, but a pilgrimage of strict austerity (*tapasya*), self-control, and spiritual excellence. Many devotees travel without shoes, with minimal belongings. Sometimes, many of them walk barefoot from a comparatively long distance for weeks or months to reach the cave to show their devotion more intensely in nature. Their participation is extremely connected to their Shaiva traditions (tradition of worshipping deity Shiva), as Amarnath is a Shaivite (follower of Shaivism) pilgrimage (Babb, 1986).

Literature Review

The anthropological study of pilgrimage has long been involved with queries of ritual, sacred space, symbolism, and collective identity. The AmarnathYatra, an important Hindu pilgrimage to the Amarnath cave shrine in Jammu and Kashmir, provides a rich ethnographic field to comprehend these themes within the wider outline of ecology, religion, politics, and cultural identity.

Initial anthropological studies on pilgrimage emphasised its contribution in making communities and reiterating social structures. Victor Turner (1969) legendarily described pilgrimage as a "liminoid" experience wherein the social orders are temporarily suspended, allowing pilgrims to practice a sense of spiritual community and equality. This theoretical framework is predominantly relevant for the AmarnathYatra, where pilgrims from different socio-economic backgrounds accept a physically and emotionally difficult journey in search of spiritual value and divine sanctification.

More current studies, however, have shifted from a purely symbolic explanation to further materialist and structural methods, positioning pilgrimages within political economies and state structures. Eade and Sallnow (1991) critique the excessively uniting vision of Turner by highlighting the controversy, diversity, and complication of pilgrimage practices. Their idea of "contested space" is particularly relevant in understanding AmarnathYatra, specifying its location in a geopolitically delicate region and the contribution of multiple participants, including the Shri Amarnath Shrine Board (SASB), central security forces, and state government tourism departments.

In the background of Indian pilgrimages, researchers such as Jacobsen (2013) and Morinis (1984) have discovered how pilgrimage routes and practices are supported by modern set-ups, state government policies, and religious institutions. The AmarnathYatra, heavily regulated through biometric registration, registration quotas, and security rules, represents, as Dubey (2019) calls the "securitisation of sacred geographies." The expedition is as much about the planned connectedness of checkpoints, permits, and military existence as it is about spiritual consecration. This amalgamation of the sacred and the secular can be conceptualised through the "sacred-secular complex," a connotation used to explain the cohabitation of religious rituals with administrative and economic opportunities.

Moreover, the ecological anthropology of pilgrimage also has importance. Contributions by Sax (1991) and Michaels (2004) have shown how pilgrim actions impact ecological regions and local communities. In the case of Amarnath, ecological apprehensions regarding handling, degradation of delicate mountain ecosystems, and pilgrim control are central to discussions on sustainable pilgrimage. The effort of Dabla (2007) precisely addresses the ecological consequences of the Yatra in the Kashmir Himalayas, inspiration for an anthropological understanding that indicates ecological susceptibilities.

The function of sādhus in Indian pilgrimages has also been a subject matter of anthropological research. Srinivas (1952) and Parry (1985) have mentioned that sādhus role is both as abstainers, carriers of religious and cultural tradition. During the AmarnathYatra, sādhus not only take part but also personify the Shaiva ethos of renunciation and spiritual excellence. Their hectic journeys often become important points for both media and pilgrims, strengthening the religious sacredness of the journey while also contributing to its symbolic and performative scopes.

The AmarnathYatra also provides a scope to study identity politics and intercommunity relationships. As Jaffrelot (2007) demonstrates, religious marches and pilgrimages in India frequently intersect with dominant nationalism and local politics. The AmarnathYatra, occurring in the conflict-driven region of Kashmir, is commonly explained through a political lens. For example, Bose (2003) and Zutshi (2014) highlight how state-sponsored arrangements around the Yatra help to develop a sense of Indian nationalism and religious validity over the conflicted area of Kashmir. This narrative is further intricated by the Kashmiri Muslim population's varied reactions—ranging from providing hospitality towards pilgrims to confrontation against militarisation.

Anthropological consideration has also been given to the commodification of the AmarnathYatra. As Coleman and Eade (2004) claim, pilgrimage in the present world is deeply entwined with consumerism, tourism, and economic pursuit. The AmarnathYatra, with its wide commercial provision of tent accommodations, pony rides, a human carrier for devotees (*palanquin*), helicopter services, and souvenir markets, dryfruit shops, illustrates this convergence. Notably, the government's promotion of religious tourism in Jammu and Kashmir supports broader neoliberal approaches of regional development branded with a spiritual flavour (Singh & Raj, 2016).

Gender roles in pilgrimage anthropology have uncovered the wide range of experiences of male and female pilgrims. Mostly, studies on the AmarnathYatra do not exclusively focus on gender, evolving ethnographic interpretations indicate that women pilgrims face discrete challenges, including security concerns, access to facilities like toilets, and gendered restrictions of ritual performance. This remains an emerging area for further anthropological study.

Another notable expansion is the digital inclusion of pilgrimage. Scholars like Helland (2007) and Campbell (2012) have highlighted the rise of "e-pilgrimage" or online pilgrim. The AmarnathYatra now includes virtual puja booking, online worship viewing (*darshan*), and digital registration; the reconstituted religious experience is governed by digital technology. Anthropological studies could further look into how technology configures sacred space and ritual performance.

The literature on AmarnathYatra attractions from multiple elements of anthropological review—political anthropology, ritual theory, identity politics, environmental concerns, and digital religion. While few ethnographic studies focus completely on the Yatra, it remains a rich, underexplored site for anthropologists interested in the connections of ecology, culture, governance, and faith. A systematic ethnography of the AmarnathYatra has the scope to contribute to the broader anthropology of pilgrimage.

Objectives of the Study

Scholarly works on the AmarnathYatra have emphasized its role as a site of symbolic interaction (Eade & Sallnow, 1991), a liminal journey (Turner, 1969), and a complex process of both sacred meanings and secular structures (Morinis, 1992). In the context of Amarnath, these perspectives gain further relevance as the pilgrimage unfolds through heavily mediated state protocols, militarized landscapes, and ecologically sensitive zones (Bose, 2003; Zutshi, 2014). This multiple-layered complexity makes the AmarnathYatra a rich ethnographic field for exploring ritual expressions, infrastructural challenges, identity formations, and the negotiations between continuity and change in contemporary Indian religiosity. Under this backdrop, the following objectives were considered:

To understand the ritual practices and symbolic meanings of the yatra.
To explore the social composition and diversity of pilgrims.
To assess the sacred and secular complex co-existence in the yatra.
To discern the impact of the pilgrimage on local communities, ecology, and infrastructure.
To analyse the cultural continuity and change in the practice of pilgrimage in contemporary India.

Methodology

This study employs an ethnographic (England, 1994) framework to examine the socio-cultural dimensions of the AmarnathYatra, with a specific focus on the involvement and experiences of diverse stakeholders (Chambers, 1997), including tribal communities. Ethnography, as a cornerstone of anthropological research, enables a nuanced understanding of cultural practices through immersion and participant observation (Geertz, 1973).

Fieldwork Strategy

The fieldwork was conducted by accompanying the AmarnathYatra as a participant-observer, following the route from Baltal Base Point to the Amarnath Cave and back, before concluding in Srinagar. This dual role allowed for both embodied experience and academic analysis, following the tradition of "being there" central to ethnographic fieldwork (Malinowski, 1922; Jackson, 1989). The researcher's movement alongside fellow pilgrims between 5 July 2024 and 8 July 2024 provided access to ritual, spatial, and interpersonal dimensions of the pilgrimage.

Data Collection Methods

Participant Observation: Prolonged engagement in campsites, religious ceremonies, and travel segments facilitated thick description of daily practices and spatial-temporal rhythms of the Yatra (Geertz, 1973). Special emphasis was placed on observing the labour and cultural expressions of tribal workers and other service providers involved in the pilgrimage infrastructure.

Semi-Structured Interviews: Guided by Bernard's (2011) framework, semi-structured interviews were conducted with pilgrims, sādhus, tribal community members, local authorities, and shopkeepers to elicit narratives on spirituality, livelihood, and identity.

Focus Group Discussions: Following Morgan (1997), group discussions were organized among tribal service providers to explore collective meanings, socio-economic shifts, and adaptation to the expanding pilgrimage economy.

Informal Conversations: Informal interactions provided spontaneous, context-rich insights into personal beliefs, communal cooperation, and challenges faced by various actors (Spradley, 1979).

Visual and Audio Documentation: With informed consent, selective photo and audio documentation was undertaken to complement textual data, following Pink's (2015) recommendations for incorporating sensory and visual methods in ethnography.

Sampling and Participant Overview

Participants were selected purposively to represent the range of stakeholders involved in and impacted by the Yatra. The researcher interacted with 33 informants across multiple points of the pilgrimage route. Informants included pilgrims, sādhus, local tribal workers, shopkeepers, security personnel, NGO volunteers, environmental activists, and administrative staff. The diversity of the sample allowed for a multi-perspective analysis of the sacred–secular complex, pilgrimage economy, and inter-community relations.

Informant Profile—AmarnathYatra Study

Sl. No.	ID code	Stakeholder type	Gender	Location of interaction	Date of interaction	Brief role/Relevance to study
1	*P*-01	Pilgrim (First-time)	M	Jammu Bus Terminal	05.07.2024	Shared expectations and motivations for Yatra before departure
2	*P*-05	Pilgrim (Female)	F	Baltal Base Camp	06.07.2024	Spoke on gendered experiences of safety and ritual during Yatra
3	*S*-01	Naga Sadhu	M	Baltal Base Camp	06.07.2024	Provided insights into ascetic practices and critique of commercialization
4	*G*-09	Gujjar Porter	M	Baltal Route	07.07.2024	Discussed economic dependency on Yatra and inter-religious cooperation
5	*K*-04	Kashmiri Shopkeeper	M	Baltal Route	07.07.2024	Spoke about seasonal economy and shifting pilgrim-local relations
…	…	…	…	…	…	…
33	*E*-03	Environmental Activist	F	Srinagar	08.07.2024	Highlighted ecological threats and sustainable pilgrimage strategies

(Full table in Appendix for detailed reference.)

Ethical Considerations

Ethical protocols followed the AAA Code of Ethics (AAA, 2012). Informed verbal consent was obtained, and anonymity was ensured through pseudonyms and coded identifiers. Cultural and religious sensitivities were respected, especially during sacred rituals and restricted spaces (Riley, 1999).

Reflexive Commentary: Walking as Pilgrims, Listening as Anthropologists

Conducting the Yatra together as both pilgrims and anthropologists allowed us to navigate the physical, emotional, and analytical dimensions of the journey in tandem. On the steep ascent from Baltal (06–07 July 2024), one of us walked in close conversation with *G-09*, a Gujjar porter who spoke of escorting pilgrims during curfews, while the other documented the rhythmic chants and exchanges along the path (Theme 1: Inter-Religious Coexistence). Sharing langar meals with *P-13* and *P-05* (Theme 2: Sacred–Secular Complex) immersed us in moments of devotion, while evening tent discussions with *S-01* and *S-04* revealed sādhus' resistance to commodification (Theme 3: Cultural Custodianship).

Our dual presence often diffused suspicion—particularly when one of us joined the aarti at the cave while the other quietly noted gestures, offerings, and security interactions. However, the visible act of note-taking occasionally prompted more guarded responses, as observed during a conversation with *K-12*, a shopkeeper cautious about speaking openly on political issues (Theme 4: Pilgrimage Economy). By alternating between devotional participation and analytical observation, we were able to gather both intimate, affective accounts—such as *F-07*'s reflections on safety as a woman pilgrim (Theme 6: Gender & Accessibility)—and structural insights from local service providers like *C-08* and *T-02* (Theme 7: Traditional–Modern Integration).

This reflexive (Davies, 2008), collaborative approach ensured that the sacred–secular interplay was captured not only in institutional frameworks and public performances but also in whispered confidences, bodily fatigue, ritual gestures, and fleeting solidarities. Our combined positionalities—insiders in faith and outsiders in analysis—shaped the co-production of these narratives, making the ethnographic record as much a product of our presence as of the pilgrimage itself.

Findings

Theme 1: Inter-Religious Coexistence and Socio-Political Dynamics

The AmarnathYatra, while rooted in Hindu devotional practice, unfolds in the Muslim-majority region of Kashmir, generating a unique space of inter-religious engagement. The pilgrimage thus becomes a living ethnographic site where pluralism, mutual dependence, and conflict coexist. Historically, Kashmiri Muslims—including the Gujjars and Bakarwals—have played vital roles in facilitating the Yatra by offering transport, shelter, and logistical support. Despite political tensions, many locals continue to regard the pilgrims as *mehmaan* (guests), embodying a moral ethic of *insaniyat* (humanity). However, recent developments—such as the abrogation of Article 370—have increased mistrust, and the heavy militarization of the route has made the Yatra feel like a state-controlled project to some local communities. This duality of cooperation and alienation reflects the complex dynamics of coexistence under the shadow of state politics and religious nationalism.

A middle-aged shopkeeper in Pahalgam recalls how his father would serve tea to pilgrims free of charge, believing it to be an act of faith. Now, while business sustains his family, the increased security presence makes every interaction feel monitored.

> *"We call them mehmaan, because in our culture, a guest is like God. But these days, God comes with so many soldiers."—K*-05, Kashmiri Muslim Shopkeeper

A Gujjar porter explains that during curfews, they have secretly escorted pilgrims to safe zones, risking their own safety. His sense of duty outweighs political pressures.

> *"Politics changes, but when a yatri is tired or hungry, I don't see Hindu or Muslim—I see a human being who needs help."*—G-09, Gujjar Porter

A young Bakarwal transporter describes how in recent years, some pilgrims avoid local services due to mistrust, affecting both income and the spirit of shared humanity that once defined the Yatra.

> *"After 2019, there is more suspicion. Some think we are here for money only, but service is also part of our deen."—B*-03, Bakarwal Transporter

Theme 2: Sacred-Secular Complex: Ritual, Governance, and Surveillance

The AmarnathYatra exemplifies the sacred-secular complex where spiritual devotion is deeply enmeshed with administrative, military, and economic systems. On the one hand, the Yatra is a site of ritual acts—chanting, puja, vows, and ascetic practices—symbolizing spiritual transformation and divine communion. On the other, it is a highly regulated event managed by the Shri Amarnath Shrine Board (SASB) and secured by the Indian Army and police forces. Pilgrims must undergo biometric registration, pass through checkpoints, and are monitored via GPS and CCTV. The juxtaposition of divine protection and military escort typifies the sacred-secular tension—where the sanctity of the space coexists with protocols of control, surveillance, and governance.

A teacher from Varanasi compares the journey to a corridor where one side is lined with chanting pilgrims and the other with armed guards, both claiming to ensure safety.

"Lord Shiva protects me, but the army checks my bag first."—P-22, Pilgrim from Uttar Pradesh

A first-time pilgrim from Gujarat is surprised at how structured the route feels—registration counters, security scanners, and queue systems—contrasting with the imagined free-flowing spiritual path he had envisioned.

"Darshan today means ID proof, biometric slip, and then the ice lingam."—P-13, Pilgrim from Gujarat

A woman pilgrim from Delhi notes that checkpoints sometimes disrupt the rhythm of prayer, yet she also feels safer knowing the route is constantly monitored.

"The Yatra feels like walking between two worlds—one of devotion, one of rules."—P-05, Female Pilgrim from Delhi

Use of digital registration, helicopters, and social media modernizes the yatra. Shift from spiritual austerity to "religious tourism". Tension between traditional values and modern conveniences.

Sacred element	Secular response
Spiritual solitude and withdrawal	Constant surveillance and crowd management
Pilgrimage as a penance or austerity	Use of helicopters, VIP passes, and mobile apps
Religious inclusiveness	Regulated access, permit systems
Divine protection belief	Military escort, weather alerts, insurance

Theme 3: Sādhus as Cultural Custodians and Liminal Figures

The presence of sādhus during the Yatra adds another layer of cultural and ritual depth. These Hindu ascetics—whether Naga Sādhus, Dasnami Sanyasis, or Aghoris—are perceived as embodiments of Shaiva tradition and carriers of mythic spirituality. Their austere practices, including barefoot journeys, symbolic attire (tridents, matted hair, ash-smeared bodies), and public blessings, evoke reverence and cultural continuity. Many sādhus reject the commodification of the Yatra, criticizing helicopter rides, VIP darshan, and bureaucratic structures for eroding the spiritual core of the pilgrimage. Anthropologically, sādhus can be seen as liminal figures who resist institutional religion while anchoring the symbolic and performative aspects of sacred practice.

A Naga Sadhu walking barefoot from Haridwar rejects food from tourist langars, relying on foraged herbs and offerings, as a way to preserve the austerity of the Yatra.

"Moksha cannot be booked on a helicopter ticket."—S-01, Naga Sadhu

A Dasnami Sanyasi describes his journey as a test of discipline, contrasting it with what he sees as a "picnic culture" emerging with tourism packages.

"The mountain tests your body before you can meet Shiva."—S-04, Dasnami Sanyasi

A Sadhu from Madhya Pradesh narrates his discomfort when pilgrims ask for selfies instead of blessings, symbolizing a shift from spiritual reverence to consumer spectacle.

"Every step on snow is a mantra—machines can't chant for you."—*S*-06, Sadhu from Madhya Pradesh

Types of Sādhus in the Yatra

Sādhu order	Features
Naga Sādhus	Naked or semi-naked, smeared with ash, symbolic of death of ego
JunaAkhara	One of the oldest Shaiva sects; many members undertake Amarnathyatra
AghoriSādhus	Extreme ascetics practicing taboo-breaking rituals
DasnamiSanyasis	Followers of AdiShankaracharya's orders, more philosophical
UdasiSādhus	Founded by Sri Chand (Guru Nanak's son), often less orthodox

Theme 4: Commodification and Pilgrimage Economy

In contemporary times, the AmarnathYatra has undergone a transformation from spiritual austerity to what many term "religious tourism." The rise of pilgrimage packages, helicopter services, and branded offerings reflects a broader trend of commodification. The economy surrounding the Yatra supports thousands of seasonal jobs, including ponywalas, porters, shopkeepers, tent providers, and transporters. Local trade in woollens, saffron, and dry fruits thrives during the pilgrimage season, creating temporary prosperity. However, this seasonal dependency also highlights the limitations of a short-term pilgrimage economy and calls for diversified, sustainable development models.

A ponywala explains that although helicopter rides cut his earnings, they also reduce the danger of accidents on steep trails, meaning fewer injured animals.

"Helicopters have taken away half our customers, but also half our income risk."—*G*-11, Ponywala

A Kashmiri shopkeeper thrives during the Yatra season but worries about dependence on this two-month income burst for the whole year.

"Selling woolens during Yatra is like Diwali for us—short, intense, and uncertain."—K-12, Shopkeeper

A dry fruit seller describes how the Yatra attracts traders from as far as Rajasthan and Punjab, creating a temporary "bazaar of India" in the valley.

"The cave may be Shiva's, but the market belongs to everyone."—K-04, Dry Fruit Seller

Theme 5: Environmental Impact and Sustainable Pilgrimage

The ecological consequences of the Yatra in the fragile Himalayan region cannot be overstated. With lakhs of pilgrims walking through sensitive terrain, the area faces challenges such as waste accumulation, deforestation, and pressure on glaciers and water bodies. Local communities and environmental activists have raised concerns about the long-term damage to sacred ecology. While the Shrine Board and state authorities provide sanitation and medical facilities, the pace of environmental degradation demands a stronger shift toward sustainable pilgrimage practices, including waste management, crowd regulation, and green infrastructure.

An NGO volunteer spends her mornings collecting plastic bottles along the Baltal route, describing the work as seva (service) to both God and nature.

"A clean path is also a prayer."—V-01, NGO Volunteer

A local guide notes that the snowline has been retreating for decades, altering water sources and affecting camp site selection.

"The glacier is shrinking. We see it year after year."—G-07, Local Guide

An activist warns that without strict waste control, the pilgrimage could harm the very environment that makes it sacred.

"If the mountains fall sick, so will the Yatra."—E-03, Environmental Activist

Theme 6: Gender and Accessibility

The increasing participation of women, elderly, and differently-abled individuals marks a shift in the demographic composition of the Yatra. Historically perceived as a masculine and physically strenuous journey, the pilgrimage is becoming more inclusive, thanks to improved infrastructure and healthcare provisions. This change also prompts new questions about the gendered experience of ritual, safety in militarized zones, and access to spiritual spaces.

A mother-daughter duo undertakes the Yatra, carefully timing their movements around sanitation stops and safe zones.

"We came as mother and daughter—our devotion is double, but so are our safety concerns."—F-07, Female Pilgrim

A volunteer recounts pushing a wheelchair pilgrim up a muddy slope, calling it both inspiring and revealing of the infrastructure gap.

"Wheelchair pilgrims inspire us, but the path still isn't ready for them."—V-02, Volunteer

A woman pilgrim emphasizes that spiritual equality also depends on physical accessibility and dignity.

"Bathrooms are as important as bhajans for women on this journey."—F-11, Female Pilgrim from Punjab

Theme 7: Indianism and National Integration

The Yatra is not just a regional event—it resonates with the broader narrative of Indianism. The pilgrimage embodies India's sacred geography and cultural pluralism. The communal langars run by diverse religious and social groups reflect the values of hospitality, inclusivity, and spiritual democracy. Pilgrims from across India, irrespective of caste, region, or language, participate in the Yatra, reinforcing a shared civilizational ethos. In this sense, the Yatra functions as both a spiritual journey and a performative act of national integration. However, this symbolism becomes politically charged in Kashmir, where it may also be seen as a tool of soft power and cultural assertion.

A langar volunteer from Punjab recounts the joy of serving tea to pilgrims from every corner of India. In his kitchen tent, people from Nagaland to Gujarat sit cross-legged on the same floor, sharing jokes and weathering the cold together. He sees the Yatra as a living example of *VasudhaivaKutumbakam* (the world is one family), a vision enacted daily through ladles of dal and piles of chapati.

"Here, in the langar line, no one asks who you are—only if you are hungry."—L-05, Langar Volunteer from Punjab

A pilgrim from Kanyakumari smiles as she recalls her tent assignment at Panchtarni. Her tent-mate was a Ladakhi farmer who had never seen the ocean, while she had never seen snow before. Over steaming cups of kahwa, they exchanged stories about coconut harvesting and barley farming, marvelling at how the same nation could hold such different worlds.

"I came from Kanyakumari, and my tent-mate is from Leh. Where else can you meet the whole country on one mountain?"—P-31, Pilgrim

A Kashmiri journalist, covering the Yatra for years, notes that while the spectacle of national integration is real, in the Valley it carries extra weight.

"When the national flag waves here," he says, "some see unity, others see assertion."—*K*-18, Kashmiri Journalist

His perspective reflects the layered meanings the Yatra acquires in a politically sensitive space. He also said that

"Unity is good, but in Kashmir, it is never just unity—it is also politics."—K-18, Kashmiri Journalist

Theme 8: Sacred-Institutional Tensions and Local Agency

The growing institutionalization of the Yatra has led to tensions between traditional religious actors (like sādhus) and modern administrative frameworks. Many sādhus express dissatisfaction with what they perceive as the over-bureaucratization and commercialization of the Yatra. Simultaneously, Kashmiri locals often feel excluded from decision-making processes. While they contribute substantially through labour, logistics, and hospitality, they lack representation in governing bodies like the SASB. This exclusion fosters alienation, despite the economic symbiosis that characterizes pilgrim-local relations.

A Sadhu from Uttarakhand, resting by the Lidderriver, voices his frustration over the SASB's growing control. For him, bureaucratic permits, ID cards, and fixed routes feel like interference in a centuries-old spiritual practice.

"Shiva gave no forms to fill,"—*S*-11, Sadhu from Uttarakhand

He says, suggesting that institutional order risks dulling sacred spontaneity.

"The board decides, but the mountain belongs to everyone."—S-11, Sadhu from Uttarakhand

A Kashmiri porter, who has worked every Yatra season for 15 years, explains how decisions about route closures, camp placements, or ration supplies are made without consulting those who know the trails intimately.

"We carry their loads, set up their tents, cook their food—but have no seat at the table."—*G*-14, Kashmiri Porter

His account reveals how economic dependence coexists with political exclusion.

A local shopkeeper in Baltal acknowledges the necessity of crowd control and sanitation rules but resents the way officials sometimes issue commands without engaging respectfully.

"A rule with explanation is guidance,"—*K*-09, Local Shopkeeper

He says,

"A rule without it is just an order."—K-09, Local Shopkeeper

His distinction hints at the fragile trust between administrators and locals.

> *"Rules are needed, but rules without respect are just orders."*—K-09, Local Shopkeeper

Theme 9: Conflict, Cooperation, and Resilience

Despite recurring political unrest (e.g., in 2010, 2016, and post-2019), the Yatra has continued, often with active cooperation from the local population. Even during curfews and heightened security, many Kashmiri Muslims have protected pilgrims and ensured their safe passage—demonstrating moral pluralism and human solidarity. This cooperation amidst conflict highlights the resilience of shared traditions and the potential for peace-building through cultural and economic interdependence.

A Kashmiri elder in Anantnag recalls the summer of 2016, when protests and curfews swept the Valley. Yet, when a group of stranded pilgrims knocked on his door, his family took them in, offering food and shelter for three days.

> "Religion is not the guest's first identity," he says. "First, they are a human seeking safety."—*K*-03, Kashmiri Elder

> *"Even in 2016, when the streets burned, the pilgrims drank tea in my house."*—K-03, Kashmiri Elder

A Gujjar guide explains that political unrest has forced him to learn alternate routes through forests and meadows, avoiding main roads when tensions run high.

> "The mountain doesn't close," he says. "We just find safer paths."—*G*-06, Gujjar Guide

His knowledge has kept many pilgrims moving even when official routes were temporarily blocked.

> *"The mountain doesn't close because of politics—we just find safer paths."*—G-06, Gujjar Guide

A Bakarwal tent owner reflects on decades of hosting yatris.

> "They know our faces,"—*B*-07, Bakarwal Tent Owner

He says,

> "We know their chants."—B-07, Bakarwal Tent Owner

For him, the relationship is reciprocal: pilgrims bring income and blessings, while locals offer shelter and guidance. This mutual recognition sustains the Yatra through crises.

> *"We know their prayers, they know our hospitality—that is why the Yatra survives."*—B-07, Bakarwal Tent Owner

Theme 10: Toward Balanced and Inclusive Growth

For the Yatra to thrive as both a spiritual and socio-economic phenomenon, inclusive policies and community participation are essential. Recommendations include promoting eco-tourism, providing micro-finance for local entrepreneurs, building year-round infrastructure, and integrating local voices into institutional decision-making. Such steps would ensure that the pilgrimage remains not only a sacred journey but also a sustainable model of coexistence and regional development.

A local guide argues for integrating community representatives into SASB planning sessions.

"We know where the snow melts first, where the wind is dangerous,"—*G*-01, Local Guide
He says,
"If we can guide pilgrims in the snow, we can guide policy too."—G-01, Local Guide

He sees formal inclusion as a way to bridge experiential knowledge with institutional decision-making.

"If we can guide pilgrims in the snow, we can guide policy too."—G-01, Local Guide

An environmental NGO worker in Srinagar envisions a model where the economic benefits of the Yatra are reinvested into maintaining trails, funding waste management, and supporting reforestation.

"Tourism should feed the mountain,"—*E*-05, Environmental NGO Worker

She says,

"not just take from it."—E-05, Environmental NGO Worker

Her approach ties ecological sustainability directly to economic planning.

"Tourism should feed the mountain, not just take from it."—E-05, Environmental NGO Worker

A young Kashmiri man, who works as a pony handler in summer, spends the rest of the year unemployed. He advocates for winter tourism, handicraft cooperatives, and skill training to make pilgrimage-related income more stable.

"We need jobs in winter too,"—*K*-21, Kashmiri Youth

He says,

"not only when Shiva calls."—K-21, Kashmiri Youth

"We need jobs in winter too, not only when Shiva calls."—K-21, Kashmiri Youth

Theme 11: Traditional Knowledge and Modern Technology Integration in Logistical Support

An important dimension of this study involves examining how tribal communities—particularly Gujjars and Bakarwals—combine traditional knowledge systems with modern technologies to manage logistical support during the AmarnathYatra. Based on qualitative interviews with tribal camp managers, porters, guides, and transport service providers, several key patterns emerged, which illuminate the resilience and adaptability of indigenous logistical systems.

A camp manager uses wind patterns, soil texture, and satellite maps to decide camp locations—calling it a marriage of old wisdom and new tools.

> *"We read the land like a book, but now the book has maps too."*—C-08, Camp Manager

A guide teaches younger workers to navigate by both GPS and natural signs, ensuring skills survive even without digital aids.

> *"When the phone dies, the moss shows the way."*—*G*-04, Guide

A transporter combines ergonomic saddles from NGOs with herbal tonics learned from his grandfather, blending safety with tradition.

> *"Modern saddles are good, but herbs keep the pony's legs strong."*—T-02, Transport Provider

Camp Management Practices

Respondents involved in camp management emphasized the use of traditional methods to select safe and elevated locations to avoid landslides and floods. Structures were often built using biodegradable tarpaulin, wooden frames, and layered ground coverings that balance insulation and drainage. Food preparation incorporated age-old practices of bulk cooking using clay ovens or open fire, supplemented by modern tools like portable stoves and gas cylinders. Several camp workers still utilized herbal remedies for minor ailments such as altitude sickness, nausea, and joint pain, drawing from intergenerational medicinal knowledge.

Traditional sanitation practices—such as designated waste pits and biodegradable cleaning agents—were supplemented by modern bio-toilets provided by state agencies. However, respondents expressed concern over the lack of training in operating modern sanitation facilities, leading to misuse or neglect in some instances.

Navigation and Safety by Guides and Porters

The porters and guides, many of whom have over a decade of experience, rely heavily on environmental cues—moss patterns, bird movement, wind direction, and changes in river colour—for navigation and risk assessment. Rituals such as lighting incense at key trail points or offering prayers at high-altitude passes were commonly performed to seek protection.

Simultaneously, many have integrated the use of GPS-enabled smartphones, although connectivity remains unreliable. Those who received basic training from NGOs or government programs expressed greater confidence in using digital navigation tools. However, some older guides voiced scepticism and discomfort, citing overreliance on battery-powered devices as a risk in remote areas.

Traditional Animal Care and Transport Services

Transport service providers, especially those handling ponies and palanquins, described traditional practices such as mixing herbal tonics into feed, resting animals at specific altitudes, and rotating service schedules based on lunar cycles. Ponies were ritually blessed before the Yatra, reinforcing their spiritual and economic role.

Modern equipment such as ergonomic saddles and reflective harnesses have improved animal safety, though many respondents noted difficulties in sourcing or maintaining this gear. GPS trackers for ponies were used in a pilot initiative but discontinued due to logistical challenges and maintenance costs. Animal welfare knowledge remained largely oral, with no formal training systems in place for most handlers.

Attitudes Toward Modern Technology

Respondents generally viewed the integration of modern technologies as beneficial —particularly for health, safety, and food storage—but also warned of potential cultural erosion. Younger respondents were more enthusiastic about digital integration, while older ones emphasized the importance of maintaining ancestral wisdom.

One camp manager noted:

> Steel beds and stoves may make us efficient, but we must not forget how to cook in the rain without gas or shelter.

This highlights the tension between efficiency and authenticity—a theme echoed by other respondents who favoured a hybrid model.

Training, Support, and Future Directions

Most participants emphasized the need for targeted, hands-on workshops in local dialects to build capacity for integrating modern tools with traditional systems. Suggestions included training on mobile navigation apps, solar-powered lighting, waste treatment, and animal care. Respondents also expressed a desire for formal recognition of their knowledge—through certifications or documented manuals—to preserve and validate tribal practices in a rapidly modernizing pilgrimage infrastructure.

Discussion

The AmarnathYatra, one of India's most renowned Shaivite pilgrimages, offers more than a spiritual ascent to a Himalayan cave. It is also a living stage where faith, governance, ecological concerns, market forces, and intercommunity relations continuously interact (Eade & Sallnow, 1991). Viewing the Yatra through the "sacred–secular complex" reveals a dynamic interplay in which rituals and myths merge with bureaucratic systems, security protocols, and commercial networks.

At its narrative heart lies the Shaivite legend of Lord Shiva revealing the secret of immortality to Parvati within the cave (Bhardwaj, 1997). Yet, today's pilgrimage is shaped by a strong secular overlay: biometric enrolment, multiple checkpoints, military escorts, and GPS tracking systems (Kaul, 2017). For some pilgrims, these measures ensure safety; for others, they highlight the way the state mediates access to the sacred. This coexistence of transcendence and regulation echoes Turner's (1969) observation that pilgrimage can dissolve social boundaries but also reproduce inequalities through structured access, helicopter services, and VIP privileges.

The Gujjar and Bakarwal service providers—porters, pony handlers, camp managers—are central to sustaining the Yatra. Their traditional expertise in navigation, animal care, and mountain survival is now intertwined with technologies such as GPS mapping and helicopter evacuations (Chatterjee, 2021). As one Bakarwal porter expressed, "The moss on the stones still tells me the way, but I also check the map on my phone." While some welcome technological integration for safety and efficiency, others worry it reduces their role from cultural custodians to service contractors, echoing concerns about the marginalisation of indigenous epistemologies (Panda, 2023).

Relations between Hindu pilgrims and the local Muslim population illustrate another facet of the sacred–secular continuum. Kashmiri Muslims provide logistical, commercial, and emergency services crucial to the Yatra's functioning (Rashid, 2016). While this economic interdependence can foster cooperation, it is also influenced by shifting political climates—sometimes celebrated as pluralism, sometimes strained by mistrust (Zutshi, 2014). As

Appadurai (1996) notes, religious events can serve both as acts of worship and as expressions of national identity, a duality clearly visible in the Yatra's political symbolism.

Environmental concerns were a recurrent theme in the field data. Tribal respondents described how earlier practices—composting, restricting night travel, using biodegradable materials—are giving way to more resource-intensive arrangements to accommodate rising pilgrim numbers. This shift reflects a broader tension between ecological stewardship and infrastructural expansion (Guha, 2000). A camp worker remarked, "Cleaning the river was part of our prayer; now it's fenced, and no one touches it."

The sādhus' presence underscores the resistance to over-institutionalisation. Their barefoot journeys and austere practices contrast sharply with the commercialised, tour-packaged versions of the Yatra. Yet even they are drawn into the spectacle, with curated routes, VIP passes, and photo opportunities transforming them into cultural icons as much as spiritual guides (Fuller, 2004).

Shifts in gender and accessibility are also noteworthy. More women, elderly participants, and differently-abled pilgrims are joining the Yatra. While infrastructural changes have enabled wider participation, issues of sanitation, safety, and dignity remain, making the intersection of devotional commitment (*bhakti*) and secular rights (*adhikar*) a critical part of the pilgrimage discourse (Chopra, 2015).

The AmarnathYatra thus emerges not as a fixed tradition but as a continuously negotiated field where sacred aspirations and secular realities collide and converge. It exemplifies India's sacred–secular entanglement as a spectrum rather than a binary—where governance coexists with devotion, ecology with ritual, and local economies with national politics.

By positioning the Yatra as a microcosm of India's cultural complexity, this research contributes to anthropological debates on pilgrimage as a site for negotiating identity, power, and belonging. It also points toward the need for *epistemic sustainability*: ensuring that future pilgrimage policy honours not only environmental and infrastructural priorities but also the lived knowledge of those who physically and culturally sustain the route to the cave. Only then can the AmarnathYatra remain both a sacred journey and a sustainable model of coexistence.

Acknowledgement We gratefully acknowledge the authorities of the University of Calcutta for granting permission and extending institutional support to conduct this study. We also extend our sincere thanks to the faculty colleagues, administrative staff, and members of the Department of Anthropology for their encouragement and cooperation. Our heartfelt gratitude goes to the participants of the study—pilgrims, tribal service providers, local residents, and officials—whose generous sharing of experiences and perspectives made this research possible. We further thank the logistical support teams and all individuals, named and unnamed, who contributed to the successful completion of the fieldwork and research.

Appendix

Informant Profile—Amarnathyatra Study

Sl. No.	Informant ID	Type/Role	Demographic details	Key relevance to study	Data collection method	Date/ Location of interaction
1	K-05	Local Shopkeeper	Male, 48, Muslim, Pahalgam	Insights on inter-religious coexistence and post-2019 changes	Semi-structured interview	05.07.2024, Jammu end route halt
2	G-09	Gujjar Porter	Male, 36, Muslim, Anantnag	Experience of cooperation during political unrest	Participant observation + interview	06.07.2024, Baltal Base Camp
3	B-03	Bakarwal Transporter	Male, 40, Muslim, nomadic	Livelihood link between pastoralism and pilgrimage	Semi-structured interview	06.07.2024, Baltal Base Camp
4	P-22	Pilgrim from UP	Male, 52, Hindu	Navigating sacred devotion under security checks	Informal conversation	07.07.2024, YatraRoute—Domel
5	P-13	Pilgrim from Gujarat	Male, 44, Hindu	View on regulated, structured Yatra	Semi-structured interview	07.07.2024, Yatra Route—Brarimarg
6	P-05	Female Pilgrim from Delhi	Female, 39, Hindu	Gendered ritual experience under surveillance	Informal conversation	07.07.2024, Yatra Route—Panchtarni
7	S-01	Naga Sadhu	Male, 60, Hindu ascetic	Critique of commodification and helicopter travel	Key informant interview	07.07.2024, Near Holy Cave
8	S-04	DasnamiSanyasi	Male, 55, Hindu ascetic	Discipline and spiritual testing in pilgrimage	Informal conversation	07.07.2024, Sheshnag Trail Section
9	S-06	Sadhu from MP	Male, 50, Hindu ascetic	Change from spiritual reverence to tourist spectacle	Key informant interview	07.07.2024, Baltal Return Camp
10	G-11	Ponywala	Male, 33, Muslim, Sonamarg	Impact of helicopters on earnings and safety	Semi-structured interview	06.07.2024, Baltal Pony Stand
11	K-12	Kashmiri Shopkeeper	Male, 45, Muslim	Seasonal economic dependence on Yatra	Semi-structured interview	06.07.2024, Baltal Market
12	K-04	Dry Fruit Seller	Male, 38, Muslim	Seasonal market expansion during Yatra	Informal conversation	06.07.2024, Baltal Market

(continued)

(continued)

Sl. No.	Informant ID	Type/Role	Demographic details	Key relevance to study	Data collection method	Date/ Location of interaction
13	V-01	NGO Volunteer	Female, 29, Hindu, Jammu	Waste management and eco-awareness	Participant observation	07.07.2024, Yatra Route—Baltal to Domel
14	G-07	Local Guide	Male, 42, Muslim, Baltal	Observations on glacier retreat	Semi-structured interview	07.07.2024, Yatra Route—Brarimarg
15	E-03	Environmental Activist	Male, 35, Kashmiri Pandit	Advocacy for sacred ecology protection	In-depth interview	08.07.2024, Srinagar
16	F-07	Female Pilgrim	Female, 48, Hindu, Maharashtra	Mother–daughter spiritual journey	Informal conversation	07.07.2024, Panchtarni
17	V-02	Volunteer	Male, 26, Sikh, Punjab	Accessibility for differently-abled pilgrims	Participant observation	07.07.2024, Yatra Route—Holy Cave
18	F-11	Female Pilgrim	Female, 55, Hindu, Punjab	Sanitation concerns for women	Semi-structured interview	07.07.2024, Baltal Return Camp
19	C-08	Camp Manager	Male, 40, Muslim, Gujjar	Traditional–modern camp site selection	In-depth interview	06.07.2024, Baltal Base Camp
20	G-04	Guide	Male, 37, Muslim, Anantnag	GPS and environmental cue navigation	Participant observation	07.07.2024, Pissu Top
21	T-02	Transport Provider	Male, 41, Muslim, Bakarwal	Modern saddles + herbal care for ponies	Semi-structured interview	06.07.2024, Baltal Pony Stand
22	L-05	Langar Volunteer	Male, 50, Sikh, Punjab	Inclusivity through langar service	Informal conversation	07.07.2024, Near Holy Cave
23	P-31	Pilgrim from Kanyakumari	Female, 34, Hindu	Cultural exchange across India	Semi-structured interview	07.07.2024, Panchtarni
24	K-18	Kashmiri Journalist	Male, 43, Muslim, Srinagar	Critical view on national integration narratives	In-depth interview	08.07.2024, Srinagar Press Enclave
25	S-11	Sadhu from Uttarakhand	Male, 58, Hindu ascetic	Concern over bureaucratization of Yatra	Key informant interview	07.07.2024, Near Holy Cave
26	G-14	Kashmiri Porter	Male, 39, Muslim	Exclusion from decision-making despite logistical role	Semi-structured interview	07.07.2024, Baltal Return Camp

(continued)

(continued)

Sl. No.	Informant ID	Type/Role	Demographic details	Key relevance to study	Data collection method	Date/ Location of interaction
27	K-09	Local Shopkeeper	Male, 46, Muslim, Baltal	Views on rule enforcement and local–administration trust	Informal conversation	06.07.2024, Baltal Market
28	K-03	Kashmiri Elder	Male, 70, Muslim, Anantnag	Stories of sheltering pilgrims during unrest	Oral history interview	08.07.2024, Srinagar
29	G-06	Gujjar Guide	Male, 44, Muslim	Adapting routes during conflict	Semi-structured interview	07.07.2024, Yatra Route—Brarimarg
30	B-07	Bakarwal Tent Owner	Male, 47, Muslim	Reciprocal pilgrim–local relationships	Semi-structured interview	06.07.2024, Baltal Camp Area
31	G-01	Local Guide	Male, 38, Muslim	Advocating local knowledge in Yatra planning	Semi-structured interview	07.07.2024, Baltal to Domel stretch
32	E-05	Environmental NGO Worker	Female, 31, Hindu, Delhi	Reinvestment of Yatra profits into ecology	In-depth interview	08.07.2024, Srinagar
33	K-21	Kashmiri Youth	Male, 25, Muslim	Calls for year-round livelihood options	Semi-structured interview	08.07.2024, Srinagar

References

American Anthropological Association. (2012). *American anthropological association code of ethics*.

Appadurai, A. (1996). *Modernity at large: Cultural dimensions of globalization*. University of Minnesota Press.

Babb, L. A. (1986). *Redemptive encounters: Three modern styles in the Hindu tradition*. University of California Press.

Bernard, H. R. (2011). *Research methods in anthropology: Qualitative and quantitative approaches*. AltaMira Press.

Bhardwaj, S. M. (1997). *Hindu places of pilgrimage in India: A study in cultural geography*. University of California Press.

Bose, S. (2003). *Kashmir: Roots of conflict, paths to peace*. Harvard University Press.

Campbell, H. A. (2012). *Digital religion: Understanding religious practice in new media worlds*. Routledge.

Chambers, R. (1997). *Whose reality counts? Putting the first last*. Intermediate Technology Publications.

Chatterjee, S. (2021). Pilgrimage, technology, and transformation: Changing practices in Himalayan yatras. *Journal of Cultural Geography*, *38*(2), 211–229. 10.xxxx/jcg.2021.xxxxx

Chidester, D., & Linenthal, E. T. (Eds.). (1995). *American sacred space*. Indiana University Press.

Chopra, R. (2015). Women, religion, and public space: Gendered encounters in pilgrimage. *Indian Anthropologist*, *45*(1), 23–40.
Coleman, S., & Eade, J. (2004). *Reframing pilgrimage: Cultures in motion*. Routledge.
Coleman, S., & Elsner, J. (2005). *Pilgrimage: Past and present in the world religions*. Harvard University Press.
Dabla, B. A. (2007). AmarnathYatra: A socio-environmental analysis. *Kashmir Journal of Social Sciences*, *1*(2), 25–34.
Davies, C. A. (2008). *Reflexive ethnography: A guide to researching selves and others*. Routledge.
Dubey, M. (2019). Securitization of sacred geographies in India: The case of Amarnath. *South Asia: Journal of South Asian Studies*, *42*(3), 450–469.
Eade, J., & Sallnow, M. J. (Eds.). (1991). *Contesting the sacred: The anthropology of Christian pilgrimage*. Routledge.
England, K. V. L. (1994). Getting personal: Reflexivity, positionality, and feminist research. *The Professional Geographer*, *46*(1), 80–89.
Fuller, C. J. (2004). *The camphor flame: Popular Hinduism and society in India*. Princeton University Press.
Geertz, C. (1973). *The interpretation of cultures*. Basic Books.
Guha, R. (2000). *Environmentalism: A global history*. Oxford University Press.
Gupta, A., & Ferguson, J. (1997). *Anthropological locations: Boundaries and grounds of a field science*. University of California Press.
Helland, C. (2007). Diaspora on the electronic frontier: Developing virtual connections with sacred sites. *Journal of Computer-Mediated Communication*, *12*(3), 956–976.
Jackson, M. (1989). *Paths toward a clearing: Radical empiricism and ethnographic inquiry*. Indiana University Press.
Jacobsen, K. A. (2013). *Pilgrimage in the Hindu tradition: Salvific space*. Routledge.
Jaffrelot, C. (2007). *Hindu nationalism: A reader*. Princeton University Press.
Kaul, N. (2017). The political economy of pilgrimage in Kashmir: Security, sovereignty, and sacred space. *South Asia: Journal of South Asian Studies*, *40*(4), 789–805. 10.xxxx/southasia.2017.xxxxx
Malinowski, B. (1922). *Argonauts of the western Pacific*. Routledge.
Michaels, A. (2004). *Hinduism: Past and present* (B. Harshav, Trans.). Princeton University Press.
Morgan, D. L. (1997). *Focus groups as qualitative research* (2nd ed.). Sage Publications.
Morinis, A. (1984). *Pilgrimage in the Hindu tradition: A case study of West Bengal*. Oxford University Press.
Morinis, A. (Ed.). (1992). *Sacred journeys: The anthropology of pilgrimage*. Greenwood Press.
Panda, S. (2023). Indigenous knowledge and infrastructural change: The case of Himalayan pilgrimage workers. *Asian Ethnology*, *82*(1), 45–68.
Parry, J. P. (1982). Sacrificial death and the necrophagous ascetic. In M. Bloch & J. Parry (Eds.), *Death and the regeneration of life* (pp. 74–110). Cambridge University Press.
Parry, J. P. (1985). *The anthropology of Hindu pilgrimage*. N. K. Bose Memorial Foundation.
Pink, S. (2015). *Doing sensory ethnography* (2nd ed.). SAGE Publications.
Rashid, S. (2016). Tourism, livelihoods, and political economy in Kashmir. *Economic and Political Weekly*, *51*(32), 69–75.
Riley, P. J. (1999). Researching pilgrimage. In J. S. Swatos Jr. (Ed.), *Encyclopedia of religion and society* (pp. 391–393). AltaMira Press.
Sax, W. S. (1991). *Mountain goddess: Gender and politics in a Himalayan pilgrimage*. Oxford University Press.
Singh, S., & Raj, R. (2016). *Religious tourism and pilgrimage: Management, marketing and development perspectives*. CABI.

Spradley, J. P. (1979). *The ethnographic interview*. Holt, Rinehart and Winston.
Srinivas, M. N. (1952). *Religion and society among the Coorgs of South India*. Clarendon Press.
Turner, V. (1969). *The ritual process: Structure and anti-structure*. Aldine Publishing Company.
Zutshi, C. (2014). *Kashmir's contested past: Narratives, sacred geographies, and the historical imagination*. Oxford University Press.

Chapter 10
Rooted Resilience: Indigenous Agricultural Practices and Traditional Farming Wisdom Among India's Tribal Communities

Rabiya Haritaksh and Abhik Ghosh

Abstract This paper revolves around the indigenous agricultural practices embedded in the traditional knowledge systems. These agricultural practices are originated from the depth of ancient cognition and emotional disposition of mind and how some of these practices has been used in contemporary world to promote sustainability in agricultural practices. The modern agricultural practices were taking a toll upon the earth's carrying capacity. Tribal people consider earth as mother goddess so their agricultural practices are near to the raw organic forms in contrast to the modern high-intensity agriculture which imposed a threat to the environment and this paper also included the case studies related to the resilience and enculturation of the indigenous knowledge. Lahari Bai has collected 150 varieties of the seeds and preserved them in her hut. Although she faced humiliation from other people, too, she never stopped to store the seeds of wisdom, knowledge of the tribes which grew wide varieties of crops, preserving the biodiversity and promoting the preservation of native species. This paper also narrates the case study of the Car Nicobarese, where the efforts from the government and the local/tribal people came forward to enhance the production of the coconut trees.

Keywords Indigenous knowledge systems · Indigenous agricultural practices · Sustainability

Introduction

The inception of agriculture in India traces its roots back to the Neolithic period, a transformative era that marked the shift from a nomadic, hunter-gatherer lifestyle to settled agricultural communities. Around 7000 BCE, in the northwestern regions of the

R. Haritaksh (✉) · A. Ghosh
Department of Anthropology, Panjab University, Chandigarh, India
e-mail: rabiyaharitksh@gmail.com
e-mail: abhikg@rediffmail.com

R. C. Sobti et al. (eds.), *Traditional Knowledge Systems in India: Sustainability, Ecology, and Resource Governance*, People, Cultures and Societies: Exploring and Documenting Diversities,
https://doi.org/10.1007/978-981-92-1346-7_10

Indian subcontinent, evidence of early cultivation of crops such as wheat and barley has been unearthed at sites like Mehrgarh in present-day Pakistan. This Neolithic settlement, dating back to approximately 6500 BCE, reveals the nascent stages of agriculture, characterized by the cultivation of plants and domestication of animals (Kenoyer, 1991). Sustainable agricultural practices such as crop rotation and irrigation are evidence of their first inkling. This meant that when humans started cultivation of plants such as wheat and barley, this set the course for their settlement, excess food production and ultimate development of complex societies like the Indus Valley Civilization in the Indian subcontinent. Thus, the Neolithic period is the most important period in the history of Indian agriculture as it marked the beginning of a new era where man's relation with nature changed, and the fruits of agricultural tradition were ripened for many generations to come (Kenoyer, 1991).

The ancient threads of these traditional practices wove a fine tapestry from India's early agricultural past. The close relationship that indigenous peoples have towards land is due to these ancient pioneers who were from Mehrgarh. Our forefathers practiced settled agriculture, which resembles to the present indigenous communities of today (Kenoyer, 1991).

We are witnessing how India's older agricultural practices blend with the sustainable traditions of its aboriginal communities. From the agricultural practices that began during the Neolithic era and continued towards the indigenous knowledge systems and threaded throughout cultures, this fabric connects many stories about human relations with nature. These indigenous agricultural practices, developed from a deep understanding of ecosystems and biodiversity are a living example demonstrating how wisdom is passed from one generation to the next through the process of enculturation. The earth-farmer relationship is remembered in a complete manner even today because it has remained as an inspiring factor for fostering contemporary ecologically-friendly farming techniques. By examining Mehrgarh's legacy and the continuum of native customs, we witness where timeless wisdom merges into current necessities so that sustainable agriculture keeps in shape for things to come (Richards, 1985).

Indigenous agriculture refers to farming practices that are native to a particular region and have been developed and passed down from generation to generation within the indigenous community. These practices are often deeply rooted in the local ecosystem, culture tradition, and tradition promoting sustainable land use and biodiversity conservation (Berkes et al., 2000).

Indigenous agriculture is closely intertwined with indigenous knowledge systems which encompass the accumulated knowledge, skills and practices developed by indigenous communities over generations, many indigenous agricultural practices have originated from the indigenous knowledge system reflecting the deep understanding of the local ecosystem, weather patterns, soil conditions, and plant diversity.

Conklin (1967b) presented a report on the integral system of shifting cultivation in the Philippines that has provided us with the model of intensity. Shifting cultivation is represented by how cultural practices influence the natural environment and how human societies adapt to and modify their surroundings based upon cultural beliefs and practices. Conklin also stated the Numerically Labelled Temporal Distinctions referring to five successive phases of Swidden Farming as

Selecting, Cutting, Burning, Cropping, and Fallowing. Vidyarthi (1963) studied the Maler tribe of Santhal Parganas and also gave similar successive phases related to shifting cultivation and worshiping and merrymaking as part of cultural paraphernalia. He emphasized this totality with Nature-Man-Spirit Complex association which embodies the network of mutual interaction and intimate interaction and intimate interrelations between nature, man, and spirit.

Objectives

1. To explore the case studies of the tribal people who excelled in the field of agriculture.
2. To delineate the practices that can provide vantage to sustainability in agricultural practices.
3. To know the importance of Indigenous Cultural Practices

Importance of Preserving Indigenous Agriculture

Before delineating the importance of preserving the indigenous culture, the narration of the tale of Gaura Devi, a tribal woman and pioneer of the Chipko Movement hailing from the Himalayan region of Uttarakhand (Pathak, 2021). Affiliated with an indigenous group, she took a stand and embraced in her arms to shield her beloved companion, i.e., the tree. She held a strong sentiment that she could lay down her life to protect her companion. This serves as a wondrous example of devotion to protecting the environment without being a part of the formal education system and without knowing the scientific basis of environmental protection. All she had was an innate instinct to protect her surroundings.

Similarly, an intimate understanding of Cultural Resource Management often aids in adopting sustainable agricultural practices. Indigenous people possess traditional knowledge of soil fertility, plant diversity, and natural cycles, allowing them to cultivate crops in harmony with the environment.

Community Forest Management is evident in tribes like the Bishnois in Rajasthan and the Khasis in Meghalaya. They have traditional systems of managing forests sustainably, including agroforestry, rotational grazing, and the protection of sacred groves(Agrawal & Gibson, 1999).

Sustainable development Goals 2030 which are a total of 17 in number; include two sustainable goals in particular that state Zero Hunger and the target 2.3. They aim to double the agricultural productivity and incomes of small-scale food producers, in particular women, indigenous peoples, family farmers, and fishers, including through secure and equal access to land, other productive resources, and inputs. So, this becomes an important issue to preserve the indigenous agricultural practices too because they sustain the diversity in the social fabric (Department of Economic and Social Affairs United Nations, 2015).

Cultural Aids in Storing Indigenous Agricultural Practices

The process of enculturation refers to how the human mind acquires knowledge. Family and community are the primary sources of information that make up a society's culture. This information is passed down from generation to generation, making the indigenous knowledge system an important part of preserving cultural heritage and simultaneously cultural heritage is preserving the indigenous agricultural practices.

Strathern (1968) explored the symbiotic relationships in cultural systems that could be applied to the symbiosis between indigenous cultures and their agricultural practices. The reciprocal relationship between cultural beliefs, rituals, and sustainable farming methods reflects a symbiotic integration of culture and ecology.

Types of Indigenous Agricultural Practices

India, with its well-preserved cultural heritage, boasts a vast diversity in seasons, each with its enticing charm. However, every season also carries biophysical echoes, and the aftermath of these echoes is the inception of agricultural practices to sustain mankind. These ancient yet well-defined and preserved practices continue to make their way into the modern technological world, and because these are environment friendly regional practices; they are a matter for research to vaunt the sustainability of the modern world. These are the following types of the indigenous Agricultural Practices.

1. **Cultivation of** *Jhum* in North-eastern states: Indigenous communities in states such as Manipur, Nagaland and Mizoram practice shifting cultivation known as *jhum* cultivation. This involves clearing the land, burning vegetation and cultivating on it for some years before moving to another plot. It demonstrates indigenous peoples' deep relationship with their territories and traditional knowledge of sustainable farming practices (Cairns, 2015).
2. **Dry land Farming in Rajasthan**: It is dryland farming which is done by indigenous communities living in dry regions like Rajasthan who utilize water harvesting techniques that are traditionally built such as check dams, contour bunding and *khadins* or underground reservoirs. All these methods make it possible for crops to be grown even when rainfall is lacking (Pokharia et al., 2014).
3. **Agro forestry (Western Ghats)**: Indigenous communities in the Western Ghats region practice agro forestry, integrating trees, crops, and livestock on their land. This system enhances soil fertility, conserves water, and provides multiple

sources of income for farmers while maintaining forest cover (Kumar & Takeuchi, 2009).

4. **Organic Farming**: Indians have been involved in organic farming for a long time, which entails avoiding artificial pesticides and manure. This is manifested through their acts that are based on traditional wisdom and cultural practices that promote coexistence with nature and the achievement of sustainable environment (Modi, 2003).
5. **Traditional seed saving and exchange (Various Regions)**: People in different parts of the country use conventional methods of seed saving to protect ancient crop varieties or region-specific food requirements Indian farmers have conserved precious seeds over countless generations to develop crops that are most adapted to their specific ecological situations for agriculture, indicating an unchanged agricultural heritage interwoven with our cultural heritage since then till date (Press Trust of India 2023).
6. **Community Seed Saving and Exchange**: Indigenous groups like Bishnois in Rajasthan and Khasis in Meghalaya have indigenous systems of forest management such as agroforestry, rotational grazing as well as conservation of sacred groves (Agrawal & Gibson, 1999).
7. **Hill Terracing:** The Lepchas who belong to Sikkim and the Khasis living in Meghalaya construct terraces on steep slopes for cultivation purposes including rice, maize and vegetables thereby controlling soil erosion and improving water retention characteristics (Giri et al., 2020).
8. **Aquaculture:** Indigenous communities within regions hosting water bodies like Mishing tribe in Assam resort to customary fish farming techniques which they integrate into their farming practices such as pond-fisheries or river-fisheries (Inaotombi and Mahanta, 2018).

From the Fields: Case Studies

PVTG Tribal Woman to Train the Ph.D. Scholars

Lahari Bai, a tribal woman and seed collector, has been making significant contributions to the preservation of biodiversity through her innovative initiatives. Her work in seed collection is a testament to the importance of individual efforts in safeguarding our planet's rich flora and sustaining agricultural diversity (Press Information Bureau, 2023a).

Lahari Bai belonging to the Baiga tribe from Madhya Pradesh was declared the Ambassador for international year of millets. Baiga tribe is a Particular Vulnerable Tribal Group. This particular tribal group is believed to possess special knowledge on biodiversity and environment. Lahari was inspired from her grandmother's word that a number of indigenous varieties were disappearing from the eyes of the world. Lahari Bai, when she was just 18 years old, was collecting seeds of indigenous varieties, and all the information about these indigenous seeds from the elderly members of the community and preserved 150 varieties of seeds. The current Prime Minister of India also acknowledged her and her efforts for the preservation of the millets or as the Prime Minister of India called them *Shree ann* (Millets). Dindori district collector Vikas Mishra who has nominated Lahari (who never went to school) for a coveted Rs 10 lakh scholarship of Jodhpur-based ICAR said, "If she succeeds in getting the scholarship, then she would be seen guiding PhD students."

Climate-Smart Agriculture in Odisha: Resilient Dongria Kondh Practices

1. Introduction: In the state of Odisha, where there are recurrent droughts and erratic rainfall, Dongria Kondh tribe has become a symbol of hope that climate-smart agriculture works. They cultivate different crops as their traditional farming method is an epitome of sustainable and resilient agriculture in the face of climate change.
2. Background: When pests attacked crops, there was a drought and untimely rainfall leading to crop failure in most parts of Odisha during 2017; however, the harvest was good for the Dongria Kondh tribe who reside in the South West part of Odisha. Their success comes from the traditional practice of cultivating many types of plants as protection against uncertainties caused by nature.
3. Diverse Cropping Practices: The farmers are well suited to their land since they grow over 50 varieties including rice, millets, sorghum, vegetables, pulses, tubers, etc. This diversity not only ensures constant food supplies but also acts as a natural shield towards climate change related risks.
4. Resilience to Climatic Challenges: The Kondh farmers of the Dongria did not lose their crops despite Odisha being in severe agricultural distress in 2017. Dongrias' sustainable agriculture hinges on genetic profiling for crops such as millets which are drought-tolerant and are recognized as climate smart cereals for arid areas, where temperatures are extreme. Furthermore, they have been adopting traditional upland paddy varieties that consume less water hence suitable even when there is drought.
5. Natural Pest Control: One remarkable attribute of the farming practices carried out by the Dongria Kondh tribe is that they bank on natural means of pest control. This group does not use chemical fertilizers or pesticides thus allowing pests like insects and their natural enemies to coexist in their fields. Consequently, this habit helps maintain a balanced ecosystem that protects crops while attracting pest-controlling birds.
6. Climate Resilience Integrated with Tradition: Climate change has seen their traditional cropping patterns evolving in line with nature, thus serving as an

organic approach towards climate resilience. By growing diverse crops mixed with traditional seeds, they have created a resilient system capable of surviving through major weather occurrences and other climatic perturbations.

7. Lessons for Sustainable Agriculture: At this time when the world is experiencing increased concerns over effects of global warming on farming activities, Dongria Kondh Practices. Patnaik and Das Patnaik (1980) provide this comprehensive ethnographic account of the Kondh tribe of Orissa.

20 Crops and Biodiversity

The Bonda community in Odisha practices a unique form of shifting cultivation known as *dangar chas*, which allows them to plant and grow more than 20 crop types, while conserving biodiversity, protecting the forest and climate. The farmers are Bonda people who belong to one of the extremely vulnerable tribes group hence adopt time-tested ways of agro-farming.

These are some important things about farming by the Bonda tribe:

1. *Dangar Chas* (Highland Cultivation): Malkangiri hills in Eastern Ghats are where the tribe carries out *dangar chas*. This type of shifting cultivation involves cultivating on tiny areas referred to as *podu* that arise due to cutting down weeds and grasses after burning them. These patches are left idle for more than three years after being cultivated because they allow natural regeneration of tree cover and shrubs.
2. Cultivation Cycle: Farming begins immediately following the initial rains towards early May. The first harvest takes place between November and January. When beginning this process, farmers perform *Buhi Parabor* ritual which is done by praying for their ancestors and Mother Earth.
3. Seedbed Preparation: The tribe enriches the soil and stores moisture by burning ash, vegetable residues, and grass stubble on the ground prepared for sowing. Instead of ploughing, they gently break the soil to sow seeds avoiding the destruction of trees.
4. Crop Diversity: Different types of indigenous paddy varieties (*dangar dhaan*), local millets and oilseeds are planted in designated seedbeds. This rich diversity makes farming system sustainable.
5. Fallow Cycle: The fallow cycle ensures that different patches are in various stages of natural vegetation, supporting a wide range of flora. This is also useful practice for plant conservation as seen in Malkangiri hills which has more than 450 plant species including 34 medicinal herb varieties.
6. Climate and Environmental Benefits: Shifting cultivation method is known to have climate-friendly characteristics, it does not involve generation of methane—powerful greenhouse gas produced in flat/inundated croplands especially rice fields; *dangar chas* also offers carbon sinks through young growing forests it supports, offsetting carbon emissions during preliminary stage burning of vegetation.

Challenges: However, with Bonda population rapidly increasing, the sustainability of *dangar chas* is being threatened. This is due to a decrease in fallow period from four to five years to two to three years as a result of increased pressure on land for more cropping seasons owing to the increase in population. It has been established that youth are moving out of the village in search of better income. However, the farming technique by the Bonda tribe presents a symbiotic relationship with nature that ensures food security and conserves biodiversity while reducing environmental damage caused by them (Sahoo et al., 2023).

Amalgamation of Good Governance and Indigenous Knowledge Practices

Take an anecdote of Nicobarese, their economy is simple which is based upon horticulture, but some of them have taken up some modern economic pursuits after coming in contact with outsiders. The government has established two farms in the Car Nicobar- A multipurpose farm at Malacca of over 1.5 ha of land and a coconut plantation at Auckchung over 22.23 ha of land, the multipurpose farm served as a knowledge centre for providing technological education to the farmers. The main cash crop was coconut and the Nicobarese had been using their local and technological knowledge imparted by the outsiders (Rath, 2006).

A new system of coconut plantation has been pioneered by the Car Nicobarese, in which new coconut plants are raised in the place of old ones as a new coconut grove declines in production after 40 years, coconut plantation is done by the joint efforts of the members of the *tuhets* (forms of the extended family). In order to protect their coconut plantations, they erect *Takoyo: a* symbol displayed by erecting two vertical parallel poles joined at the top by a horizontal stick. Jaws of pigs are hung on the horizontal stick, along with a few tender coconut leaves and a red white cotton cloth. A hole was punched in unripe coconuts, which were perched on poles. It acted as a warning to members of a large extended family residing within a *Tuhet* not to pluck any nuts from the forbidden plantation. With this practice, the coconut production increased (Rath, 2006).

Conclusion

The Human Development Report (1997) defined absolute poverty as some absolute standard of minimum requirement, and relative poverty as that which fell behind most others in the community.

They have no treasure as the world counts gain,
Some starving cattle; a small bin of grain
Torn scraps of dirty cloth; a string of beads,

A mat, a broken bed, a pot of seeds,
The basket of roots, a little meat,
The bows and arrows and a wooden seat,
Is all their low-roofed hovel boast of store
Such a sad accounting of the poor (Rustomji, 1989:54–55)

The above-given poem projects the poverty amongst the Kuttia Kond tribe of Orissa. Tribal poverty puts a huge question mark on the grand idea of the Nehruvian Model for the eradication of tribal poverty (Rath, 2006). However, in all the adversaries, tribal people still preserved their traditional cultural knowledge in contrast to the modern society which is affected by the constant pressure of materialism. As we all say, necessity is the mother of all invention (Sah, 2009). Madhya Pradesh, situated in the Western tribal belt, is characterized by extensive nitrogen deficiency. Ergo, the consumption of Nitrogen is the agronomic choice that is why non-tribal people in the region use fertilizers and other facilities provided by the state like high-yielding variety and intensity irrigation but in the tribal region these modern practices were still new, the spread of HYV was also less and seed fertilizer technology has bypassed the belt. However, overcoming all the odds, Lahari Bai preserved the seeds of 150 Crop varieties in the form of traditional knowledge and was now providing the insights of her wisdom to the Ph.D. students.

In Cop28, it is stated that globally only 11% of policies on nature prioritize the rights of indigenous people and local communities yet safeguard the world's 80% cultural and biological diversity. The major tool that these indigenous people have is their knowledge.

Indigenous knowledge is, "an invaluable basis for developing adaptation and natural resource management strategies in response to environmental and other forms of change," as underlined by the Intergovernmental Panel on Climate Change in 2007 (Parry et al., 2007).

Traditional knowledge tends to be more holistic and intuitive and has a spiritual dimension also and doesn't make a distinction between sacred and empirical. Traditional Knowledge is Qualitative and passed on orally from one generation to another.

Written words provide permanence in contrast to oral literature, written literature won't vanish into thin air. Ergo, for safekeeping, it should be preserved in the literary papers and there should be constant dissemination of this knowledge to the masses.

The government was imparting modern technological knowledge but a stronger emphasis should be on the Cultural Knowledge Exchange, since from the above example it is clear that by the joint effort of society and the government, one may provide food security to mankind.

Finally, the author would like to put emphasis upon the syncretism of the technology, traditional knowledge systems and good governance for attaining true sustainable development in agriculture and the need for preserving Indigenous agricultural practices for future use (Fig. 10.1).

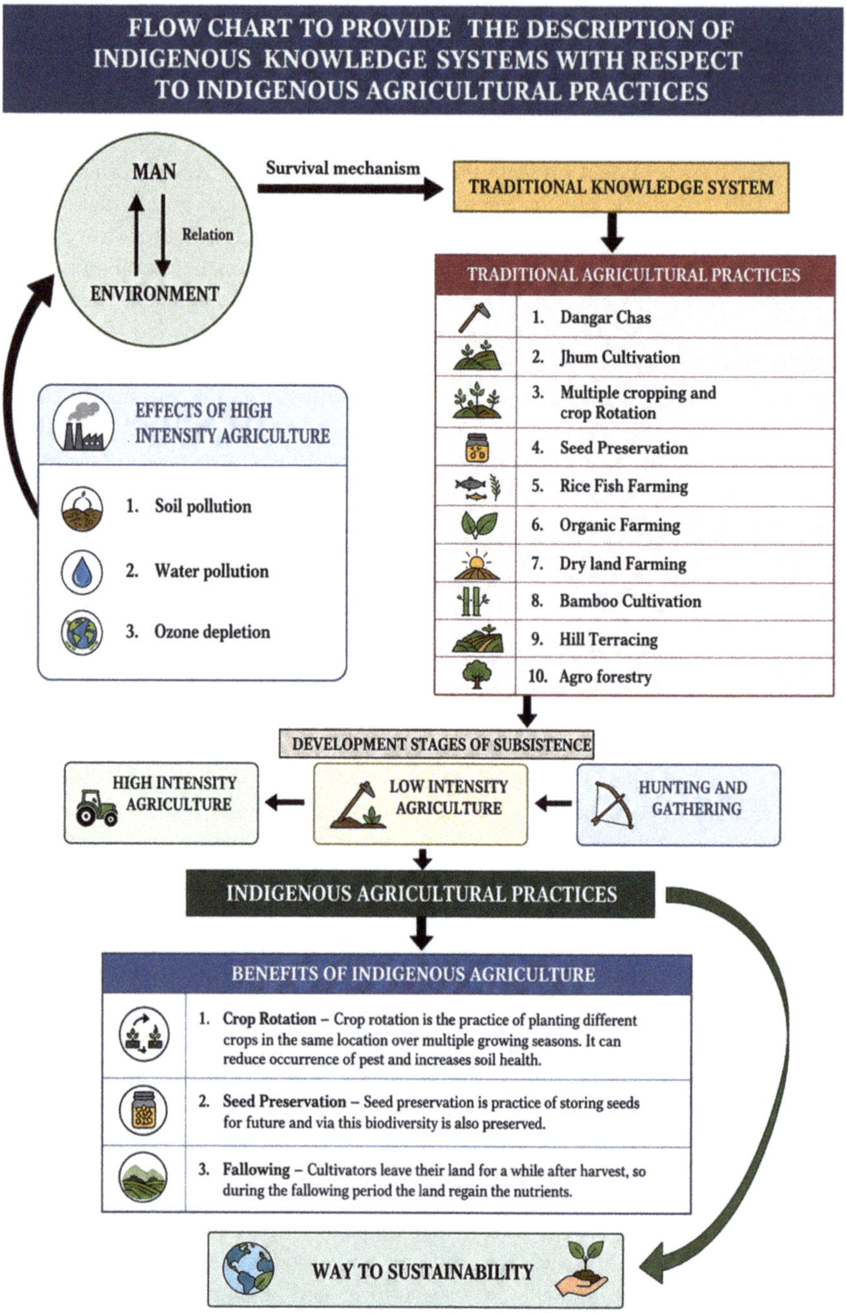

Fig. 10.1 Flow chart to provide the description of indigenous knowledge systems with respect to indigenous agricultural practices

References

Agrawal, A., & Gibson, C. H. (1999). Enchantment and Disenchantment: The role of community in Natural resource Conservation. *World Development*, *27*(4), 629–649. https://doi.org/10.1016/s0305-750x(98)00161-2

Berkes, F., Colding, J., & Folke, C. (2000). Rediscovery of traditional ecological knowledge as adaptive management. *Ecological Applications*, *10*(5), 1251. https://doi.org/10.2307/2641280

Cairns, M. F. (Ed.). (2015). *Shifting cultivation and environmental change: Indigenous people, agriculture and forest conservation*. Routledge.

Conklin, H. C. (1967b). Division of anthropology: Some aspects of ethnographic research in ifugao*,†. *Transactions of the New York Academy of Sciences*, *30*(1 Series II), 99–121. https://doi.org/10.1111/j.2164-0947.1967.tb02455.x

Department of Economic and Social Affairs (Director). (2015). *Sustainable Development Goals-The 17 Goals*. Retrieved March 16, 2024, from https://sdgs.un.org/goals

Giri, K., Mishra, G., Rawat, M., Pandey, S., Bhattacharyya, R., Bora, N., & N, J. P. (2020). Traditional farming systems and agro-biodiversity in eastern Himalayan region of India. *Rhizosphere Biology*, 71–89. https://doi.org/10.1007/978-981-15-1902-4_5

Inaotombi, S., & Mahanta, P. C. (2018). Pathways of socio-ecological resilience to climate change for fisheries through indigenous knowledge. *Human and Ecological Risk Assessment*, *25*(8), 2032–2044. https://doi.org/10.1080/10807039.2018.1482197

Kenoyer, J. M. (1991). The indus valley tradition of Pakistan and Western India. *Journal of World Prehistory*, *5*(4), 331–385. http://www.jstor.org/stable/25800603

Kumar, B. M., & Takeuchi, K. (2009). Agro forestry in the Western Ghats of peninsular India and the satoyama landscapes of Japan: A comparison of two sustainable land use systems. *Sustainability Science*, *4*(2), 215–232. https://doi.org/10.1007/s11625-009-0086-0

Modi, A. T. (2003). What do subsistence farmers know about indigenous crops and organic farming? Preliminary experience in KwaZulu-Natal. *Development Southern Africa*, *20*(5), 675–684. https://doi.org/10.1080/0376835032000149306

Parry, M. L., Canziani, O., Palutikof, J., Van Der Linden, P., & Hanson, C. (2007). *Climate change 2007: Impacts, adaptation and vulnerability*. https://assets.cambridge.org/97805217/05974/copyright/9780521705974_copyright_info.pdf

Pathak, S. (2021). *The chipko movement:a people's history*. Permanent Black.

Patnaik, N., & Das Patnaik, P. S. (1980). *The Kondh of Orissa*. Tribal and Harijan Research-cum-Training Institute. BHUBANESWAR, Orissa. https://repository.tribal.gov.in/bitstream/123456789/73910/1/SCST_1982_book_0091.pdf

Pokharia, A. K., Kharakwal, J. S., & Srivastava, A. (2014). Archaeobotanical evidence of millets in the Indian subcontinent with some observations on their role in the Indus civilization. *Journal of Archaeological Science*, *42*, 442–455. https://doi.org/10.1016/j.jas.2013.11.029

Press Information Bureau. (2023a). *PM lauds the efforts of Lahari Bai for preserving more than 150 varieties of millet seeds*. https://pib.gov.in/PressReleseDetailm.aspx?PRID=1897559

Press Trust of India. (2023). *G-20 first spouses enjoy the farm-to-fork millet experience*. The Hindu. https://www.thehindu.com/news/national/g-20-first-spouses-see-breakthroughs-in-indian-agriculture-at-iari-enjoy-farm-to-fork-millet-experience/article67288532.ece

Rath, G. C. (2006). *Tribal development in India: the contemporary debate*. http://ci.nii.ac.jp/ncid/BA79036515

Richards, P. (1985). *Indigenous Agricultural Revolution: Ecology and food production in West Africa*. I). Routledge.

Rustomji, N., & Elwin, V. (Eds.). (1989). *Verrier Elwin philanthropist selected writing* (1st). Oxford University Press.

Sah, D. C. (2009). *Tribal development in India: The contemporary debate*. Rawat Publication.

Sahoo, M., Kuiti, B. K., & Patra, P. K. (2023). Livelihood Practices of the Bonda Tribe of Odisha. *The Researchers' International Research Journal*, *9*(1), 36–48.

Strathern, M. (1968). [Review of the book *Pigs for the Ancestors: Ritual in the Ecology of a New Guinea People*, by R. A. Rappaport]. *Man, 3*(4), 687–688.
UNDP. (1997). *Human Development Report*. Oxford University Press.
Vidyarthi, L. P. (1963). *The maler: A study in nature-man-spirit complex of a hill tribe*. Bookland Private Limited.

Chapter 11
Harmony of Heights: Harnessing Himalayan Indigenous Knowledge for Sustainable Pastoralism

Muzamil Rashid, Irfan Ali Banka, and Abhik Ghosh

Abstract Indigenous Knowledge Systems (IKS) represent invaluable reservoirs of ancestral wisdom in indigenous communities worldwide. In the unique context of Himalayan pastoralism in Jammu and Kashmir, India, the relevance and resilience of IKS become particularly pronounced. This chapter presents a comprehensive case study exploring the practical applications of IKS among two prominent pastoral groups: the transhumant Gujjar-Bakarwal and the Chopan shepherds. Through ethnographic fieldwork from 2022 to 2023, focusing on pastoral mobility, weather prediction, ethnomedicines, and sustainable practices, the study sheds light on the intricate ways traditional knowledge systems contribute to livelihood security and ecological resilience in challenging environments.

The findings underscore the importance of recognising and preserving IKS as vital wisdom and adaptive capacity repositories. By integrating traditional knowledge into policymaking and development initiatives, policymakers can harness the full potential of indigenous communities in addressing contemporary challenges and fostering sustainable development. The study advocates for a holistic approach to resource management that incorporates scientific and indigenous perspectives, emphasising the need for participatory research and emic perspectives in policy formulation to ensure the success and relevance of interventions. Efforts to document and preserve traditional knowledge should be accompanied by initiatives to empower indigenous communities and protect their rights to land, resources, and

M. Rashid (✉)
Anthro International, Srinagar, India
e-mail: burzahama@gmail.com

I. A. Banka
Department of Anthropology, Panjab University, Chandigarh, India
e-mail: anthroirfan@gmail.com

A. Ghosh
Department of Anthropology, Panjab University, Chandigarh, India
e-mail: abhikg@rediffmail.com

R. C. Sobti et al. (eds.), *Traditional Knowledge Systems in India: Sustainability, Ecology, and Resource Governance*, People, Cultures and Societies: Exploring and Documenting Diversities,
https://doi.org/10.1007/978-981-92-1346-7_11

cultural heritage. We can work towards a more equitable and resilient future for all by fostering partnerships between scientists, policymakers, and indigenous communities.

Keywords Indigenous knowledge · Pastoralism · Gujjar and bakarwal · Ethnomedicine

Introduction

Indigenous Knowledge Systems (IKS) represent the rich reservoirs of ancestral wisdom in indigenous communities worldwide (Barnhardt & Kawagley, 2008; Macusi et al., 2023). These systems, encompassing of knowledge domains, spanning agriculture, medicine, ecology, education, social customs, and environmental stewardship, serve as vital conduits for societal cohesion, environmental stewardship, and resource management. Rooted deeply in cultural heritage and environmental awareness, indigenous communities continue to rely on indigenous knowledge due to the absence of modern scientific systems or the inaccessibility of such systems. Within the unique context of Himalayan pastoralism in Jammu and Kashmir, India, the relevance and resilience of IKS become particularly pronounced. Drawing from ethnographic studies conducted among pastoral communities in the Himalayas of Jammu and Kashmir (India), this chapter examines the intricate relationship between IKS and sustainable pastoralism, shedding light on the practical wisdom that sustains livelihoods and fosters resilience in challenging environments. It underscores the ongoing relevance of IKS, serving as either an alternative or the sole option where modern knowledge systems are lacking or ineffective.

Jammu and Kashmir hosts two prominent pastoral groups: the transhumant *Gujjar-Bakarwal* and the *Chopan* shepherds. The former migrate between summer pastures in Kashmir and winter pastures in Jammu, while the latter are local Kashmiri shepherds spending extended periods in the highland pastures during summers. Ethnographic fieldwork conducted in 2022 and 2023 focused on these groups, revealing their profound understanding of their environment and resources, cultivated over centuries of experiential learning. Within these communities, diverse age groups, genders, and roles possess various forms of knowledge, contributing to the intricate fabric of IKS, which contrast with globally generated knowledge systems.

The significance of IKS in addressing contemporary challenges, such as environmental degradation and socioeconomic disparities, cannot be overstated. IKS, developed within indigenous societies, offers solutions to socioeconomic problems and has proven successful throughout history (Rizqi, 2022). Despite facing historical injustices and ongoing marginalisation, indigenous wisdom offers invaluable insights applicable to modern-day concerns (Turnbull & Poelina, 2022). By blending traditional knowledge with contemporary needs, IKS presents pathways toward a more equitable, balanced, and sustainable future.

Objectives

1. To examine the relevance and resilience of IKS within the context of Himalayan pastoralism in Jammu and Kashmir.
2. To explore the practical applications of IKS in addressing challenges related to pastoral mobility, weather prediction, ethnomedicines, and sustainable practices.
3. To assess the effectiveness of IKS compared to modern scientific knowledge systems in sustaining livelihoods and fostering resilience among pastoral communities in the Himalayas.

This chapter unfolds in three main sections, beginning with a discussion of the relevance of IKS. Through a detailed examination of Himalayan pastoralism in Jammu and Kashmir, the authors explore the practical applications of IKS, focusing on pastoral mobility, weather prediction, ethnomedicines, and sustainable practices. The authors conclude by discussing the challenges facing IKS and opportunities for recognising, revitalising, and empowering indigenous wisdom within sustainable development.

By explaining the relevance of IKS in sustaining livelihoods and fostering resilience among Himalayan pastoral communities, this chapter seeks to contribute to the ongoing discourse on preserving and integrating traditional wisdom into contemporary frameworks for sustainable development.

At specific junctures and locations, IKS outweigh the Scientific Knowledge System in importance. This holds particularly true for pastoralists in the Himalayas, whose nomadic lifestyles or residence in remote highlands limit access to modern scientific resources. Hence, fostering participatory research becomes imperative to safeguard indigenous knowledge and cultivate robust IKS. Moreover, IKS often emerges as the sole viable alternative beyond being a mere option, as demonstrated by the case study in this chapter.

Relevance of Indigenous Knowledge Systems (IKS)

Indigenous Knowledge Systems serve as vital conduits for societal cohesion, facilitating the transmission, preservation, and acquisition of knowledge within indigenous communities. These systems are the cornerstone for local agricultural decision-making, pastoralism, healthcare, and natural resource management, and they are deeply connected to their environment (Farooquee et al., 2004). These systems (IKSs) are knowledge shaped by acculturation and kinship ties, transmitted orally and through cultural rituals. IKS embodies implicit and explicit knowledge, reflecting a holistic understanding of traditional lifestyles and sustainable resource utilisation (Chiwanza et al., 2013). Particularly relevant in sustainable development, it offers valuable insights into resource management and environmental conservation.

IKS are characterised by their holistic approach, resilience, and adaptability, recognising the interconnectedness of spiritual, cultural, ecological, and social elements (see Fig. 11.1). These systems demonstrate adaptability, evolving over generations through observation and experimentation to suit changing environments and societal contexts. IKS are characterised by their deep-time perspective on complex systems, encompassing both local and nonlocal aspects, and are guided by Indigenous protocols (Sakapaji, 2022). These Knowledge Systems exhibit resilience, drawing upon traditional wisdom and collective strategies to endure adversity and recover from crises. The practices in IKS are culturally enshrined and embedded in myths, taboos, values, folklore, traditional beliefs, rituals, and institutions.

Contrary to the misconception of indigenous knowledge being outdated or static, these systems are recognised as dynamic and scientific, blending traditional

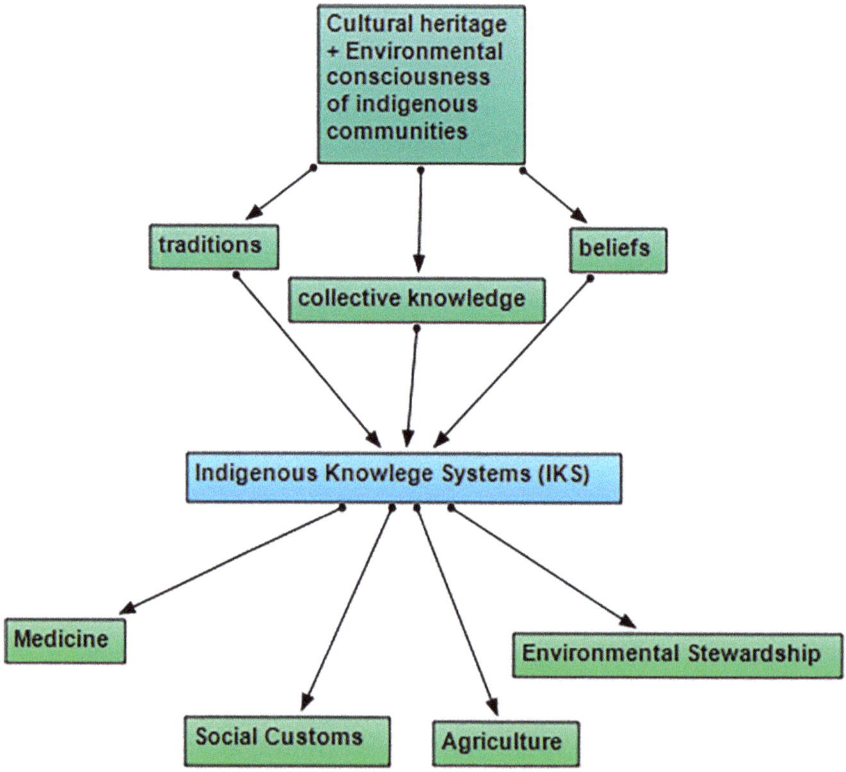

Fig. 11.1 Indigenous knowledge systems

and global perspectives while focusing on the future. They embrace radical, inventive, creative, and intuitive approaches to address present and future challenges, demonstrating remarkable adaptability and flexibility. Indigenous knowledge is the foundation for shaping our identity and maintaining the cohesion of social structures within communities. Evolving over millennia, IKS are grounded in ancient philosophies and offer invaluable insights applicable to contemporary challenges. These IKS not only enlighten historical and practical approaches but also have the potential to address contemporary challenges and benefit society as a whole.

IKS hold profound relevance in modern society, particularly concerning resource management, environmental sustainability, and health and well-being. These systems embody a wealth of intergenerational and transgenerational wisdom that has sustained indigenous communities and empowered them to adapt to environmental changes (Shava, 2019). In many countries, indigenous knowledge influences many areas of life; its role in the social and economic well-being of the nation and management of its resources is immense. It is widely accepted that indigenous knowledge is crucial for the survival of many indigenous communities. This knowledge forms the foundation for decision-making at the grassroots level, usually within indigenous organisations, where they identify problems and find solutions. IKS offer innovative solutions across diverse domains, including agriculture, animal husbandry, healthcare, water management, community development, poverty alleviation, and climate resilience. For instance, ethno-medicine leverages natural remedies for healing, while sustainable farming methods like permaculture and agroforestry enhance soil fertility and crop productivity. Moreover, indigenous communities utilise traditional techniques such as pitcher-watering systems and rainwater harvesting to conserve water resources.

Indigenous Approaches to Resource Management and Conservation

Indigenous approaches emphasise holistic approach to safeguarding species and ecological communities of cultural significance, aligning with international obligations such as the Convention on Biological Diversity and the United Nations Declaration on the Rights of Indigenous Peoples (Goolmeer et al., 2022). By nurturing a profound reverence for nature and facilitating intergenerational knowledge transmission, indigenous resource management practices uphold ecological integrity and foster resilience and cultural continuity within communities. Indigenous people are the custodians and stewards of their forests, land, and environment, and they have been entrusted by ancestral charters to care for them through successive generations. Indigenous food production and preservation systems contribute significantly to food production. Techniques such as rotational farming, agroforestry, and seasonal harvesting manifest indigenous peoples' intimate understanding of local ecosystems and their adeptness in adapting to evolving environmental conditions. For thousands of years, such techniques have proven superior in many cases to alien technologies. These technologies and know-how rely on locally available skills and materials and thus are often more cost-effective than exotic technologies.

The sustainable practices showcase the ingenuity of IKS in harmonising human activities with the natural environment, offering valuable lessons for promoting sustainability and resilience in modern agriculture. For example, agroforestry practices like the Taungya system, which integrates tree planting with agriculture, home gardens, and small-scale systems of diverse crops and trees, illustrate sustainable land management. These approaches foster biodiversity, conserve soil fertility, and generate multiple income sources. Similarly, rotational farming methods such as Three-Sister farming and crop rotation promote ecological resilience and food security. By leveraging IKS, farmers utilise intercropping and alternating planting sequences to enhance soil health, reduce pest pressure, and maximise yields, ensuring long-term agricultural productivity.

IKS of water management have empowered communities to endure drought and drinking water crises. Historical structures like *johads* (small earthen check dams that capture and conserve rainwater), once prevalent in Rajasthan but neglected over time, are now being revitalised through community mobilisation led by organisations like Tarun Bharat Sangh (TBS) (Everard, 2015). This restoration of traditional water harvesting structures and the construction of new ones underscores the significance of IKS in tackling contemporary water management challenges.

Health and Well-being

IKS play a crucial role in advancing health and well-being by integrating cultural practices and traditional wisdom into healthcare policies and initiatives, offering holistic approaches to healing and wellness rooted in centuries of accumulated knowledge and profound insights into the interconnectedness between human well-being and the environment (Milroy et al., 2023). Practices such as acupuncture, Ayurveda, chiropractic care, herbal medicine, meditation, and yoga are esteemed for their effectiveness in managing chronic conditions and enhancing quality of life while preserving cultural heritage.

Rooted in ancient philosophies, traditional medicine is studied and applied across various disciplines such as anthropology, pharmacology, ethnobotany, and the pharmaceutical industry. Indigenous healers harness the power of medicinal plants, herbs, and therapies to address health's physical, emotional, mental, and spiritual aspects. Emphasising the interconnectedness of mind, body, and spirit, complementary therapies strive to restore balance and foster holistic well-being. Many of these practices have been passed down through generations within traditional healing systems and have garnered recognition for their efficacy in alleviating symptoms, managing chronic conditions, and improving overall quality of life. For instance, indigenous communities in the Morni Hills and Raipur Rani regions of Haryana, India, rely heavily on traditional herbal remedies for treating various health issues, highlighting the practicality and effectiveness of their knowledge (Balkrishna et al., 2022). Similarly, the Gujjar and Bakarwal, pastoral communities of Jammu and Kashmir, spend most of their time in the higher reaches of the Himalayas, where it becomes difficult for them to reach modern medicine. So they rely more on these IKS for their health and well-being.

Herbal medicine is commonly practised, with indigenous communities utilising medicinal plants to address cold, fever, and skin conditions. Acupuncture, originating from traditional Chinese medicine, is also present in many indigenous communities and involves thin needle insertion to alleviate pain and promote well-being. Massage therapy techniques incorporate deep tissue manipulation and pressure points to relieve muscle tension and enhance overall physical and mental health. Dietary therapy focuses on nutrition, with traditional practices including consuming specific foods or herbs like turmeric and ginger for their medicinal properties. Spiritual healing rituals, such as smudging and sweat lodges, aim to address health's spiritual, emotional, and mental aspects by connecting individuals with their cultural heritage. Bone setting methods, passed down through generations, treat musculoskeletal injuries through manual manipulation and herbal remedies. Meditation and mindfulness promote mental clarity, reduce stress, and enhance spiritual connection. These examples illustrate the breadth and depth of IKS in addressing health conditions and promoting holistic well-being.

Social Cohesion and Resilience

IKS promote social cohesion and resilience (Hilchey, 2022). Rooted in centuries-old traditions and passed down through generations, indigenous wisdom fosters strong bonds of solidarity and interdependence within communities. Indigenous communities often face vulnerabilities due to climate risks, but their traditional ecological knowledge and practices contribute to their ability to adapt and recover from disasters (Lee et al., 2021). Indigenous institutions and culturally specific spatial regulations help maintain stability and sustainability in local spatial management, enhancing resilience and adaptability (Fernández et al., 2021). This knowledge serves as a unifying force, providing a shared framework for understanding the world and navigating life's challenges.

Indigenous communities strengthen their social fabric and promote cultural continuity through collective decision-making processes and intergenerational learning. Through transmitting ancestral wisdom from elders to younger generations, these communities uphold their cultural heritage and reinforce a sense of collective identity and belonging. For instance, the Navajo people have numerous ceremonies for healing, coming of age, marriages, and honouring the dead and

Blessing way (honouring the mother), reinforcing cultural identity (Mika, 2022). Similarly, the Maasai people's *mabaraza* councils ensure inclusive decision-making, while the Inuit's expertise in sea ice conditions showcase deep ecological knowledge. Moreover, the Māori *marae* centres promote cultural revitalisation. The Dene people uphold their cultural traditions and language through diverse methods, such as oral storytelling, intergenerational knowledge transfer, active participation in cultural ceremonies and events, and involvement in initiatives to revitalise their culture (The Dene, 2023). These examples highlight how IKS nurture strong social bonds, preserve cultural heritage, and facilitate sustainable living practices, ultimately enhancing resilience within indigenous communities.

Practical Applications of IKS in Himalayan Pastoralism: A Case Study

In this case study, the authors delve into the practical applications of IKS within the context of Himalayan pastoralism in Jammu and Kashmir, India. Focusing on pastoral mobility, weather prediction, ethno-medicines, and sustainable practices, the authors aim to explore the profound relevance and resilience of IKS in sustaining livelihoods and fostering resilience in challenging environments.

Ethnographic fieldwork commenced in 2022 and remains ongoing in Jammu and Kashmir, primarily focusing on two prominent pastoral communities: the transhumant Gujjar-Bakarwal and the Chopan shepherds. Data collection involved participant observation, interviews, and document analysis to capture the intricate nuances of indigenous knowledge practices. Participant observation allowed researchers to immerse themselves in the daily lives of pastoral communities, gaining first-hand insights into their traditional practices and beliefs. Sampling techniques involved purposive sampling, targeting individuals with expertise in traditional knowledge systems and extensive experience in pastoralism. Snowball sampling was also utilised to identify additional informants within the community network. In India's Jammu and Kashmir region, a survey conducted by the Jammu and Kashmir Tribal Affairs Department in 2021 revealed that it hosts the highest transhumant population globally. Among these transhumants are the Gujjar and Bakarwal communities, who migrate between winter pastures in Jammu and summer pastures in Kashmir in search of grazing lands and water sources for their livestock.

Additionally, the local shepherds of Kashmir, known as the Chopan or *Pohol*, are employed by agricultural landowners to tend their sheep in the highland pastures for around four to eight months. This pattern of pastoral mobility has garnered increasing recognition in Western academia as an environmental adaptation and a sustainable management practice. Through participant observation during fieldwork, it was observed that indigenous knowledge proved to be more beneficial than scientific knowledge among these pastoralists. This was evident during the traversal of the traditional transhumance route (called *Rasto*) from winter to summer pasture alongside a group of Gujjar-Bakarwal pastoralists, as depicted in Fig. 11.2.

The authors utilized the fieldwork observation of 2022 and 2023 in this chapter. One migratory group of pastoralists (called *Kafila*) was observed from winter pasture A (Rajouri, Jammu) up to summer pasture H (Gurez, Kashmir). The Kafila halted at B (Thanamandi), C (Noori Chamb), D (Behrangala), E (Mughal Road), and F (Shupiyan). The Kafila used intervening pastures to feed their livestock at B (Thanamandi), C (Bafliaz), and D (Behrangala Sunderkote) as shown in Fig. 11.3.

Indigenous wisdom proved invaluable in organising the journey of the Kafila from the winter pasture to the summer pasture. About a month before departure, the caravan members diligently prepared and packed essential items for the migration, such as utensils, torches, a radio, *tambo* (shelter tents), and sharpened tools, and settled any outstanding matters in the winter pastures. This process, known as packing of *Dera*, draws upon centuries-old indigenous knowledge and traditions passed down through generations. Despite checking weather updates at A (Dadarsan Bala), their faith in God did not deter them from embarking on the migration journey from Dadarsan Bala, Rajouri (Jammu) to Gurez (Kashmir) following *Baisakhi* in April.

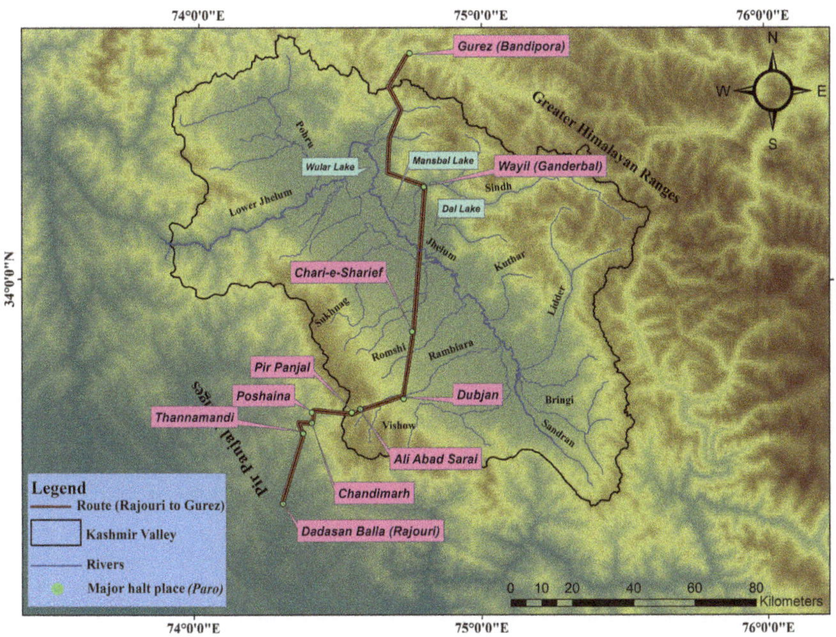

Fig. 11.2 Traditional route of transhumanism followed by the Kafila under observation

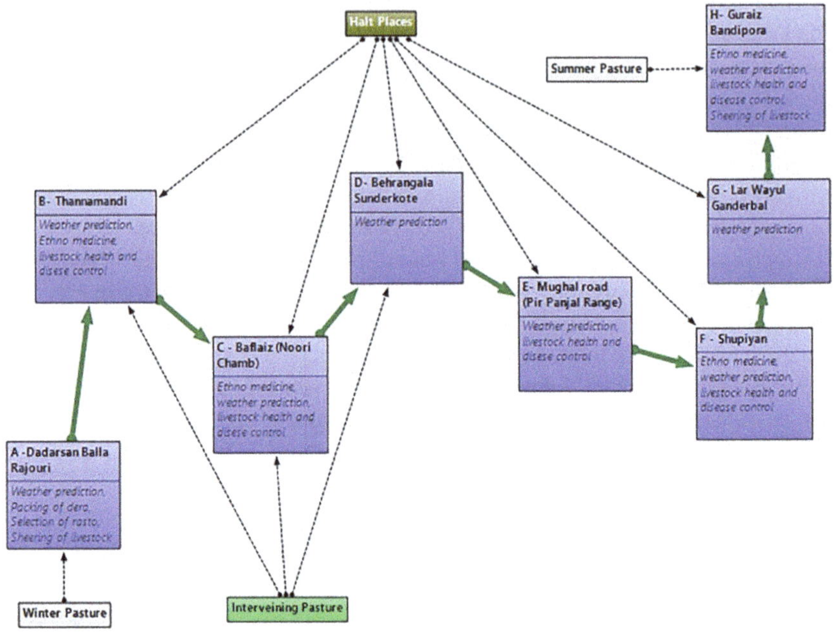

Fig. 11.3 Use of Indigenous Knowledge along Rasto (migratory route)

In examining the pastoralists' approach to mobility, the authors explored their perspectives and motivations for migration through their practices. The pastoralists uniformly cited the search for pasture and water as their primary reasons for migration. While there were some variations in their motivations regarding pasture—some seeking better quality (pull effect) while others due to insufficient quantity (push effect)—from an ecological standpoint, these reasons reflect the pastoralists' efforts to manage the variability of natural resources and uphold sustainable development practices.

However, it is noteworthy that the pastoralists did not frame their mobility considerations regarding sustainability or environmental concerns. Instead, their discussions centred on meeting the needs of their livestock. This distinction highlights a disparity between ecological rationality and the pastoralists' priorities. For the Gujjar and Bakarwal pastoralists, mobility ensures the survival and well-being of their livestock, rather than solely an ecologically driven decision.

The Rasto was chosen based on traditional and ancestral knowledge passed down through generations. Along this route, the caravan halted at various locations selected through their accumulated experience. They typically avoided sites already occupied by other groups to prevent conflicts within the community, particularly concerning women and children. However, they prioritised night halting near water sources. In the Pir Panjal region of the Himalayas, the authors resided in tents known as tambo alongside the caravan during the fieldwork.

During one instance, authors observed an interesting phenomenon while camping at Noori Chamb along the migratory route. Around 4 pm, the women and the leader of the caravan (Kafila head) began setting up the camp (*Dera*), including the traditional tambo, when suddenly, within half an hour, a heavy rainfall accompanied by strong winds ensued. Authors hastily erected their modern tent, assisted by the pastoralists who chuckled at the fragility of modern tents in such conditions. Earlier that day, the Kafila leader had predicted the rain and emphasised the urgency of reaching the campsite (halt place) to set up shelter. When asked about their forecasting method, he explained that they rely on observing the mountains and the sky. Unlike modern weather updates unavailable in remote areas like the higher Himalayan pastures, their predictions were remarkably accurate, drawing on their experiential wisdom and ancestral knowledge.

This episode highlighted the reliance of Himalayan pastoralists on their IKS for weather prediction and related decision-making processes, such as determining travel speed and selecting campgrounds in areas where access to modern facilities is challenging. Despite discrepancies with modern weather forecasts available online, the pastoralists' indigenous methods proved more effective and reliable, underscoring the enduring value of traditional knowledge in their way of life.

In the Himalayas, where the environment was highly variable and unpredictable, transporting many animals at once using vehicles is impractical. During heavy snow or rainfall, pastoralists employ two main strategies to mitigate the risk to their livestock: mobility, where people move with their herds, or utilising indigenous knowledge to protect the animals from danger. The preferred response among Himalayan pastoralists is mobility, which involves halting at a safe location before heavy rains occur. This approach reflects what some have termed 'ecological wisdom,' as it aligns with ecological rationality. However, not all pastoralists may share this preference.

With increased intervention from policymakers, providing vehicles for transporting livestock becomes an attractive option, while mobility may be perceived as burdensome, this intervention was associated with risks and without emic perspectives, it could not work. As observed in a case study, this was evident among pastoralists in the Rajouri district of Jammu and Kashmir. Despite government intervention, pastoral mobility remained a preferred option, as waiting for government vehicles often resulted in fatalities and damage to livestock. In one instance, a Kafila's livestock sustained damage during transportation in a vehicle, while another Kafila experienced delays and livestock loss due to temperature increases and vehicle allotment delays in the Jammu region. One of the respondents from the government departments called the intervention symbolic.

Through participant observation, it was noted that some Kafilas opted to proceed without waiting for government vehicles, relying on indigenous knowledge rather than modern mobility methods influenced by government intervention. Additionally, Kafilas strategically delayed their arrival to Gurez by staying for several days in Shopiyan and Ganderbal, assessing the weather conditions by observing the mountains and sky before traversing to the highland pastures.

Based on these observations, it is argued that modern knowledge systems did not effectively serve these transhumant pastoralists in the Himalayas. Instead, IKS, encompassing various practices and beliefs, have proven crucial for the survival and well-being of societies, including pastoralists in Jammu and Kashmir.

The authors also accompanied the local Kashmiri pastoral shepherds called as Chopans or pohol in the Kashmiri language, on their journey from Delina (village in Baramulla district) to Lienmarg, Gulmarg (summer pasture in Baramulla district), during the summer season. The Chopan group under observation began moving towards the Lienmarg in the last week of April. However, the group did not take their women and children along, unlike the Gujjar and Bakarwals. Chopans are entrusted with herding sheep belonging to local farmers to highland pastures for four to eight months, attracted by the promise of better quality grazing lands (pull effect). Initially, the Chopan heavily relied on meteorological weather updates before departing for the highland pastures. However, upon reaching Lienmarg, they encountered communication difficulties as their mobile phones ceased functioning. Consequently, their sheep fell ill and faced challenges accessing timely animal husbandry services. Despite the absence of support from animal husbandry employees during the authors' fieldwork, the Chopan relied on their traditional knowledge to care for their sheep.

The traditional shearing practices, timing, and sheep identification methods were all rooted in the Chopans' extensive experience and indigenous shepherding knowledge. The term "*pohol*" reflects their shepherding heritage, passed down through generations, signifying their mastery of skills and indigenous knowledge acquired from their way of life. The Chopans' expertise in ethno-medicines and healing practices, particularly in the highland pastures, is renowned throughout Jammu and Kashmir. Even individuals who typically rely on modern allopathic medicines frequently seek out the Chopan for ethno-medicines sourced from the mountains and guidance on their usage. These ethno-medicines not only benefit the Chopan themselves in the mountains but also prove invaluable to local farmers in the plains of the Kashmir valley. The widespread belief in the efficacy of their indigenous knowledge of ethno-medicines and healing practices underscores their significance in the region. A selection of ethno-medicines documented during the researchers' fieldwork is presented in Table 11.1.

Table 11.1 Ethnomedicines used by the Chopan Community

Indigenous Name	Botanical name	Part used	Ethnomedical uses
Shanger	*Glycyrrhiza glabra* L.	Root	Cough, cold, skin, beauty, and ageing
Tethwan	*Artemisia absinthium*	Whole plant	Obesity, diabetes, liver infection
Kahzaban	*Arnebia benthamii*	Root	Common cold, cough, fever, blood purification
Kulhak	*Nasturtium officinale*	Leaf	Stomach pain
Kuth	*Saussurea costus*	Rhizome	Joint pain, back pain, sore ulcer, dysentery
Tsok-tsen	*Oxalis corniculata*	Whole plant and leaf	Toothache, blood purification
Doan kul	*Juglans regia*	Leaf and bark	Tooth infection, scrofula, rickets
Kukliporte	*Cuscuta reflexa*	Whole plant and leaf	Joint pain, wound healing, hair loss
Sotal	*Malva sylvestris*	Seeds	Cough, fever, eyesight
Datur	*Datura stramonium*	Seeds	Rheumatism, frostbite, toothache

Note Indigenous names were collected in fieldwork, and the scientific names were taken from secondary sources

Discussion

Pastoralists residing in the Himalayas possess a profound understanding of treating their livestock using traditional methods and herbal remedies indigenous to the region. They are well-versed in the traditional preparation and application of these ethno-medicines, holding invaluable knowledge that warrants preservation. It is imperative to safeguard this indigenous knowledge from extinction, as it constitutes a vital cultural heritage that should be cherished. Typically, elderly individuals possess a wealth of indigenous knowledge about ethno-medicines, which they endeavour to pass down to younger generations, albeit often without formal preservation efforts.

This case study explores the intricate nexus between pastoral mobility, livestock management, and indigenous knowledge. By illuminating the practical wisdom inherent in pastoral communities, authors contend that IKS play a pivotal role in sustaining livelihoods and fostering resilience in challenging environments, surpassing the efficacy of existing Scientific Knowledge Systems in Jammu and Kashmir, India. Integrating these two knowledge systems can yield more sustainable outcomes, necessitating actions such as preserving pastoralists' indigenous knowledge and incorporating emic perspectives into policy formulation and implementing scientific knowledge systems.

Efforts to preserve indigenous knowledge could leverage biodiversity registers, yet challenges persist in effectively implementing biodiversity laws without considering emic perspectives. Despite obstacles, indigenous wisdom offers invaluable insights across various domains, including sustainable pastoral practices, holistic healthcare, and livestock management approaches. However, the integrity and relevance of IKS are jeopardised by historical injustices, ongoing marginalisation, and misconceptions. The legacy of colonisation, globalisation, and cultural shifts has further exacerbated these challenges, eroding traditional practices and disrupting the intergenerational transmission of ancestral knowledge.

Nevertheless, opportunities exist to recognise, revitalise, and empower IKS. Through community-driven initiatives and inclusive policies, the resurgence of indigenous wisdom holds promise for fostering positive change on a global scale. By amalgamating traditional wisdom with contemporary needs, IKS offer pathways toward a more equitable, balanced, and sustainable future.

IKS encounter various obstacles in the contemporary world, such as the rapid advancement of modernisation and the weight of historical injustices. These hurdles jeopardise the conservation of indigenous knowledge, predominantly passed down orally through generations and embedded within the collective wisdom of communities. Despite its demonstrated holistic, adaptable, and empirical nature, indigenous knowledge often suffers from misconceptions of being archaic or unchanging, leading to its marginalisation and undervaluation in modern contexts (Leonard, 2022; Shava & Togo, 2020).

Historically, colonisation has played a significant role in eroding IKS, disrupting cultural norms and values (Kugara & Mokgoatšana, 2022). Globalisation exacerbates this by favouring Western viewpoints and practices over indigenous ones, resulting in cultural standardisation and marginalisation. External parties frequently misinterpret or misrepresent indigenous knowledge, perpetuating damaging stereotypes (Fernández et al., 2021). Development processes and socioeconomic disparities further contribute to the erosion of indigenous knowledge.

However, amid these challenges, opportunities arise for acknowledging, rejuvenating, and empowering IKS. By amalgamating traditional wisdom with contemporary needs, IKS can tackle global challenges and promote sustainable development. Initiatives to integrate indigenous knowledge into policymaking, education, and corporate practices are essential for leveraging its full potential (Tharakan, 2017; de F Afonso, 2013). Recognising and honouring indigenous self-governance over their knowledge is imperative (Fletcher et al., 2023).

Documenting indigenous knowledge is pivotal in safeguarding traditional practices and fostering cooperation for sustainable development. Active research, governmental backing, and community-driven endeavours are vital in conserving, revitalising, and embedding traditional wisdom into contemporary frameworks. Upholding the rights of indigenous peoples to safeguard their cultural heritage enables society to embrace a more inclusive and comprehensive approach to addressing present-day challenges and shaping an equitable and sustainable future.

In certain circumstances, IKS prove more relevant than other knowledge systems. Access to the latter can be problematic, as evidenced by the case of Himalayan pastoralists. Therefore, participatory research is essential to preserve indigenous knowledge and construct an effective IKS. In various scenarios, IKS should not be seen as merely an option but as the primary alternative, as exemplified in the aforementioned instances.

Conclusions

The IKS observed among the Gujjar-Bakarwal and Chopan pastoral communities have demonstrated their significant contribution to sustainable livelihoods in the challenging environments of the Himalayas. Unlike Modern Knowledge Systems (MKS), which often struggle to be practically applicable in the high-altitude seasonal pastures where these communities reside, IKS has proven to be resilient and effective in meeting the needs of these communities where modern knowledge falls short. However, the current initiatives taken by most policymakers often overlook the emic perspective of indigenous communities, failing to integrate their traditional wisdom into policymaking and development initiatives. There is a pressing need to amalgamate IKS with MKS, recognising these indigenous communities' invaluable experience and knowledge. By leveraging the strengths of both knowledge systems, we can better support these communities in sustaining their traditional way of life while also achieving broader sustainable development goals.

This study underscores that indigenous knowledge is essential and often the only knowledge that enables these communities to survive and thrive in their harsh environments. For example, the traditional knowledge of weather-related predictions by the Himalayan pastoralists and their knowledge of ethno-medicines and traditional healing practices are the only methods one can rely on during mobility and interaction with the environment in the Highlands. The study does not argue that IKS are universally more effective than MKS for all pastoral groups. This is because cultural and social organisation varies widely among nomadic communities, with no singular set of features standard to all or exclusive to nomads (Spooner, 1973). However, in the specific context of the Gujjar-Bakarwal and Chopan communities, IKS has proven to be indispensable for their sustainable livelihoods. Integrating indigenous knowledge into policy and development frameworks is essential for ensuring the well-being and resilience of pastoral communities in the Himalayas. By recognising the value of indigenous wisdom and working collaboratively with these communities, we can pave the way for a more inclusive, balanced, and sustainable future for all.

Acknowledgements The authors gratefully acknowledge the financial support received through the ICSSR Doctoral Fellowship awarded to Irfan Ali Banka. The authors also acknowledge the

Department of Science & Technology, JKST&IC, Government of Jammu and Kashmir, India, for sponsoring and supporting the project entitled *Indigenous Knowledge Systems & Sustainable Pastoralism* under the Sponsored Research and Extension Programme (SREP) grant through the Tribal Research Institute, J&K. The authors further extend their sincere gratitude to all participants and contributors whose support, experiences, and cooperation enriched this work.

References

Balkrishna, A., Gupta, N., Gond, D. K., Sharma, I. P., Bhandari, R., & Arya, V. (2022). Survey of indigenous knowledge of medicinal plants in India. In *Phytochemicals and Medicinal Plants in Food Design* (pp. 103–121). Apple Academic Press.

Barnhardt, R., & Kawagley, A. O. (2008). Indigenous knowledge systems and education. *Teachers College Record*, *110*(13), 223–241.

Chiwanza, K., Musingafi, M. C., & Mupa, P. (2013). Challenges in preserving indigenous knowledge systems: Learning from past experiences. In *Information and knowledge management* (Vol. *3*, No. 2, pp. 19–25).

de F Afonsonhalevilo, E. Z. (2013). Rethinking the history of inclusion of IKS in school curricula: Endeavoring to legitimate the subject. *International Journal of Science and Mathematics Education*, *11*, 23–42.

The Dene: Tracing their Indigenous Roots—The Ya-Native Blog. (2023, September 19). Ya-Native. Retrieved February 16, 2024, from https://www.ya-native.com/blog/the-dene-tracing-their-indigenous-roots/

Everard, M. (2015). Community-based groundwater and ecosystem restoration in semi-arid north Rajasthan (1): Socioeconomic progress and lessons for groundwater-dependent areas. *Ecosystem Services*, *16*, 125–135.

Farooquee, N. A., Majila, B. S., & Kala, C. P. (2004). Indigenous knowledge systems and sustainable management of natural resources in a high altitude society in Kumaun Himalaya, India. *Journal of Human Ecology*, *16*(1), 33–42.

Fernández-Llamazares, Á., Lepofsky, D., Lertzman, K., Armstrong, C. G., Brondizio, E. S., Gavin, M. C., & Vaughan, M. B. (2021). Scientists' warning to humanity on threats to indigenous and local knowledge systems. *Journal of Ethnobiology*, *41*(2), 144–169.

Fletcher, G., Waters, J., Yunkaporta, T., Marshall, C., Davis, J., & Manning Bancroft, J. (2023). *Indigenous systems knowledge applied to protocols for governance and inquiry. Systems Research and Behavioral Science.*

Goolmeer, T., Skroblin, A., Grant, C., van Leeuwen, S., Archer, R., Gore-Birch, C., & Wintle, B. A. (2022). Recognisingculturally significant species and Indigenous-led management is key to meeting international biodiversity obligations. *Conservation Letters*, *15*(6), e12899.

Hilchey, D. (2022). Weaving Western and Indigenous Knowledge for resilience. *Journal of Agriculture, Food Systems, and Community Development*, *11*(2), 1–3.

Kugara, S. L., & Mokgoatšana, S. (2022). Challenges presented by digitisation of VhaVenda oral tradition: An African indigenous knowledge systems perspective. *HTS Teologiese Studies/ Theological Studies*, *78*(3), 7428.

Lee, S. H., & Chen, Y. J. (2021). Indigenous knowledge and endogenous actions for building tribal resilience after Typhoon Soudelor in Northern Taiwan. *Sustainability*, *13*(2), 506.

Leonard, B. (2022). Indigenous knowledges and methodologies in higher education. In *Oxford Research Encyclopedia of Education.*

Macusi, E. S., Sales, A. C., Macusi, E. D., Bong-as, H. P., Cayacay, M. A., Omandam, J. L., & Vidal, C. (2023). Indigenous knowledge systems and practices (IKSP), livelihood resources and aspirations of the matigsalog and ata tribes.

Mika (2022). *The Navajo people: A culture rich in ceremonies*. Indian Country Extensions. Retrieved February 16, 2024, from https://www.indiancountryextension.org/the-navajo-people-a-culture-rich-in-ceremonies

Milroy, H., Derry, K., Kashyap, S., Platell, M., Alexi, J., Chang, E. P., & Dudgeon, P. (2023). Indigenous Australian understandings of holistic health and social and emotional well-being. In *Toward an integrated science of well-being* (pp. 158–177). Oxford University Press.

Rizqi, R. M. (2022). Islamic economics answers to a wide range of contemporary socioeconomic challenges. *Journal of Islamic Economics Lariba*, *8*(1), 143–156.

Sakapaji, S. C. (2022). Integrating local and indigenous ecological knowledge (IEK) systems into climate adaptation policy for resilience building, and sustainability in agriculture. *International Journal of Sustainable Development Research*, *8*(1), 9–24.

Shava, S. (2019). Indigenous/tribal knowledges definition and relevance in the modern era. In *Traditional and Indigenous knowledge for the modern era* (pp. 1–15). CRC Press.

Shava, S., & Togo, C. (2020). Applications of indigenous knowledges in the twenty-first Century. In D. R. Katerere, W. Applequist, & A.. (Eds), *Traditional and indigenous knowledge for the modern era*, 311–325.

Spooner, B. (1973). *The cultural ecology of pastoral nomads*. Addison-Wesley

Tharakan, J. (2017). Indigenous knowledge systems for appropriate technology development. *Indigenous People*, *123*, 123–134.

Turnbull, S., & Poelina, A. (2022). How Indigenous wisdom can sustain humanity. *Journal of Behavioural Economics and Social Systems*, *4*(1), 10–36.

VanBavel, B., MacDonald, J. P., & Dorough, D. S. (2022). Indigenous knowledge systems. In K. D. P. M. Hulme (Eds.), *A critical assessment of the intergovernmental panel on climate change* (pp. 116–125). Cambridge University Press.

Chapter 12
Preserving Tribal Wisdom: Integrating Western Himalayan Tribal Knowledge for Accelerating Sustainable Development Goals

Shalvi

Abstract Growing climatic instability and ecological stress have intensified global debates on how development can proceed within environmental limits. In fragile mountain regions, questions of sustainability are not abstract policy concerns but lived realities shaped by geography, seasonality and resource scarcity. This chapter examines tribal knowledge systems in the Western Himalayan regions of Kinnaur, Lahaul and Spiti in Himachal Pradesh, India, drawing on field-based research in Rangrik village of Spiti Valley. Adopting an anthropological perspective, the study conceptualises tribal knowledge as a locally embedded socio-ecological framework formed through long-term interaction with high- altitude cold desert environments. Rather than treating indigenous practices as static tradition, the chapter analyses them as adaptive, institutionalised and intergenerational systems of environmental regulation. It situates these systems within the broader discourse of the Sustainable Development Goals (SDGs), arguing that tribal knowledge offers context-specific insights into integrated approaches to food security, water governance, ecosystem stability and climate adaptation. By foregrounding participation and applied anthropology, the chapter highlights the need for development planning in mountain regions to move beyond uniform templates and engage collaboratively with locally grounded knowledge systems. In doing so, it contributes to ongoing debates on sustainability, indigenous knowledge and the politics of development in ecologically fragile landscapes.

Keywords Tribal knowledge · Western Himalaya · Socio-ecological systems · Sustainable development · Applied anthropology · Mountain ecology

Shalvi (✉)
Department of Anthropology, Panjab University, Chandigarh, India
e-mail: shalvi1805@gmail.com

R. C. Sobti et al. (eds.), *Traditional Knowledge Systems in India: Sustainability, Ecology, and Resource Governance*, People, Cultures and Societies: Exploring and Documenting Diversities,
https://doi.org/10.1007/978-981-92-1346-7_12

Introduction

The importance of the Western Himalayas is not limited to its geographical boundaries; it resides in the intricate relationships it sustains between ecology, water systems and human livelihoods across interconnected regions. As part of the greater Himalayan mountain system, it serves as a crucial hydrological source, feeding snow- and glacier-fed river networks that support agricultural production, biodiversity and settlement far beyond the upland terrain. Ecological transformations, mainly as a result of rampant climate change in this high-altitude region, therefore, carry implications that extend well beyond mountain communities. As climate change accelerates, shifts in snowmelt patterns and glacier retreat in the Western Himalaya have implications for water security, ecological stability and socio-economic resilience across vast downstream populations.

At the same time, the Western Himalaya represents one of the most ecologically fragile and climate-sensitive zones in South Asia. In India, the Western Himalaya includes high-altitude districts of Himachal Pradesh such as Kinnaur, Lahaul and Spiti, where cold desert landscapes and snow-fed hydrology shape patterns of settlement and livelihood. Short agricultural cycles, steep gradients, sparse vegetation and prolonged winters impose visible environmental limits. In districts such as Kinnaur, Lahaul and Spiti, particularly within cold desert landscapes like Spiti Valley, livelihoods are shaped by dependence on seasonal snowmelt, constrained arable land and restricted mobility during winter months. These conditions require systems of resource management that operate within narrow ecological margins. In villages such as Rangrik, survival has historically been dependent on knowledge systems that regulate water distribution, maintain soil fertility, diversify crops and preserve produce across long periods of isolation. These practices are not merely technical adjustments to environment; they are embedded within kinship structures, community institutions and religious worldviews that emphasise reciprocity with nature. The environmental significance of the Western Himalaya, therefore, cannot be understood without attention to the social and cultural frameworks through which its resources are managed.

The environmental conditions that shape life in the Western Himalaya bring into focus a broader concern that has come to define contemporary development discourse—the challenge of sustaining human wellbeing without putting the ecological systems at constant danger. In such sensitive mountain regions, this challenge is experienced daily rather than debated theoretically. When arable land is limited, when water depends on seasonal snowmelt, and when winters isolate communities for months, what does "development" truly mean? How should growth be imagined in landscapes where ecological margins are narrow and regeneration is slow? In such settings, decisions about cultivation, water use and resource allocation are never purely economic; they are inherently ecological. The idea of sustainability, therefore, must be understood as a lived practice before it is framed as a developmental goal. It is in response to such questions that global sustainability frameworks have emerged over the past decades. As environmental degradation, climate variability and socio-economic inequality became increasingly visible at international scales, development discourse began to shift from sector-specific targets toward more integrated approaches.

The Sustainable Development Goals (SDGs), adopted by the United Nations in 2015, represent one of the most comprehensive articulations of this shift (United Nations, 2015). Comprising seventeen interlinked goals, the framework seeks to integrate poverty reduction, food security, water access, climate action and ecological conservation within a shared global agenda. A defining feature of the SDGs is their emphasis on interdependence: progress in one domain is understood to depend upon advances in others. Environmental sustainability is not treated as a peripheral concern but as foundational to long-term social and economic stability.

This integrated architecture has been visually conceptualised in sustainability research as a nested model in which ecological systems underpin social and economic objectives (Stockholm Resilience Centre, 2016). While the SDGs operate at a universal scale, their implementation ultimately depends upon context-specific strategies capable of translating broad objectives into locally meaningful practices. In ecologically fragile regions such as the Western Himalaya, where livelihoods are directly shaped by water cycles, soil stability and seasonal variability, sustainability is negotiated through everyday resource management rather than abstract policy formulation.

It is within this intersection between global sustainability frameworks and locally embedded adaptive systems that Western Himalayan tribal knowledge becomes analytically significant. Rather than viewing tribal practices as remnants of tradition, this chapter approaches them as structured socio-ecological responses capable of informing contemporary development discourse. To understand how tribal knowledge intersects with sustainable development, it is useful to conceptualise the relationship between ecological limits, institutional regulation and development outcomes.

As illustrated in Fig. 12.1, sustainability in the Western Himalayan context emerges from the interaction between ecological constraints, institutional norms and broader development objectives. The concept of tribal knowledge has been widely discussed within anthropological and environmental domains as a form of locally embedded, experience-based understanding that evolves through sustained interaction with specific ecosystems. Rather than constituting isolated techniques, such knowledge systems represent cumulative processes of observation, adaptation and intergenerational transmission (Berkes, 2018). They are embedded within social institutions, ritual practices and systems of authority, linking ecological management with cultural norms.

In mountain environments, where ecological margins are narrow and climatic variability is pronounced, the adaptive dimension of local knowledge becomes particularly visible. Highaltitude societies must negotiate short agricultural seasons, fragile soils, fluctuating water supply and extended winters. As a result, agricultural practices, irrigation systems and storage mechanisms are shaped by continuous experimentation and refinement. These systems reflect what Toledo (2002, 2002) describes as biocultural adaptation—the co-evolution of cultural practices and ecological conditions.

In the Western Himalaya, tribal communities have historically relied on integrated agro-pastoral systems that regulate extraction and promote regeneration. Crop rotation, organic manure recycling, collective or community-based irrigation management and regulated grazing are not simply economic strategies but

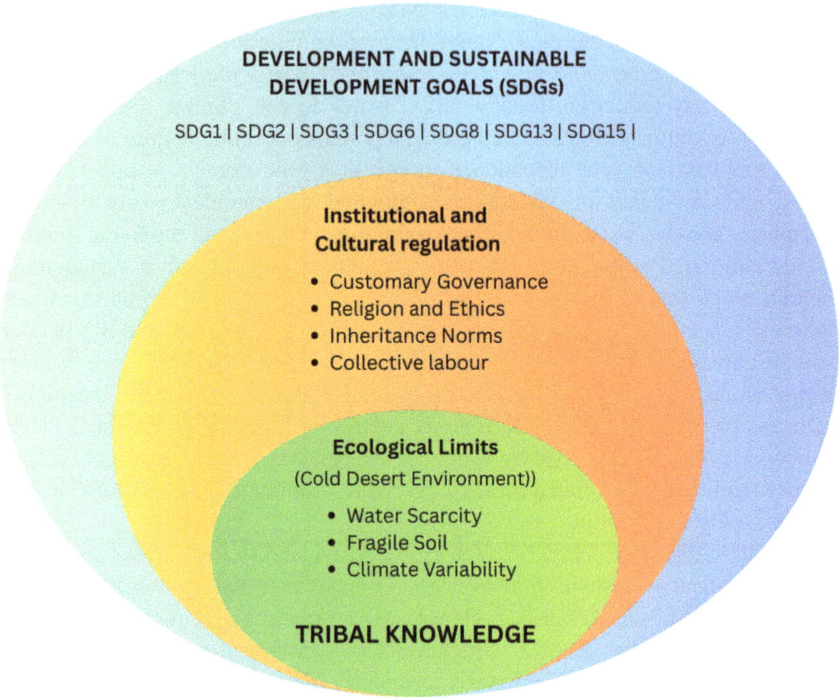

Fig. 12.1 Conceptual framework illustrating the relationship between ecological limits, institutional regulation and sustainable development outcomes in Western Himalayan tribal systems

components of a broader socio-ecological framework. Ecological management operates through customary norms and community institutions that mediate access to land and water. As Berkes (2018) argues, resilience in small-scale societies is often grounded in institutional arrangements that connect ecological knowledge with social organisation.

At the same time, tribal knowledge should not be understood as static or immune to change. Processes of state intervention, market integration and infrastructural development have altered livelihood patterns across many Himalayan regions. Yet even within these transformations, foundational principles of ecological constraint and collective regulation continue to shape local decision-making. The persistence of such principles suggests that tribal knowledge remains an active and adaptive framework rather than a residual cultural form.

In approaching Rangrik village as a field site, this chapter treats tribal knowledge as a living socio-ecological system—one that mediates the relationship between environment, livelihood and community governance. The following sections examine how this knowledge operates in practice, particularly in relation to water regulation, agricultural cycles and soil management within the constraints of a cold desert ecosystem.

Rangrik Village: Field Context and Research Orientation

This chapter examines the functioning and contemporary relevance of tribal knowledge within the context of climatic variability in the Western Himalaya. While its primary empirical focus is on field-based research conducted in Rangrik village of Spiti Valley, the analysis is situated within a broader regional context through engagement with secondary literature on Kinnaur, Lahaul and Spiti districts. By combining ethnographic insights with existing literature, the chapter evaluates how tribal knowledge systems operate across comparable cold desert environments and how they relate to contemporary sustainability discourse.

The high-altitude districts of Kinnaur, Lahaul and Spiti in Himachal Pradesh are frequently characterised as cold desert regions within the Western Himalaya. Marked by sparse vegetation, limited cultivable land and prolonged winter isolation, these regions depend heavily on seasonal snowmelt for irrigation and domestic water supply. Climatic variability—particularly changes in snowfall and glacial behaviour—directly affects agricultural cycles and livelihood stability. Secondary studies documenting agro-pastoral systems and irrigation practices in these districts provide an important comparative framework for understanding the adaptive character of local knowledge systems.

Primary data for this chapter is drawn from fieldwork conducted in Rangrik village, involving participant observation and semi-structured interviews with the local population engaged in agricultural and water management practices. Particular attention was given to irrigation systems, crop rotation, manure usage, seed preservation and storage techniques. These observations are analysed alongside documented practices from neighbouring regions to assess the broader efficacy and adaptability of tribal knowledge in cold desert ecologies.

The methodological orientation is interpretive and ethnographic, focusing on how ecological knowledge is embedded within social organisation and everyday practice. Rather than treating secondary sources as descriptive catalogues, the chapter engages them comparatively, allowing field-based findings from Rangrik to be situated within wider patterns of Western Himalayan resource management.

Tribal Knowledge in Practice: Socio-Ecological Adaptation in the Western Himalaya

Tribal knowledge in the Western Himalayan cold desert is not expressed as environmental theory; it is enacted through everyday practices that regulate cultivation, processing, storage and seasonal risk management. In Rangrik village, these

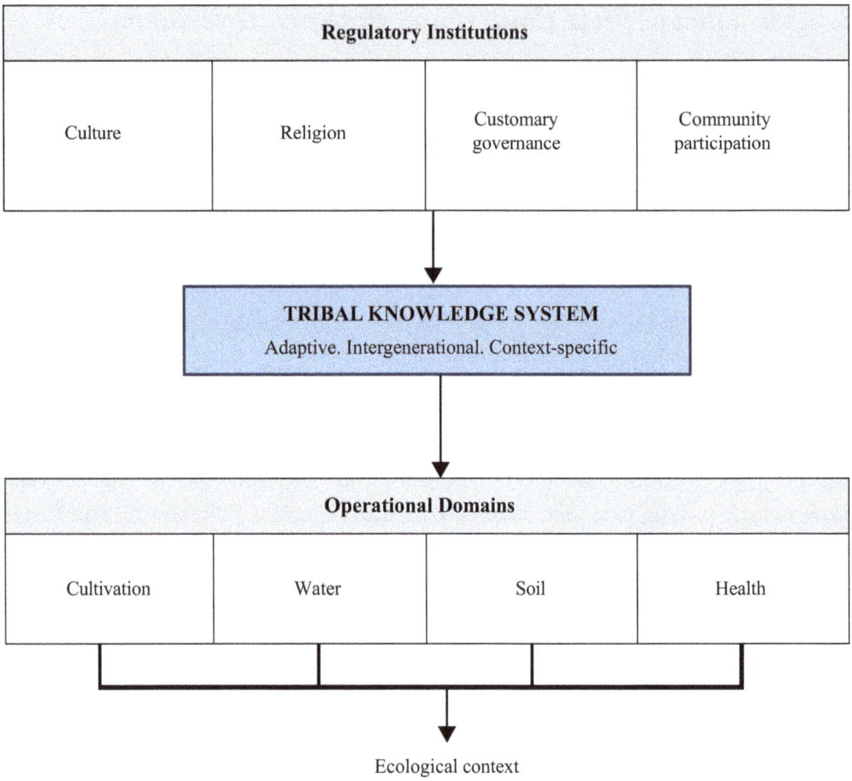

Fig. 12.2 Structural relationship between regulatory institutions, tribal knowledge systems and operational domains in Western Himalayan communities

practices emerge from sustained engagement with altitude, limited precipitation and dependence on seasonal snowmelt. Rather than functioning as isolated techniques, they form an integrated socio-ecological system through which communities negotiate environmental constraint and economic necessity. The following section examines how this adaptive framework operates in practice, beginning with cultivation systems that structure agrarian life.

As illustrated in Fig. 12.2, tribal knowledge in the Western Himalayan context operates through institutional regulation and manifests across interconnected operational domains.

Cultivation Systems in the Cold Desert Ecology

Agriculture in Rangrik unfolds within a narrow seasonal window shaped by snowmelt and temperature variability. The cultivation cycle begins only after winter thaw renders the soil workable, and sowing is guided less by calendar dates than by observation of soil moisture and meltwater flow. The cultivation window, however, has undergone observable shifts due to changes in the intensity and timing of snowfall, patterns that are consistent with documented climatic variability across the Himalayan region (IPCC, 2021). Farmers, however, repeatedly emphasised that timing determines yield; a delay of even a few weeks can shorten the growing period significantly. Such attentiveness reflects experiential ecological knowledge embedded in mountain societies (Berkes, 2018).

Field Preparation and Agricultural Calendar

Field preparation begins in late spring around April–May, once the land becomes accessible. Landholdings are generally small and situated along relatively level valley stretches where irrigation is feasible. Ploughing is conducted manually or with limited mechanised assistance, depending on household resources. Labour is largely family-based, though reciprocal support among neighbours during sowing and harvest periods remains common. Larger landholders also employ seasonal labourers who come from outside the region.

Across the Western Himalayan districts of Kinnaur and Lahaul, similar environmental constraints shape agricultural calendars, with variations in elevation influencing the timing and duration of cultivation cycles. In colder zones of Spiti, agricultural success depends heavily on synchronising cultivation with snowmelt patterns rather than fixed agricultural schedules.

Crop Selection and Rotation

Barley continues to dominate cultivation in Rangrik due to its resilience to cold temperatures and relatively short maturation period. It has formed part of local agriculture for generations, though in the last 10–15 years peas have gained prominence due to their commercial value. Peas—particularly green pea varieties—have gained prominence because of their economic value and adaptability to high-altitude conditions. Comparable cereal–legume combinations have been documented in trans-Himalayan agro-pastoral systems, where crop diversity functions as a risk-buffering strategy under climatic uncertainty (Berkes, 2018).

Crop rotation is widely practised, often alternating cereals with legumes across successive seasons. Farmers describe this as allowing the soil to "regain strength," reflecting an understanding of nutrient balance framed through lived experience rather than scientific terminology. This cyclical pattern corresponds to what Toledo (2002, 2002) describes as biocultural adaptation, wherein ecological regulation is embedded within culturally transmitted practice.

Seed preservation further strengthens resilience. Households retain seeds from previous harvests, selecting grains perceived as robust under local climatic conditions. Indigenous seed conservation practices have been recognised as important for maintaining crop diversity and strengthening resilience in Himalayan farming systems under changing climatic conditions (Sannegowda & Garkoti, 2022).

Post-Harvest Processing: Khaliyan and Chum

Harvested crops are first dried in open courtyards known locally as *khaliyan*. Once sufficiently dried, threshing is conducted using yak-cattle hybrids referred to as *chum*. Sheaves are arranged in circular formations, and pairs of *chum* are guided to trample the crop repeatedly. Farmers expressed preference for animal-assisted threshing not only due to limited mechanisation but because it preserves straw texture, which is essential for winter fodder.

Mechanical threshers, when available, are often viewed by farmers as reducing straw to a finer consistency that is less suitable for livestock consumption, reinforcing a continued preference for animal-assisted threshing as observed on the field. The persistence of *chum*-based threshing thus reflects the interdependence of crop production and pastoral survival rather than technological stagnation.

Beyond ploughing and rotation, cultivators across Ladakh, Lahaul-Spiti and Kinnaur employ specific implements designed to support root development in vegetable crops. One such wooden tool, locally known as *tokhre*, is used to loosen and aerate soil around cabbage and similar crops. Unlike heavier iron implements, the *tokhre* allows soil to be gently opened without damaging delicate root systems.

Farmers describe this practice as helping roots "spread freely" within the soil. By facilitating horizontal expansion, the loosened soil enhances moisture absorption and nutrient uptake, particularly important in high-altitude regions where soil depth and fertility are limited. The process simultaneously aids in weed removal, reducing competition for nutrients. The use of wooden implements for such practices reflects a preference for minimal disturbance and greater control over soil structure in fragile mountain environments.

The continued use of *tokhre* reflects an understanding of soil–root interaction shaped by practical experience rather than formal agronomic instruction. In fragile cold desert soils, excessive mechanical disturbance can reduce structure and moisture retention. The choice of tool therefore reveals ecological sensitivity embedded within everyday cultivation decisions.

Storage Structures: Kothar and Underground Pits

Storage practices form a critical component of seasonal risk management in Rangrik and neighbouring districts. Elevated wooden grain structures known as *kothar* are constructed to protect produce from moisture while maintaining ventilation. Their raised design prevents direct ground contact, reducing the risk of rot during snowmelt periods. Field observations and regional studies suggest that such practices are not unique to Rangrik but form part of a broader pattern of adaptation in the Western Himalayan cold desert.

Underground pit storage is commonly used for potatoes and certain vegetables. Farmers select relatively dry sites within fields to minimise seepage and line pits with dry grass locally called *booj* to provide insulation. After filling the pit with harvested tubers, layers of organic material and soil are placed above, utilising the earth's natural thermal buffering to maintain stable conditions throughout winter. In cold desert climates, where pest incidence is relatively low, such methods enable effective long-term storage without reliance on external energy inputs (FAO, 2019).

Winter Sowing and Adaptive Market Integration

In some parts of Kinnaur and Spiti, peas are sown before heavy snowfall, allowing seeds to absorb moisture gradually under snow cover and germinate early in spring. Farmers describe this practice as giving the seed time to "settle" before growth begins, indicating an experiential understanding of seasonal moisture and temperature conditions. While agronomic explanations refer to hydrothermal regulation, local interpretation remains experiential.

The expansion of apple cultivation into higher altitudes in parts of Kinnaur and Lahaul-Spiti illustrates the adaptive capacity of tribal agrarian systems. Orchard development has required modifications in storage and transport strategies, indicating that these knowledge systems are demonstrating that tribal knowledge systems are capable of accommodating commercial crops while retaining ecological sensibilities shaped by local environmental conditions.

Taken together, cultivation and post-harvest practices in Rangrik and comparable Western Himalayan regions reveal an agro-ecological system oriented toward stability, regeneration and seasonal risk mitigation. Structures such as the *kothar*, practices like *chum*-assisted threshing and seed preservation strategies illustrate how local terminology, material culture and environmental reasoning intersect. Rather than maximising short-term output, cultivation in the cold desert reflects long-term adaptation to ecological constraint and winter isolation.

Soil Regeneration and Conservation Practices

In the cold desert ecology of Spiti, soil is both limited and ecologically fragile. Arable land is confined to narrow valley stretches, and productivity depends not only on nutrient input but on protection against wind erosion, runoff and freeze–thaw cycles. In Rangrik and neighbouring districts, soil management practices reflect sustained efforts to preserve the productive top layer while maintaining fertility across generations. These practices illustrate what scholars describe as community-based land stewardship, where ecological processes are regulated through culturally embedded norms (Berkes, 2018).

Maintenance of Field Boundaries and Soil Redistribution

With the onset of spring in March–April, cultivators begin preparing fields by cleaning bunds and field corners. Weeds and grasses are removed using spades and wooden implements, and soil attached to uprooted clods is carefully separated and redistributed into lower fields.

This practice performs multiple ecological functions. Clearing field edges prevents weed encroachment and preserves cultivable area. At the same time, transferring soil from upper ridges—often enriched with organic residues—subtly enhances soil depth and nutrient concentration downslope. Such practices reflect a locally embedded understanding of soil management adapted to the constraints of mountain environments.

From a theoretical perspective, such practices contribute to regulating ecosystem services—particularly soil fertility maintenance and erosion control. Rather than relying on synthetic fertilisers, farmers redistribute locally available soil resources, reinforcing a cyclical rather than extractive model of land use.

Moisture Retention and Wind Erosion Control

Wind erosion is a persistent feature of cold desert landscapes. To protect the topsoil layer, fields are sometimes irrigated in autumn before winter sets in. The moist upper layer subsequently freezes, forming a protective crust that reduces soil displacement during winter winds. Farmers note that this also facilitates easier ploughing in spring.

This anticipatory strategy demonstrates adaptive management of seasonal climatic conditions. By using natural freeze cycles as a protective mechanism, cultivators convert environmental constraint into ecological advantage. Such practices illustrate how environmental limitations are reworked into productive strategies through locally embedded ecological knowledge. Such techniques align with principles of climate-resilient agriculture, where land management is synchronised with seasonal variability.

Pine Leaf Mulching (*Suttar*) and Runoff Regulation

In parts of Kinnaur and Lahaul-Spiti, pine leaves (*Pinus wallichiana*), locally known as *suttar*, are spread across cultivated fields, particularly on sloped terrain. The needle-like leaves slow irrigation runoff, increase water percolation and reduce soil erosion. By decreasing the velocity of surface water flow, the practice enhances moisture retention and prevents nutrient washout.

Pine leaf munching is widely used in Himalayan agro-ecosystems as a means of conserving soil moisture, reducing erosion, and enhancing crop productivity, a practice also observed across villages in Lahaul-Spiti. The use of locally available forest material reflects a low-cost, ecosystem-based adaptation strategy. Rather than importing external soil stabilisers, farmers integrate surrounding ecological resources into agricultural management.

Biofencing and Soil Stabilisation: Seabuckthorn

Biofencing with seabuckthorn (*Hippophae rhamnoides*) is widely practised in Spiti, including Rangrik village. Planted along field boundaries, this thorny shrub protects crops from stray animals while simultaneously stabilising soil through its dense root system.

As a nitrogen-fixing species, seabuckthorn contributes to soil fertility and enhances nutrient availability, reinforcing its ecological and economic significance (Zeb, 2004; Gâtlan & Gutt, 2021). Its multipurpose utility—as fuelwood, medicinal resource and nutritional supplement—further integrates ecological and economic dimensions. The plant exemplifies multifunctional land use, where soil conservation, livelihood security and biodiversity maintenance intersect.

Taken together, soil conservation practices in Rangrik reveal a locally grounded system of ecological regulation. Maintenance of field boundaries, moisture management, pine-leaf mulching and biofencing collectively sustain soil productivity within environmental limits. These strategies represent forms of ecosystem-based adaptation, in which communities manage land through low-input, regenerative techniques rather than intensive external supplementation. In this sense, soil stewardship in the Western Himalayan cold desert embodies principles of sustainability long before they are codified within global policy frameworks.

Water Governance: Kuhls, Chu *and* Zings

In the cold desert landscape of Spiti, the management of *chu* (water) is central to agricultural survival. Snowmelt serves as the primary source of irrigation, and its movement from glacier to field is mediated through gravity-fed channels known as *kuhls*. These channels are not merely physical channels; they represent long-standing systems of tribal knowledge through which communities regulate seasonal water flow under conditions of scarcity.

Kuhls as Community-Managed Irrigation Systems

Kuhls are constructed along natural gradients to allow meltwater to flow by gravity from higher elevations to cultivated fields. In Rangrik, the head of the *kuhl*—where glacier or upstream snowmelt is tapped—is considered its most critical point. Farmers emphasise that this head must remain free from debris to prevent obstruction. Stone lining is often used along sections of the channel to reduce seepage and clogging, reflecting practical adaptation to fragile mountain terrain.

Where terrain or pathways interrupt the flow, variations in construction appear: some channels are dug directly into the ground, while others are covered with stone slabs or constructed using hollowed wooden trunks to guide water across difficult stretches, reflecting adaptation to local topography and material availability. These material adaptations demonstrate flexibility within a consistent gravitational logic of irrigation. Recently, the Irrigation and Public Health Department (IPH) has also built cemented *Kuhls* to bring water faster from long distances. However, the locals feel these *Kuhls* are not that efficient when it comes to bringing water for irrigation as the water doesn't gradually seeps along the channel to keep the soil under it moisturised. These channels also clog more frequently than traditional dug-out channels and often carry greater debris into agricultural fields.

In Rangrik, these *kuhl* frequently feed into a circular tank or small regulating pond from which water is distributed into fields. This intermediary structure allows cultivators to moderate flow before directing water into individual plots. The first irrigation typically occurs several weeks after sowing, reflecting synchronisation between crop cycles and meltwater availability.

Zings and Seasonal Water Storage

In addition to flowing channels, water harvesting is undertaken through small ponds locally known as *zings*. These structures collect meltwater diverted from *kuhls* or natural runoff and function as temporary storage points within the village landscape. *Zings* allow regulated release of water, particularly when melt flow fluctuates.

Community members participate in maintaining these ponds and ensuring their functionality. In Rangrik, *spang* grass is used as an internal lining material within both *zings* and certain sections of irrigation channels. Farmers attribute to this grass strong non-permeability properties, reducing percolation losses compared to untreated soil, indicating a practical understanding of water retention in a water-scarce environment.

The use of locally available vegetation reflects ecological resourcefulness and adaptation to material constraints.

Churay: Customary Allocation of *Chu* (Water)

While *kuhls* and *zings* regulate the physical movement of water, its distribution within the village is governed through rotational turns known as *churay*. These turns are not allocated uniformly but follow established customary norms tied to social structure.

In Rangrik, eighteen households identified as *Bada Ghar*—regarded as original or higher-status families with comparatively larger landholdings—receive access to *chu* for six days of the week. These households divide water among themselves according to internal arrangements. The remaining households, referred to as *Chota Ghar*, collectively receive water access for one designated day per week.

Members of *Bada Ghar* families traditionally oversee the regulation and maintenance of the *kuhl* system. Responsibility for monitoring flow, resolving minor disputes and coordinating maintenance is embedded within these households. The legitimacy of this arrangement rests on inherited norms rather than formal state regulation.

Within the framework of tribal knowledge, such allocation is understood as continuity of established order. Water governance thus operates through customary authority, intergenerational memory and collective recognition of roles. The *churay* system demonstrates that irrigation in Rangrik is not simply technical distribution but socially embedded regulation.

Taken together, *kuhls*, *zings* and *churay* illustrate an integrated socio-ecological system in which environmental management and social organisation are inseparable. Snowmelt is channelled through gravity, moderated through storage, and allocated through customary hierarchy. In the Western Himalayan cold desert, sustainability of irrigation depends as much on institutional continuity as on hydrological flow.

Health, Healing and Environmental Knowledge

In the Western Himalayan cold desert, health practices are closely intertwined with ecological familiarity and seasonal subsistence. Limited accessibility during winter months, historical isolation and dependence on local biodiversity have shaped a medical knowledge system grounded in experiential learning. In Rangrik and neighbouring regions of Lahaul-Spiti and Kinnaur, healing practices draw upon cultivated crops, wild medicinal species and the expertise of traditional practitioners. Rather than existing as a separate domain, health knowledge forms part of the broader socio-ecological framework that structures everyday life.

***Amchi*: Local Medical Authority and Ecological Expertise**

Although Rangrik does not currently have a resident *Amchi*, villagers actively consult Amchi practitioners in neighbouring villages when health concerns arise. The *Amchi* system, rooted in Tibetan medical traditions, continues to command trust within the community. Residents particularly rely on *Amchi* for chronic ailments, seasonal imbalances and preventive care.

Amchi prescriptions are rarely limited to a single plant or preparation. Treatments often involve combinations of dried herbs, roots, minerals and dietary adjustments tailored to individual conditions. Their diagnostic approach is based on long-standing familiarity with local ecology and plant properties. Community members emphasise that *Amchi* possess detailed knowledge of the surrounding landscape and its therapeutic resources.

The continued reliance on *Amchi* illustrates that tribal knowledge in the Western Himalayan cold desert extends beyond agriculture and irrigation into structured systems of healing. Even where biomedical facilities are available in district centres, traditional consultation remains a respected and parallel form of care, reflecting confidence in locally grounded expertise.

Dietary and Cultivated Therapeutics

Health regulation in Rangrik is also embedded in everyday dietary practices. Buckwheat (*Fagopyrum tataricum*), locally known as *fafdha*, is widely regarded as beneficial for maintaining bodily balance and controlling sugar levels. Community members associate regular consumption of buckwheat with prevention of diabetes and related conditions.

Scientific studies have reported hypoglycaemic properties in Tartary buckwheat, identifying compounds such as D-chiro-inositol with insulin-like bioactivity (Kawa et al., 2003; Yao et al., 2008). While biomedical research provides one explanatory dimension, within tribal knowledge systems buckwheat is not treated as a medicinal supplement but as a staple crop with therapeutic value embedded in routine consumption.

Seabuckthorn (*Hippophae rhamnoides*), locally referred to as *tiruk*, similarly occupies both nutritional and medicinal significance. Growing along riverbanks and field margins, the bright orange berries are consumed in various forms and are valued for their restorative properties.

Scientific studies have highlighted their richness in vitamins, fatty acids and antioxidants (Zeb, 2004; Gâtlan & Gutt, 2021), contributing to broader recognition of the plant beyond the region.

At the same time, seabuckthorn has gained economic importance in recent decades, providing supplementary income during harvest months. Despite commercialisation, its role within Spiti remains embedded in local hospitality and therapeutic practice. The plant's multifunctionality—as food source, medicinal resource, biofencing species and soil stabiliser—demonstrates the interconnected nature of ecological and health knowledge.

Ethnomedicinal Plant Knowledge

Beyond cultivated crops, communities in Lahaul-Spiti possess extensive knowledge of wild medicinal species. Seasonal isolation due to heavy snowfall historically required reliance on locally available flora for treatment of common ailments. Knowledge of plant identification, harvesting seasons and preparation methods has been transmitted intergenerationally.

Species such as *Angelica glauca*, *Arnebia euchroma* (locally *ratanjot* or *khomig*), *Artemisia maritima*, *Taraxacum officinale*, and *Picrorhiza kurroa* (locally *karu* or *kutki*) hold significant use-value within the region. Dried roots of *Arnebia euchroma* are used for treating coughs, wounds and skin conditions, while *Artemisia maritima* is associated with respiratory and digestive ailments. *Taraxacum officinale* is utilised for liver and kidney-related treatments, and *Picrorhiza kurroa*—a threatened and commercially exploited species—retains both cultural and economic importance (Kala, 2005).

These practices reflect adaptive medicinal knowledge systems in which biodiversity familiarity compensates for infrastructural limitations. At the same time, increasing commercial exploitation of certain species raises questions regarding sustainability and conservation, particularly for high-demand plants such as *Picrorhiza kurroa*.

Taken together, health-related tribal knowledge in Rangrik demonstrates that healing practices are inseparable from ecological understanding. Dietary crops, wild medicinal species and the institutional authority of Amchi collectively form a locally grounded health ecology. Rather than distinguishing sharply between food and medicine, cultivation and cure, tribal knowledge systems integrate them within a single socio-environmental continuum shaped by landscape.

Discussion

The Western Himalayan cold desert is an environment structured by limits. Scarce meltwater, short agricultural seasons, fragile topsoil, steep gradients and long winters impose ecological constraints that cannot be negotiated through abundance. The practices described in Rangrik—crop rotation, seed preservation, organic soil regeneration, pine-leaf mulching, biofencing, *kuhl*-based irrigation, *churay* allocation and biodiversity-based healing—are not isolated customs or vestiges of the past. They represent historically sedimented responses to environmental precarity. Their persistence suggests that environmental protection in this region is neither incidental nor externally imposed; it is constitutive of survival.

In Rangrik, environmental protection operates through everyday regulation. Soil is redistributed from bunds because fertility cannot be wasted. Fields are irrigated in autumn to protect the top layer from wind erosion because once soil is lost, regeneration is slow. Water is channelled through gravity-fed *kuhls* because pumping is neither feasible nor sustainable. Medicinal plants are identified, harvested and prepared with knowledge of seasonal rhythms because formal health infrastructure has historically been distant. Even reverence for water sources and mountains—embedded in local cosmologies—functions as a moral mechanism discouraging pollution and reckless extraction (Berkes, 2018). Ecology here is managed through practice, institution and belief simultaneously.

Anthropologically, such systems emerge through long-term dwelling in landscape. Tribal knowledge is formed not as abstract theory but as iterative engagement: observation of snowfall patterns, experimentation with crop timing, recognition of soil fatigue, adjustment to water scarcity, refinement of medicinal plant combinations. Knowledge accumulates through repetition and is stabilised through social transmission—within families, through gendered labour, and through institutions such as inheritance norms and Amchi practice. It is this process of situated learning that allows the system to work. Because knowledge is embedded in ecological specificity, it aligns closely with the material realities of the region.

Bourdieu's concept of cultural capital illuminates this dynamic. The ecological competence embodied in Rangrik's practices constitutes a form of cultural capital that is neither monetary nor bureaucratic but experiential and relational (Bourdieu, 1986). It is acquired through participation and reinforced through habitus—the dispositions shaped by growing up within a landscape of constraint. This explains why tribal knowledge cannot be easily transplanted into other contexts, and why externally designed interventions often struggle to replicate its coherence. Its effectiveness lies precisely in its embeddedness.

When interpreted through the lens of the Sustainable Development Goals, the structural sophistication of this knowledge system becomes evident. Sustainable agriculture (SDG 2) is not achieved through yield maximisation but through risk buffering—diverse cropping, seed retention, seasonal synchronisation and soil regeneration. Clean water governance (SDG 6) is realised through collectively maintained *kuhls* and customary allocation of *churay*, ensuring that meltwater is distributed in accordance with institutional norms. Land degradation neutrality and biodiversity conservation (SDG 15) are supported through biofencing, organic manure application and protection of fragile slopes. Health resilience (SDG 3) emerges from the integration of dietary crops and *Amchi*-guided plant therapeutics. Climate action (SDG 13) is embedded in adaptive strategies responsive to snowfall variability and hydrological change.

Crucially, these goals do not operate independently. Poverty alleviation (SDG 1) depends upon agricultural productivity; economic participation (SDG 8) increasingly relies on market crops such as peas and seabuckthorn; responsible consumption (SDG 12) is reflected in the cyclical use of organic waste and restrained water use; sustainable communities (SDG 11) are sustained through settlement continuity and ecological adaptation; and partnerships (SDG 17) become central when considering how indigenous systems might engage with state and scientific institutions. The Rangrik case demonstrates that sustainability is relational. Disruption in one domain—water, soil, biodiversity—cascades into others.

However, convergence between tribal knowledge and SDGs must be interpreted carefully. These practices were not formulated to satisfy global benchmarks. They evolved because ecological miscalculation carries immediate consequences. The Western Himalayan cold desert enforces accountability. Overextraction of water leads to crop failure. Soil mismanagement results in erosion that cannot easily be reversed. In this sense, sustainability is not an aspirational agenda but a practical condition.

The difficulty arises when development is conceptualised as linear progress defined by increased output, technological substitution and market integration. Experiences from Himalayan regions, including Ladakh, suggest that the introduction of chemical fertilisers and high-yielding varieties can displace traditional agro-ecological systems, often leading to unintended ecological and social consequences (Berkes, 2018).

Similarly, commercial pressures on medicinal plants risk transforming health resources into extractive commodities. Development models that treat land as input and culture as obstacle may erode the institutional architectures that sustain ecological balance.

Such interventions expose a deeper problem within dominant development paradigms: the presumption of universality. Policies are often designed as transferable packages, assuming that sustainability can be engineered through uniform technological solutions. Yet the Western Himalayan case underscores that sustainability is culturally mediated. Institutional arrangements such as *churay* governance, inheritance norms limiting land fragmentation, and gendered agricultural coordination are not incidental—they stabilise ecological use. Disrupting them without understanding their systemic role can generate vulnerability rather than resilience.

This does not imply that tribal knowledge is static, immune to critique or sufficient in isolation. Hierarchical water distribution reflects embedded inequalities; climate variability (IPCC, 2021) intensifies stress on established systems; younger generations may be drawn toward alternative livelihoods; and integration into broader markets may alter subsistence patterns. Indigenous systems require adaptation, dialogue and support. The point is not preservation as museum artefact, but recognition as dynamic socio-ecological intelligence.

In this context, the importance of tribal knowledge within sustainable development planning becomes unmistakable. Indigenous communities in regions such as the Western Himalaya are not merely populations vulnerable to climate change or beneficiaries of state welfare; they are custodians of long-evolved systems of ecological regulation. Their practices demonstrate that sustainability is most durable when embedded in social norms, institutional memory and moral frameworks that regulate human–environment relations. Development planning that overlooks these embedded systems risks displacing precisely those mechanisms that maintain long-term resilience.

The Western Himalayan cold desert therefore compels a reconsideration of the very idea of development. If development is understood primarily as expansion—of production, infrastructure or market integration—it may conflict with ecological limits that have historically structured survival in such landscapes. If, however, development is reinterpreted as the strengthening of existing capacities for ecological balance, social continuity and adaptive flexibility, then tribal knowledge becomes central rather than peripheral to sustainability agendas.

The Sustainable Development Goals propose a universal vision for planetary well-being. Rangrik reveals that such visions must remain attentive to place. Sustainability cannot be engineered solely through technical intervention or uniform policy design; it must be negotiated within the cultural and ecological logics that organise everyday life. In landscapes shaped by scarcity, sustainability has long been practised as restraint, reciprocity and regulation.

The enduring lesson of the Western Himalayan experience is therefore not that tribal knowledge offers a blueprint for the world, but that sustainable futures require listening to the knowledge formed in environments where miscalculation carries immediate consequence. Development detached from landscape, memory and community risks becoming extraction in another form. Development

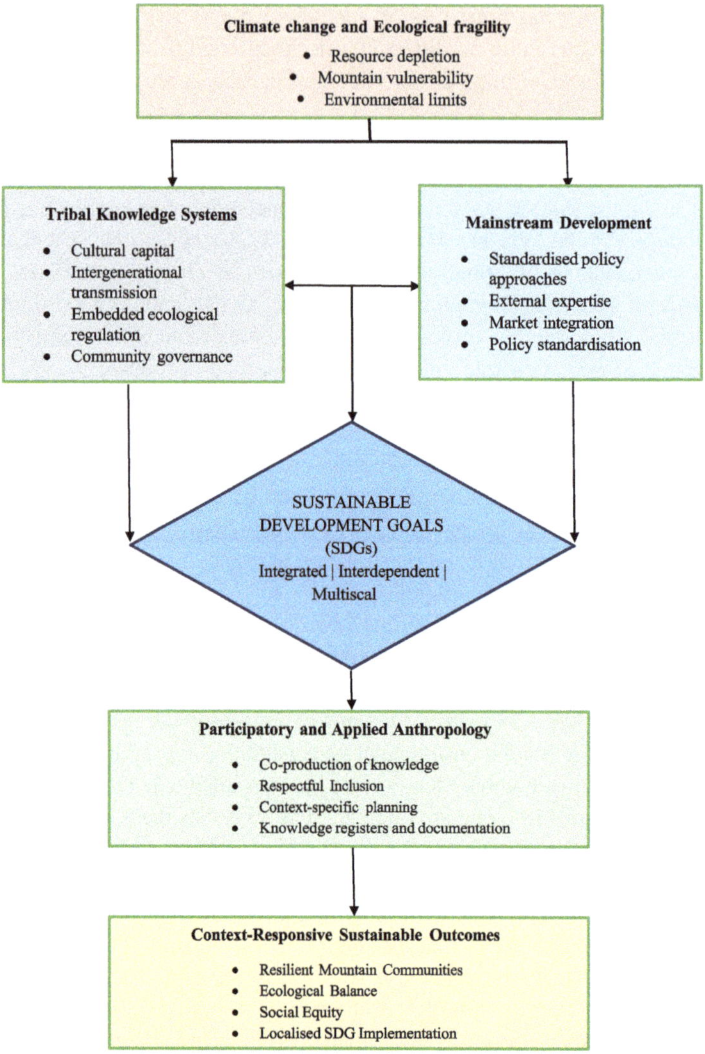

Fig. 12.3 Integrated socio-ecological sustainability model for mountain development

grounded in lived ecological intelligence holds greater promise for resilience in an era of accelerating climate uncertainty. The question is not whether tribal knowledge should be preserved unchanged, but whether development can learn to engage with place-based intelligence without dismantling the institutional structures that sustain it. The integrated relationship between ecological vulnerability, knowledge systems, development planning and sustainable outcomes is summarised in Fig. 12.3.

Conclusions and Recommendations

This chapter has demonstrated that tribal knowledge systems in Kinnaur, Lahaul and Spiti function as integrated socio-ecological arrangements shaped by environmental limits, cultural continuity and institutional regulation. In fragile mountain ecosystems marked by scarce water, vulnerable soils and climatic uncertainty, sustainability is not a policy aspiration but a condition of survival. The practices of cultivation, water governance, soil conservation and medicinal knowledge examined in this study reveal how environmental protection is embedded within social norms, inheritance systems, religious values and collective labour.

These systems, however, are context-specific and subject to change. Regions such as Kinnaur, which experience relatively greater rainfall and connectivity, are not as exclusively dependent on *kuhl* and *zing* systems as Spiti. Climatic variability, infrastructural expansion and market integration may gradually alter existing practices. The erosion of tribal knowledge—whether through climate change or externally imposed development models—represents not only cultural loss but the weakening of locally adapted sustainability frameworks that align closely with the integrated vision of the Sustainable Development Goals (SDGs).

In the light of these findings, the following recommendations are proposed. Among these recommendations, some, but not all may be scaled up for other similar communities.

Participatory Documentation of Tribal Knowledge

Region-specific practices related to agroecology, water governance, soil management and medicinal systems should be systematically documented through participatory Tribal Knowledge Registers. Such documentation must ensure community ownership and safeguard against extractive appropriation. Preserving this ecological intelligence strengthens long-term resilience and contributes to SDGs 13 (Climate Action) and 15 (Life on Land).

Institutionalising Action and Applied Anthropology in Development Planning

Development initiatives in mountain regions must adopt participatory and collaborative frameworks that recognise tribal communities as knowledge holders. Community members should be involved in the design, implementation and evaluation of interventions. The action- oriented tradition of Applied Anthropology emphasises reciprocal learning, where researchers engage with indigenous knowledge systems as contextually grounded expertise. Sustainable outcomes are more likely when development strengthens local institutional autonomy rather than displacing it through externally designed templates.

Integrated Socio-Ecological Research Approaches

Mountain sustainability challenges require interdisciplinary frameworks that address ecological processes and social institutions simultaneously. Research teams should integrate anthropology, ecology, hydrology and policy analysis to ensure that interventions aligned with SDGs 1, 2, 3, 6 and 15 do not unintentionally disrupt embedded governance systems such as customary water allocation or culturally grounded land-use practices.

Environmentally Grounded Economic Strengthening

Economic initiatives should build upon locally available ecological resources while maintaining environmental balance. Support for small-scale enterprises, value addition to traditional crops and medicinal plants, and regulated market access can contribute to SDG 8 (Decent Work and Economic Growth) without undermining biodiversity or soil stability. Commercialisation must remain aligned with ecological limits.

Conservation of Biodiversity Beyond Agroecosystems

While this study has focused primarily on agroecosystems, the protection of naturally occurring plant species, grazing lands and mountain forests is equally critical. Biodiversity conservation underpins health resilience (SDG 3), ecosystem stability (SDG 15) and climate adaptation (SDG 13). Sustainable development strategies must therefore address both cultivated and uncultivated landscapes.

In conclusion, tribal knowledge in the Western Himalayan cold desert demonstrates that sustainability is most effective when grounded in ecological specificity, cultural continuity and collective regulation. These systems have evolved not through abstract environmental policy but through lived negotiation with scarcity and climatic uncertainty. Their alignment with multiple Sustainable Development Goals is not coincidental; it reflects an integrated logic of resource management that predates global sustainability discourse.

The challenge before policymakers and researchers is not to replace these systems with uniform development templates, but to strengthen them through respectful collaboration. Action-oriented, participatory approaches rooted in applied anthropology offer a pathway toward development that is both context-sensitive and socially just. In fragile mountain ecosystems, sustainable futures will depend not on imposing external models, but on working with the knowledge systems that have long enabled survival within environmental limits.

References

Berkes, F. (2018). *Sacred ecology* (4th ed.). Routledge.

Bourdieu, P. (1986). The forms of capital. In J. G. Richardson (Ed.), *Handbook of theory and research for the sociology of education* (pp. 241–258). Greenwood.

Food and Agriculture Organization of the United Nations. (2019). *Mountain agriculture: Opportunities for harnessing Zero Hunger in Asia.* FAO.

Gâtlan, A. M., & Gutt, G. (2021). Sea buckthorn in plant-based diets: An analytical approach of sea buckthorn fruits composition, nutritional value, applications, and health benefits. *International Journal of Environmental Research and Public Health, 18*(17), 8986. https://doi.org/10.3390/ijerph18178986

Intergovernmental Panel on Climate Change. (2021). *Climate change 2021: The physical science basis*. Cambridge University Press.

Kala, C. P. (2005). Indigenous uses, population density and conservation of threatened medicinal plants in the Indian Himalaya. *Conservation Biology, 19*(2), 368–378.

Kawa, J. M., Taylor, C. G., & Przybylski, R. (2003). Buckwheat concentrate reduces serum glucose in streptozotocin-induced diabetic rats. *Journal of Agricultural and Food Chemistry, 51*(25), 7287–7291.

Sannegowda, U. B. M., & Garkoti, S. C. (2022). Traditional community-led seed system for maintaining crop vigour, diversity and socio-cultural network in view of the changing climate: A case study from western Himalaya, India. *Climate Action, 1*, Article 19. https://doi.org/10.1007/s44168-022-00020-7

Stockholm Resilience Centre. (2016). *How food connects all the SDGs: The wedding cake model.*

Toledo, V. M. (2002). Ethnoecology: A conceptual framework for the study of indigenous knowledge systems. In J. R. Stepp et al. (Eds.), *Ethnobiology and biocultural diversity* (pp. 511–522). University of Georgia Press.

Toledo, V. M. (2002). Indigenous peoples and biodiversity. In *Encyclopedia of biodiversity* (Vol. 3, pp. 451–463). Academic Press.

United Nations. (2015). *Transforming our world: The 2030 agenda for sustainable development.* United Nations. https://sdgs.un.org/2030agenda

Yao, Y., Shan, F., Bian, J., Chen, F., Wang, M., & Ren, G. (2008). D-chiro-inositol-enriched tartary buckwheat bran extract lowers the blood glucose level in KK-Ay mice. *Journal of Agricultural and Food Chemistry, 56*(21), 10027–10031.

Zeb, A. (2004). Important therapeutic uses of sea buckthorn (*Hippophae*): A review. *Journal of Biological Sciences, 4*(5), 687–693. https://doi.org/10.3923/jbs.2004.687.693

Part II
Tribal Indigenous Knowledge—Food Security and Health

Chapter 13
Indigenous Fermented Foods of the Ethnic Meitei's of Manipur and Its Nutritional Benefits

Khwairakpam Sharmila, Ningthoujam Samarendra Singh, and Khwairakpam Bembem

Abstract The wide variety of fermented foods found across North-East India highlights the region's rich cultural heritage and the use of locally available resources. By fermenting ingredients such as soybeans, bamboo shoots, and seasonal vegetables, various tribes have created distinctive flavours and textures shaped by their surroundings, history, and traditions. Fermentation, one of the oldest and most economical techniques for food processing and preservation, has been traditionally practiced by the people of Manipur in North-East India. These fermented products are deeply embedded in everyday life, as well as in festivals and rituals, contributing significantly to both physical nourishment and maintaining cultural heritage. In Manipur, commonly consumed fermented foods include *Hawaijar* (fermented soybean), *Ngari* (fermented fish), and *Soiboom* and *Soidon* (fermented bamboo shoots), which are appreciated not only for their taste but also for their nutritional and health benefits. Food plays an essential role not only in sustaining life but also in expressing cultural identity. Fermented foods, in particular, serve as markers of ethnic traditions, reflecting the histories and ecological contexts of different communities. In addition to their cultural importance, these foods are valued for their health-promoting properties, such as improving digestion, supplying probiotics, and enhancing overall well-being. These benefits further reinforce their enduring popularity and cultural relevance.

K. Sharmila (✉)
Department of Human Development and Family Studies, School of Home Science, Babasaheb Bhimrao Ambedkar University, Lucknow, Uttar Pradesh, India
e-mail: khwairakpamsharmila@gmail.com

N. Samarendra Singh
Division of Post-Harvest Management, Central Institute of Sub-Tropical Horticulture, Lucknow, Uttar Pradesh, India
e-mail: tairenki@gmail.com

K. Bembem
Division of Transfer of Technology, Central Institute of Post-Harvest Engineering and Technology, Ludhiana, Punjab, India
e-mail: Bembemkhwai@gmail.com

R. C. Sobti et al. (eds.), *Traditional Knowledge Systems in India: Sustainability, Ecology, and Resource Governance*, People, Cultures and Societies: Exploring and Documenting Diversities,
https://doi.org/10.1007/978-981-92-1346-7_13

Keywords Fermentation · Fermented foods · Indigenous foods · Nutritional benefits

Introduction

The ethnic tribal communities of North-East India, particularly those in Manipur, are renowned for their rich cultural traditions, distinctive food habits, and the harmonious presence of both tribal and non-tribal populations (Wahengbam et al., 2020). The diverse cultural practices and traditional lifestyles of these groups have led to the development of a wide variety of indigenous food items. Among the Meitei community of Manipur, an impressive assortment of vegetables is consumed, including more than a hundred varieties of leafy greens. A range of traditional dishes is prepared using locally available fish, vegetables, herbs, and spices. Prior to the introduction of modern preservation techniques, conventional methods such as smoking, drying, salting, fermentation, and marination were widely used, especially for fish preservation. These techniques help prolong shelf life by lowering moisture levels or modifying pH, thereby restricting microbial activity and preventing spoilage (Majumdar et al., 2015).

The Meitei community of Manipur has a long-standing tradition of preparing and consuming a diverse range of homemade fermented foods and beverages (Singh et al., 2018). From ancient times, they have developed specialized methods for fermenting fish, bamboo shoots, and soybeans (Devi & Suresh, 2012). This traditional knowledge of preservation is still practiced within households and is passed down from one generation to the next through daily activities guided by elders.

Fermentation is regarded as one of the oldest and most economical techniques for ensuring food safety and quality. It improves digestibility, taste, aroma, and nutritional content by enhancing beneficial nutrients while reducing or eliminating antinutritional substances (Das et al., 2016; Soibam & Ayam, 2018). Moreover, the process can detoxify certain harmful components found in raw materials. Each fermented product contains a distinct group of microorganisms that contribute to increased levels of proteins, vitamins, essential amino acids, and fatty acids.

The word "fermentation" comes from the Latin term *Fervere*, which means "to boil," which refers to the bubbling effect observed during yeast activity in fruit juices or malted grains used in the production of alcoholic beverages (Haokip et al., 2020). Indigenous fermented foods are produced from various raw materials using naturally present or selected beneficial microorganisms as starter cultures. These products are widely available in local markets for community consumption and play a vital role in post-harvest management by enhancing product value and reducing post-harvest losses (Soibam & Ayam, 2018) (Fig. 13.1).

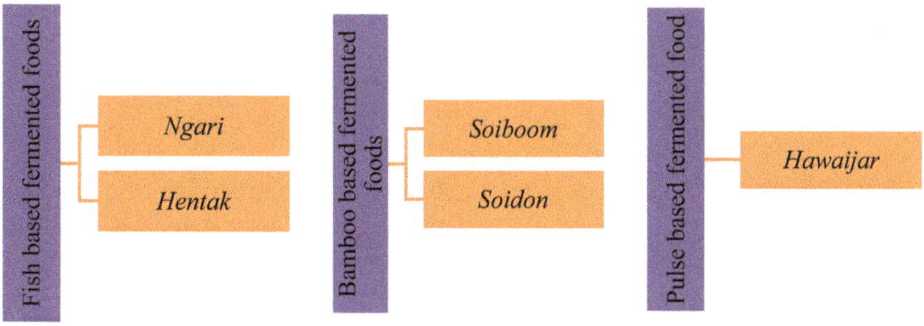

Fig. 13.1 Classification of fermented foods

Fish-Based Fermented Foods

The Meitei community of Manipur, India, possesses a rich culinary heritage that features a variety of fermented fish products. Among the most well-known traditional fermented fish items are:

(i) *Ngari* and
(ii) *Hentak*

(i) ***Ngari*** is a traditional fermented fish delicacy that occupies a significant position in Manipuri cuisine, particularly among the Meitei community. Given that fish is protein-rich and highly prone to spoilage, effective preservation methods are necessary (Majumdar et al., 2015). Although it is a staple in Manipur, Ngari is also valued as an important culinary ingredient in several other North-Eastern states of India.

Ngari is made from small freshwater fish, *Puntius sophore*, locally called *phabou nga*, which are either procured locally or imported from states such as Odisha and West Bengal for commercial purposes. For the Manipuri population, especially the Meitei community, a meal is often regarded as incomplete without ngari. It is generally added in small amounts as a seasoning to various dishes such as stews (*kangsoi*), curries, chutneys, and sauces (Devi & Suresh, 2012). Its unique flavour deepens the taste and overall appeal of these dishes, making it an essential component of traditional Manipuri cuisine.

In addition to its culinary significance, ngari holds considerable cultural value among the Meiteis. It is commonly incorporated into rituals, festivals, and ceremonial events, where it represents prosperity and abundance.

The process of preparing *ngari* involves several traditional steps:

- The fish are first sorted, cleaned, and thoroughly washed with water.
- They are then sun-dried until they become crisp.
- The heads of the dried fish are crushed carefully, often using a hammer, to facilitate quicker and more uniform fermentation.
- The dried fish are arranged systematically and tightly packed into an earthen pot coated with mustard oil. This pot, known locally as *ngari kharung*, is used specifically for fermentation.
- The container is tightly sealed and left untouched to enable natural fermentation.
- The fermentation period generally ranges from three to six months, depending on environmental conditions such as temperature and the preferred strength of flavour.

A distinct, strong aroma signals that the fermentation process is complete and the product is ready for use. During this period, naturally occurring microorganisms decompose the fish proteins and generate compounds such as lactic acid, which contribute to *ngari's* unique taste and characteristic flavour.

(ii) ***Hentak:*** It is a traditional fermented fish paste commonly consumed in Manipur, particularly by the Meitei community. It is prepared from selected local freshwater fish species, including "*Puntius sophore* (Phabou nga), *Esomus danricus* (Ngasang), *Amblypharyngodon mola* (Mukanga), *Puntius manipurensis* (Ngakha), and *Colisa fasciatus* (Ngapemma)" (Soibam & Ayam, 2018).

The preparation of *hentak* involves a series of traditional steps:

- The fish are first sorted, cleaned, and thoroughly washed.
- They are then sun-dried until they become crisp.
- The dried fish are pounded into powder using a wooden mortar (*Sumban*) and pestle (*Suk*).
- Cleaned slices of the petioles from wild *Alocasia macrorrhiza* (Hongu) or *Colocasia esculenta* (Lampal) are then added and pounded along with the powdered fish to create a smooth, thoroughly blended paste.
- The paste is shaped into round, spherical, or elliptical forms as desired, and stored in an airtight condition for one to two weeks.
- It is then pounded again, tightly pressed into earthen pots known as *hentak chaphu*, sealed, and left to ferment for at least another two weeks.

The pounding process is repeated periodically to enhance shelf life and improve flavour (Soibam & Ayam, 2018). During fermentation, natural enzymes and microorganisms degrade fish proteins, producing a pungent and richly flavoured paste. *Hentak* is characterized by a strong umami flavour with earthy undertones. Like *ngari*, it is used sparingly in Manipuri cuisine as a flavour enhancer in soups, stews, curries, chutneys, and *iromba*, adding depth and distinctiveness to these dishes.

Health and Nutritional Advantages of Fermented Fish Products

Fermented fish products serve as excellent sources of protein, bioactive peptides, and amino acids. The proteins and peptides generated during fermentation are not only nutritionally significant but also biologically active, contributing to various regulatory functions in the human body (Keishing & Banu, 2015). Furthermore, these products supply substantial amounts of minerals, especially calcium, since the bones and scales are fermented and consumed along with the fish.

In addition, fermented fish products are known for their antioxidant activity and possible probiotic effects that support overall health (Rapsang & Joshi, 2013). Ngari, for example, contains a diverse microbial population, including "*Lactococcus lactis* subsp. *cremoris*, *Lactococcus plantarum*, *Enterococcus faecium*, *Lactobacillus fructosus*, *L. amylophilus*, *L. corneformis* subsp. *torquens*, *Lactobacillus plantarum*, *Bacillus subtilis*, *B. pumilus*, *Micrococcus*, and fungi such as *Candida* and *Saccharomycopsis*" (Singh et al., 2012). Several *Lactobacillus* strains isolated from ngari demonstrate strong cell surface hydrophobicity, indicating promising probiotic potential (Singh et al., 2012).

Bamboo-Based Fermented Foods

Bamboo shoots, the young sprouts of bamboo plants, are consumed in various forms, including fresh, dried, pickled, and fermented. While they are eaten worldwide, their use is particularly common in Asian countries. From a nutritional perspective, bamboo shoots are low in fat and simple sugars but are rich sources of potassium and dietary fibre. They also contain essential vitamins, amino acids, and antioxidant compounds such as flavones, phenols, and steroids (Chaudhary et al., 2011).

In Manipur, the Meitei community includes both cultivated and wild bamboo shoot varieties in their diet, consuming them fresh, fermented, or dried. Among these, certain fermented bamboo shoot preparations are especially popular among the local population (Nongdam, 2015).

(i) *Soiboom*
(ii) *Soidon*

(i) ***Soiboom*** is among the most popular fermented foods in Manipur and can be found in almost all local vegetable markets. It usually appears off-white and has a characteristic sour taste. This preparation is made from tender bamboo shoots of selected varieties, including "*Bambusa tulda* (Utang), *Bambusa balcooa* (Ching Saniebi), Wanap, Unap, *Dendrocalamus hamiltonii* (Pecha), and *Melocanna bambusoides* (Moubi/Muli)" (Behera & Balaji, 2021).

The preparation of fermented bamboo shoots is commonly practiced in the Kwatha region of Manipur. However, *soiboom* produced in the Kakching region is particularly renowned among local consumers for its superior taste and distinctive aroma.

Now the product is ready to consume.

Steps involved in preparation of *Soiboom*
Stage I
Cleaning the tender bamboo shoots thoroughly and cutting them into slices
Allowing the sliced bamboo shoots to ferment in a large plastic container filled with water for about 7–8 days
The partially fermented bamboo shoots are then securely sealed in plastic bags and stored upside down for another 10–15 days to allow fermentation to continue
Stage II
The next stage of preparation uses a specially made bamboo basket, roughly 4–5 feet long, 3–4 feet wide, and 4–5 feet high
The interior of the basket is fully lined with polythene sheets, with small perforations provided at the base
The bamboo shoots that have undergone partial fermentation are removed from the plastic bags and transferred into the basket, ensuring they are tightly packed by pressing them firmly during the process
The basket is then completely covered with a plastic sheet
Heavy stones or similar weights are placed on top to apply sufficient pressure, ensuring compact packing of the bamboo shoots
After about 3–4 days, liquid released from the bamboo shoots begins to drain out through the perforations at the bottom of the basket
The plastic covering is removed after 3–4 months, once the exudation of liquid has significantly decreased

(ii). ***Soidon:*** Soidon is a partially fermented bamboo shoot delicacy that is commonly consumed across Manipur. It has a light green colour and is typically formed into circular coils. Common bamboo species used for its preparation include "*Bambusa tulda* Roxb., *Dendrocalamus giganteus* Munro, *Melocanna bambusoides* Trin., and *Teinostachyum wightii*" (Behera & Balaji, 2021).

The method for preparing soidon is largely comparable to that of soiboom. However, unlike soiboom, where the shoots are cut into thin slices, soidon shoots are shaped into slender coils. The fermentation of soidon is associated with the bacterium "*Cephalostachyum capitatum*" (Singh et al., 2003).

Fermented bamboo shoots are generally consumed with steamed rice and incorporated into various dishes, particularly iromba, a traditional Manipuri chutney made from mashed boiled vegetables. They are also used as essential ingredients in other preparations, such as stir-frying with fish or cooking in stews along with vegetables and smoked fish (Jeyaram et al., 2009).

Health and Nutritional Benefits of Fermented Bamboo Shoots

Fermented bamboo shoots are a vital part of the tribal diet. They are high in dietary fibre and minerals while being low in fat (Behera & Balaji, 2021). These shoots are rich sources of potassium, vitamins such as Vitamin E (α-tocopherol), Vitamin C, Vitamin B6, thiamine, riboflavin, and niacin, as well as minerals like iron and phosphorus. They also contain various dietary fibres, including hemicellulose, cellulose, pectin, and lignin (Choudhury et al., 2012). Incorporating fermented bamboo shoots into the diet can help combat micronutrient deficiencies and support the immune system in resisting infectious diseases (Singhal et al., 2021).

Fresh bamboo shoots naturally contain certain anti-nutritional compounds, such as phytates, tannins, alkaloids, saponins, and cyanogenic glycosides (HCN). Fermentation substantially reduces most of these compounds, although tannin levels may remain relatively unchanged (Sarangthem & Singh, 2013).

In summary, fermented bamboo shoots are highly nutritious and offer a range of health benefits. Some of their key advantages include (Behera & Balaji, 2021):

Pulse Based Fermented Foods

Hawaijar: Hawaijar is a customary fermented soybean delicacy native to Manipur, prized for its unique flavour and aroma. Among the Meitei community, it is regularly consumed as an affordable and nutritious source of protein (Devi & Suresh, 2012). The production of hawaijar relies on spontaneous or natural fermentation techniques that have been passed down through generations.

Health Benefits of Fermented Bamboo shoot	
	Anti-oxidant
	Anti-cancer
	Anti-senescence
	Free radical-neutralizing
	Weight loss
	Protect against cardiovascular diseases
	Improves digestion
	Antimicrobial effect attributed to various glycosides
	Flavones
	Lowers blood pressure

This product is popular not only in Manipur but also across other regions of North-East India and South-East Asia. A comparable fermented soybean product in Japan, known as natto, also involves the activity of *Bacillus subtilis*, similar to hawaijar. While there is a recognized need to modernize and standardize the traditional fermentation process for consistent quality (Devi et al., 2024), most local producers still adhere to the traditional methods that have been used for centuries (Jeyaram et al., 2009).

Preparation
Soybean seeds of medium or small size are chosen, cleaned, and sorted
The sorted seeds are soaked overnight in water, using approximately twice the volume of water to seeds
After soaking, the seeds are rinsed thoroughly two to three times under running water.
The soybeans are then pressure-cooked until they become soft
The excess water is drained, and the cooked beans are placed in a bamboo basket. They are rinsed again with lukewarm water until the surface is no longer greasy
The beans are gently turned over once or twice to ensure proper draining
A coarse bamboo basket is lined with a clean muslin cloth folded three times

(continued)

(continued)

Preparation
A dense layer of *Ficus hispida* leaves (locally called Asse heibong) or banana (*Musa* spp.) leaves is laid over the cloth
The cooked and cleaned soybeans are arranged inside the basket in layers.
A second clean muslin cloth, folded two or three times, is laid on top, and the whole bundle is tightly tied to ensure an airtight seal
During the fermentation period of 3–5 days, the bundle is kept in the sun during the daytime and placed near the fireplace at night to maintain warmth and ensure proper fermentation while preventing spoilage

The fermentation of soybeans usually takes about three days during the summer, while in winter it can extend up to five days. Once fermentation is complete, the product is called Hawaijar (Devi & Suresh, 2012). However, its shelf life is relatively short—typically three to four days—so it is often sun-dried to prolong storage (Somishon & Thahira, 2013).

Hawaijar is prized for its unique sensory properties, including its distinctive flavour and aroma. It is a staple in Manipuri households and features prominently in many traditional Meitei dishes. It is commonly used as a seasoning in vegetable preparations, eaten directly as a paste mixed with chili and salt (known as Ametpa), or added to various vegetable dishes to enhance taste and texture (Premarani & Chhetry, 2008).

One well-known Meitei delicacy, chagempomba, is made from rice, hawaijar, and locally sourced vegetables, with fermented fish (ngari) or smoked fish occasionally added to intensify the flavour (Premarani & Chhetry, 2008). The dried fermented beans can also be prepared with meat or used to create a chutney with chili, tomato, and salt (Mao & Odyou, 2007). More recently, hawaijar has also been prepared as a pickle. In this process, it is sautéed in oil with spices and salt, then placed in bottles with the remaining oil and sealed to prolong shelf life. (Devi & Kumar, 2012). "To make fermented soy chutney, whole dried hawaijar is soaked, deep-fried in vegetable oil, and then blended with salt and chilies"(Tamang, 1996).

Nutritional Benefits of Fermented Soybeans

Fermented soybeans provide numerous nutritional advantages and can serve as a highly beneficial component of a balanced diet. Compared to unfermented soybeans, they offer enhanced nutrient availability and improved support for gut health.

Improved Digestibility: Fermentation breaks down complex carbohydrates and proteins into simpler forms that are easier to digest, enhancing nutrient absorption and protein utilization.

Protein Quality: Soybeans provides a complete source of protein, containing all essential amino acids. Fermentation further improves protein digestibility and absorption. Proteolytic enzymes produced by microorganisms during fermentation hydrolyse proteins, releasing free amino acids. Notably, levels of essential amino acids, particularly arginine, increase significantly (Ketnawa & Ogawa, 2019).

Vitamin Content: Fermentation can enhance the concentration of certain vitamins, especially B-complex vitamins such as B12 (in some fermented soy products like tempeh and natto) and vitamin K2 (notably present in natto).

Mineral Bioavailability: The process also improves the absorption of minerals such as iron, calcium, and magnesium found in soybeans.

Probiotic Benefits: Some fermented soy foods, including miso and natto, contain beneficial probiotic microorganisms that support digestive health and maintain gut microbial balance.

Antioxidant Properties: Fermented soybeans may exhibit higher antioxidant levels than unfermented varieties, helping to reduce oxidative stress in the body.

Reduction of Antinutritional Factors: Fermentation decreases or eliminates antinutrients including phytic acid, protease inhibitors, oxalic acid and urease, which can otherwise hinder mineral absorption.

Potential Health Effects: During fermentation, isoflavone glycosides are converted into free isoflavone aglycones, which demonstrate greater biological activity (Liu et al., 2022). Regular intake of fermented soy products has been linked to improved cardiovascular health, stronger bones (partly due to vitamin K2), and enhanced immune function.

Conclusion

The traditional fermented foods of the Meitei community in Manipur represent a rich culinary heritage deeply connected to both cultural identity and nutritional health. Items such as Hawaijar (fermented soybean), Ngari (fermented fish), and Soiboom and Soidon (fermented bamboo shoots) not only add unique flavours to Manipuri cuisine but also offer significant health benefits. Fermentation enhances the nutritional value of these foods, increasing the bioavailability of proteins, vitamins, and minerals. Additionally, they provide beneficial probiotics that support digestive health and aid in nutrient absorption. Preserving and promoting these traditional practices helps safeguard cultural heritage while fostering a nutritious and health-promoting diet for the Meitei community of Manipur.

Summary

- The Meitei community of Manipur includes a diverse array of fresh and preserved foods in their diet, with fermented food products and drinks playing a significant role.
- In nearly every Meitei household, the traditional practice of fermenting fish, bamboo shoots, and soybeans is carried out and transmitted from one generation to the next.
- Fermentation involves the action of beneficial microorganisms and serves to extend the shelf life of food products.
- *Ngari* and *Hentak* are the principal fermented fish products in Meitei cuisine and hold notable cultural importance within the community.
- The methods used in preparing *ngari* and *hentak* follow carefully defined steps that are essential for developing their characteristic flavours.
- Fermented fish products provide abundant protein, amino acids, and minerals like calcium, and they also offer probiotics and antioxidants that support overall health.
- *Soiboom* and *Soidon* are widely consumed fermented bamboo shoot products made from specific bamboo varieties and are highly popular among the Meiteis.
- Fermented bamboo shoots are highly nutritious, supplying dietary fibre, vitamins, minerals, and antioxidants. The fermentation process diminishes most of the antinutritional compounds found in fresh shoots—such as phytates, alkaloids, saponins, and cyanogenic glycosides (HCN)—while tannin content may remain largely unchanged.
- *Hawaijar* is a customary fermented soybean food from Manipur, known for its unique flavour and aroma.

References

Behera, P., & Balaji, S. (2021). Health benefits of fermented bamboo shoots: The twenty first century green gold of northeast India. *Applied Biochemistry and Biotechnology*, *193*, 1800–1812.

Chaudhary, D., Sahu, J. K., & Sharma, G. D. (2011). Bamboo shoot based fermented food products: A review. *Journal of Scientific & Industrial Research*, *70*, 199–203.

Choudhury, D., Sahu, J. K., & Sharma, G. D. (2012, August). Value addition to bamboo shoots: A review. *Journal of Food Science and Technology*, *49*, 407–414.

Das, G., Patra, J. K., Singdevsachan, S. K., Gouda, S., & Shin, H. S. (2016). Diversity of traditional and fermented foods of the Seven Sister states of India and their nutritional and nutraceutical potential: A review. *Fronties in Life Science*, *9*(4), 292–312.

Devi, A. N., Devi, L. G., & Devi, C. B. (2024). Traditional soybean fermentation of Meitei of Manipur. *International Journal of Agriculture Extension and Social Development*, *7*(1), 75–78.

Devi, P. A., & Suresh, K. P. (2012). Traditional, ethnic and feremented foods of different tribes of Manipur. *Indian Journal of Traditional Knowledge*, *11*(1), 70–77.

Devi, P., & Kumar, S. P. (2012). Traditional ethnic and fermented foods of different tribes of Manipur. *Indian Journal of Traditional Knowledge*, *11*(1), 70–77.

Haokip, S. W., Shankar, K., & Sheikh, K. A. (2020) Traditional fermented foods and their health benefits of the North Eastern states of India. Available from https://www.researchgate.net/publication/359158881_Traditional_fermented_foods_and_their_Health_benefits_of_the_North_Eastern_states_of_India. Accessed February 24, 2024.

Jeyaram, K., Singh, T. A., Romi, W., Devi, A. R., Singh, W. M., Dayanidhi, H., Singh, R., & Tamang, J. P. (2009). Traditional fermented foods of Manipur. *Indian Journal of Traditional Knowledge*, *8*(1), 115–121.

Keishing, S., & Banu, T. (2015). Fermented fish (Ngari) of Manipur—Preparation technique and its potential as a functional food ingredient. *ELixir Food Science*, *8*, 34502–34507.

Ketnawa, S., & Ogawa, Y. (2019). Evaluation of protein digestibility of fermented soybeans and changes in biochemical characteristics of digested fractions. *Journal of Functional Foods*, *52*, 640–647.

Liu, L., Chen, X., Hao, L., Zhang, G., Jin, Z., Li, C., Yang, Y., Rao, J., & Chen, B. (2022). Traditional fermented soybean products: Processing, flavor formation, nutritional and biological activities. *Critical Reviews in Food Science and Nutrition*, *62*(7), 1971–1989.

Majumdar, R. K., Bejjanki, S. K., Roy, D., Shitole, S., Saha, A., & Narayan, B. (2015). Biochemical and microbial characterization of Ngari and Hentaak-traditional fermented fish products of India. *Journal of Food Science and Technology*, *52*, 8284–8291.

Mao, A. A., & Odyou, N. (2007). Traditional fermented foods of the Naga tribes of Northeastern, India. *Indian Journal of Traditional Knowledge*, *6*(1), 37–41.

Nongdam, P. (2015). Traditional fermented bamboo shoot foods of North-East India and their characteristic natural microbial flora. In *10th world bamboo congress, Korea 2015*.

Premarani, T., & Chhetry, G. K. N. (2008). Microbiota associated with natural fermentation of Hawaijar (an indegeneous fermented soyabean product) of Manipur and their enzymatic activity. *Journal of Food Science and Technology*, *45*(6), 516–519.

Rapsang, G. F., & Joshi, S. R. (2013). Molecular and probiotic functional characterization of Lactobacillus spp. associated with traditionally fermented fish, Tungtap of Meghalaya in northeast India. *Proceedings of the National Academy of Sciences, India Section B: Biological Sciences*, *85*, 923–933.

Sarangthem, K., & Singh, T. N. (2013). Fermentation decreases the antinutritional content in bamboo shoots. *International Journal of Current Microbiology and Applied Sciences*, *2*(11), 361–369.

Singh, A. S., Singh, A. A., & Devi, M. S. (2012). Ngari—a traditional fish product of Manipur, India. *INFOFISH International*, *1*, 41–45.

Singh, R. K. P., Satapathy, K. K., & Singh, K. S. (2003). Common fermented food products of Manipur. *Indian Journal of Hill Farming*, *16*(1&2), 113–115.

Singh, T. A., Sarangi, P. K., & Singh, N. J. (2018). Traditional process foods of the ethnic tribes of western hills of Manipur, India. *International Journal of Current Microbiology and Applied Sciences*, *7*(10), 1100–1110.

Singhal, P., Satya, S., & Naik, S. N. (2021). Fermented bamboo shoots: A complete nutritional, anti-nutritional and antioxidant profile of the sustainable and functional food to food security. *Food Chemistry: Molecular Sciences*, *3*.

Soibam, H., & Ayam, V. S. (2018). The traditional fermented foods of Meities in Manipur, India: A case study. *Journal of Pharmacognosy and Phytochemistry*, *7*(4), 535–539.

Somishon, K., & Thahira, B. A. (2013). Hawaijar—A fermented soya of Manipur, India: A review. *IOSR Journal of Environmental Science, Toxicology and Food Technology*, *4*(2), 29–33.

Tamang, J. P. (1996). Fermented soyabean products in India. In A. Buchanan (Ed.), *Proceedings of the Second International Soyabean Processing and Utilization Conference* (pp. 8–13).

Wahengbam, R., Thangjam, A. S., Keisam, S., Asem, I. D., Ningthoujam, D. S., & Jeyaram, K. (2020). Ethnic fermented foods and alcoholic beverages of Manipur. In J. Tamang (Ed.), *Ethnic fermented foods and beverages of India: Science History and culture*. Springer. https://doi.org/10.1007/978-981-15-1486-9_14

Chapter 14
Fermenting Heritage: The Traditional Alcoholic Beverages of Kinnaura Tribes in Kinnaur, Himachal Pradesh

Chanchal Yadav and Pushp Lata

Abstract The tribal communities in the Indian Himalayan region have long preserved complex traditional practices that connect closely with their environmental knowledge and cultural rituals. Among these practices, fermentation of alcoholic beverages emerge as a key part of identity, livelihood, and spirituality. This chapter examines the indigenous alcoholic beverages among the Kinnaura people of Kinnaur, Himachal Pradesh, focusing on Angoori (from Ribba), Nesang brandy (from Nesang), and Mixed Fruit Rak (from Kanam). Each of these beverages shows how the community utilizes local materials such as grapes, barley, and seasonal fruits, and their specific preparation methods. The ancient modes of preparation, handed down from generation to generation, highlight the complex nexus between traditional knowledge and environmental adaptation. These practices contribute to enhancing community identity and resilience, particularly in the context of remote Himalayan livelihoods. We document the methods of preparation, of these beverages, their cultural significance, and the science behind it, emphasizing the necessity of preserving this cultural heritage amidst modernization.

Keywords Indigenous knowledge · Kinnaura tribe · Fermentation · Traditional alcoholic beverages · Ethnic culture · Himalayan tribes · Angoori · Nesang brandy · Mixed fruit rak

Introduction

India is referred to have the second largest tribal community in the world possessing a rich cultural heritage (Roy, 2020). The ethnic tribes of Indian Himalayan region, inhabiting the high-altitude areas are prominent for their indigenous or traditional

C. Yadav · P. Lata (✉)
Department of Zoology, University of Delhi, Delhi 110007, India
e-mail: plata@zoology.du.ac.in

R. C. Sobti et al. (eds.), *Traditional Knowledge Systems in India: Sustainability, Ecology, and Resource Governance*, People, Cultures and Societies: Exploring and Documenting Diversities,
https://doi.org/10.1007/978-981-92-1346-7_14

knowledge and intricate life system around the globe. Their geographical condition, culinary preference, and lifestyle make them distinct (Rawat et al., 2021). The tribal region of Kinnaur, Himachal Pradesh is a secluded district which encompasses both sides of the river Satluj. Kinnaur district forms the North-eastern region of Himachal Pradesh and western region of great Himalayan Ranges (Negi, 2017). This mountainous district of Himachal Pradesh was formed on 1 May, 1960 which has an altitude ranging from 1600 to 6816 meters (Kumar et al., 2023). The north side of Kinnaur district is bordered by Spiti area, and on the south it is surrounded by Tehri Garhwal and Uttarkashi district of Uttaranchal, on the eastern region by Tibet and on the western side by Shimla district of Himachal Pradesh. The tribes that inhabit Kinnaur region are widely known as Kinnauras or Kinnaurese. Kinnaura tribe is a heterogenous community that displays variation in caste, religion, race, and dialects (Negi, 2017). The tribal people of Kinnaur district in Himachal Pradesh constitute the economically backward community due to region's remote location and adverse geo-climatic conditions. Since ancient times, these communities have been preserving cultural and biological diversity through different indigenous conservation practices. These tribal people have gathered their unique traditional knowledge and formed deep-rooted connections with the locally available biological resources and varied geographical and climatic conditions, thereby establishing a harmonious balance with nature (Lal & Prasher, 2016). Indigenous and tribal communities worldwide are firm believers of nature and accordingly employ them in different traditional rituals, worshiping, and preservation of natural resources (Lata et al., 2024). One of the major indigenous practice associated with tribal communities is preparation of fermented foods which remains closely linked with their customs and rituals (Roy, 2020). The practice of fermenting beverages in India dates back 5000 years. The fermentation practices for the preparation of ethnic beverages by tribal communities are very prominent and outlined. Over 350 varieties of major traditional beverages are believed to be produced using indigenous knowledge. These fermented beverages are homemade, locally crafted and prepared at family or local level. These drinks are region-specific and prepared by mixing an old starter culture with raw ingredients (Rawat et al., 2021). The consumption of locally fermented and distilled alcoholic drinks is a common social practice among many ethnic communities in the Himalayan region (Tamang et al., 2010). In this chapter, we focus on the famous beverages of the Kinnaura tribe: Angoori from Ribba village, Brandy from Nesang village, and Mix Fruit, which is available in all villages of Kinnaur but is studied here specifically from Kanam village.

Study Area

Kinnaur, located between 31°55'50'' and 32°05'15'' North latitudes and 77° 45'00'' and 79°00'35'' East longitudes, is one of the secluded tribal districts of Himachal Pradesh (Lata et al., 2021). Kinnaur district covers an area of about 6400 km^2 in the western Himalayas. The total population of this region is believed to be approximately 84,121 in accordance to 2011 census. The community lives exclusively in rural areas and the members predominantly follow Buddhism and

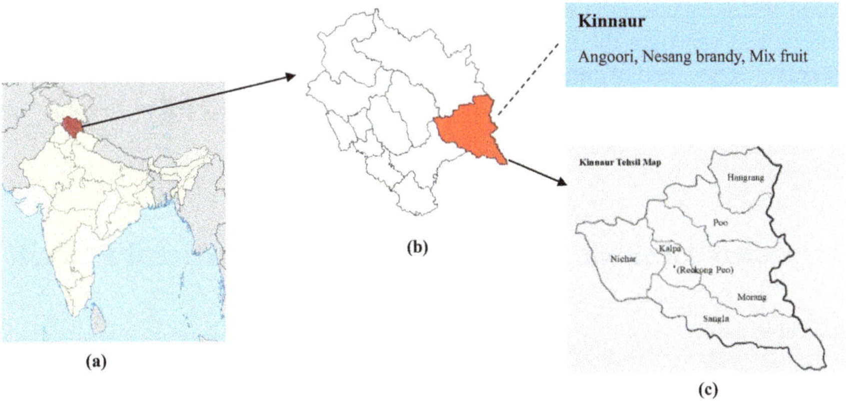

Fig. 14.1 Location of district Kinnaur (study area) in Himachal Pradesh and India

Hinduism, while a small proportion follows Christianity. (Kumari, 2025). It features three distinct climate zones: wet, semi-arid, and arid. The region's minimum temperature varies between – 6.9 °C and 14.6 °C, while the maximum ranges from 7.5 °C to 23.1 °C (Chawla et al., 2012). Kalpa, Sangla, Nichar, Morang, and Pooh are the five administrative tehsils of Kinnaur district (Lata et al., 2021) (Fig. 14.1).

Diversity and Preparation of Traditional Alcoholic Beverages

The Indian Himalayan Region is considered the epicentre of a diverse food culture, which includes both indigenous food as well as alcoholic beverages. The tribes of the Indian Himalayan Region (IHR) produce a diverse range of traditional alcoholic beverages using a variety of fruits, cereals, and grains in the form of raw material. These indigenous alcoholic beverages are grouped into different types, such as palm wine, cereal and rice wine, and distilled alcoholic beverages on the basis of used raw materials and preparation methods (Rawat et al., 2021).

The high-lander tribes of the Himalayan region have developed their own cultures on the basis of available native resources for the preparation of indigenous fermented alcoholic beverages (Roy et al., 2004). Traditional alcohol brewing is a local household industry which is led by tribal women by utilizing their indigenous knowledge of fermentation. Most of the native fermented alcoholic beverages are recognized by various local names, each distinguished by its unique taste, aroma, and preparation method. Almost all the tribes across the Indian Himalayan Region follow identical method for production of fermented alcoholic beverages that includes cooking the raw ingredients, drying them, adding a starter culture, fermenting, and lastly extracting the end product. (Rawat et al., 2021).

A variety of indigenous alcoholic beverages have been prepared and consumed in the tribal communities of Himachal Pradesh, often with limited awareness of the role of microorganisms in fermentation process (Kanwar et al., 2011). Lactic acid bacteria and yeast are the microorganisms primarily involved in fermenting these beverages. They tend to enhance the taste, flavour, acidity, digestibility, and texture of these beverages. The quality of fermented products is influenced by the presence of these microbial communities (Madilo et al., 2024).

Fermented Alcoholic Beverages of Kinnaur, Himachal Pradesh

See Table 14.1.

Table 14.1 Description of fermented alcoholic beverages of Kinnaura tribe (Lata et al., 2021)

S. no.	Ethnic beverages	Substrate	Common name	Part used
1.	Angoori	*Vitis vinifera*	*Dakhang*	Fruit
2.	Nesang Brandy	*Hordeum vulgare*	*Doo*	Grains
3.	Zamashang Phasur	*Arisaema jacquemontii*	*Zamashang*	Roots
4.	Chawa Phasur	*Fagopyrum tataricum, Hordeum vulgare, Triticum aestivum*	*Phaphra, Cha, Zod,* respectively	Grains
5.	Guram Phasur	Raw jiggery		
6.	Chhang	*Oryza sativa or Triticum aestivum*		Grains
7.	Chul Rak	*Prunus armeniaca*	*Chulli or Chul*	Fruit
8.	Seo Rak	*Maluspumila*	*Pale or Seo*	Fruit
9.	Regu Phasur	*Prunusmira*	*Behmi or Reg*	Fruit
10.	Naspoti Rak	*Pyrus communis*	*Nashpati or Nahpoti*	Fruit
11.	Aadu Rak	*Prunus persica*	*Aadu*	Fruit
12.	Plumu Rak	*Prunus ceracifera*		Fruit
13.	Luchau Rak	*Prunus domestica*	*Lucha*	Fruit
14.	Japani Pholangu Rak	*Diospyros kaki*	*Japani Pholang*	Fruit
15.	Mixed Fruit Rak	*Vitis vinifera, Prunus armeniaca, Prunus mira, Malus pumila, Prunus persica, Prunus domestica, Prunus ceracifera* and *Pyrus pyrifolia*		Fruit

Angoori

Angoori, also known as Kinnauri is a liquid dark brown alcoholic beverage of Kinnaur region. Cultivation of variety of grapes is prominently done in Kinnaur district which is later used in the production of alcoholic beverages (Bhalla, 2007).

Ripened grapes (red and green) are plucked and cleaned thoroughly. The plucked grapes are dried in sun for 2–3 days. Later, these grapes are separated from bunches and crushed. The crushed fruit material is then placed inside an underground drum known as "***Bragdum***", which is made of slate stone. In some households, people put the crushed grapes inside a wooden drum which is locally called as "***Jochak***". In present time, mostly plastic drums are being used in place of large wooden and underground drums to ease the process of fermentation. The crushed material is allowed to undergo fermentation for about 6–12 months. Lastly, the fermented material is poured inside a large vessel called "***Lamthu***" and prepared to extract *Angoori*.

The alcoholic beverage is preserved by local people using indigenous methods. Some people fill the Angoori drink in bottles and seal the lid using mud pack, so that air cannot enter it. These bottles are then stored in granaries. These granaries are locally called "***Urch***". These drinks are kept stored in between grains for 5–10 years. When these bottles are opened after years, the beverage inside them becomes very thick just like honey. Angoori is sold locally at Rs 1500/litre. People also use this strong wine for medicinal purposes. Two spoons of *Angoori* along with two spoons of ghee is used as a medicine to cure cough, chest pain, and back pain. It is also used for massage, which provides relief in knees and joints (Figs. 14.2 and 14.3).

Fig. 14.2 **a–b** Ripened grapes used for the preparation of *Angoori*. **c** Steam distillation method practiced by Kinnaura tribe of Kinnaur, Himachal Pradesh for the preparation of fermented alcoholic beverage *Angoori*. **d** Final distillation product of *Angoori* [Image credit: **a–c** Kalyan Bhagti Bhandari, Ribba]

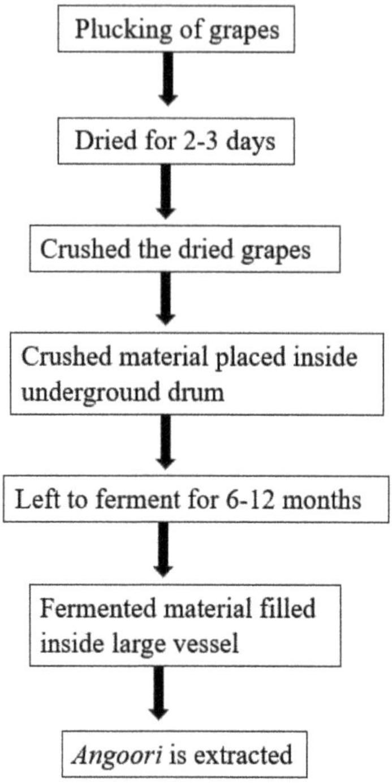

Fig. 14.3 Stepwise representation of the processes involved in the preparation of indigenous alcoholic beverage Angoori

Nesang Brandy

Nesang brandy is a popular alcoholic beverage which is predominantly prepared in Nesang village of Kinnaur district and is named after its place of origin. The village of Nesang is renowned for its unique customs and traditions, and the production of this brandy is a symbol to the villagers' connection to their land and heritage. The substrate used for the preparation of this famous beverage is barley grains (*Hordeum vulgare*). Barley is viable in high altitude and dry-temperate climate of Kinnaur region and thus, it is considered one of the staple crops and commonly utilized as primary ingredient in local food and beverages. Barley grains are boiled in large vessel for about 1–2 hours. After boiling, these grains are withdrawn from the vessel and dispersed on a woollen sari, which is locally called as *Dohru.* The spreading of grains helps in absorption of water from grains.

Fig. 14.4 **a** Barley grains used for the preparation of Nesang brandy (Image credit: Rahul Shom, Nesang) **b–c** Steam distillation method practiced by Kinnaura tribe of Kinnaur, Himachal Pradesh for the preparation of Nesang brandy (Image credit: Ashok Chaaras, Nesang) **d** Final distillation product of Nesang brandy

Later, *phab* is added as an inoculum, which is made up of millets and rice powder. It is a very popular traditional starter to ferment various alcoholic beverages in Trans-Himalayan region. The preparation of this indigenous starter primarily depends on the temperature and driven by microorganisms. It is easily accessible in the form of dried white cake (Kanwar & Bhushan, 2020). *Phab* is blend well with barley grains. This mixture of *phab* and barley grains is put inside jute bags. These bags are covered with the help of stones and allow to undergo fermentation at room temperature for one week.

On decomposition, jute bags release juice of barley grains. This mixture is then shifted to big drum. Lastly, the mixture gets ready for distillation after 12–15 days. The nutritive as well as therapeutic values of these products are elevated through the process of fermentation, which is crucial in boosting human health. Brandy is sold locally at Rs 1000/litre (Figs. 14.4 and 14.5).

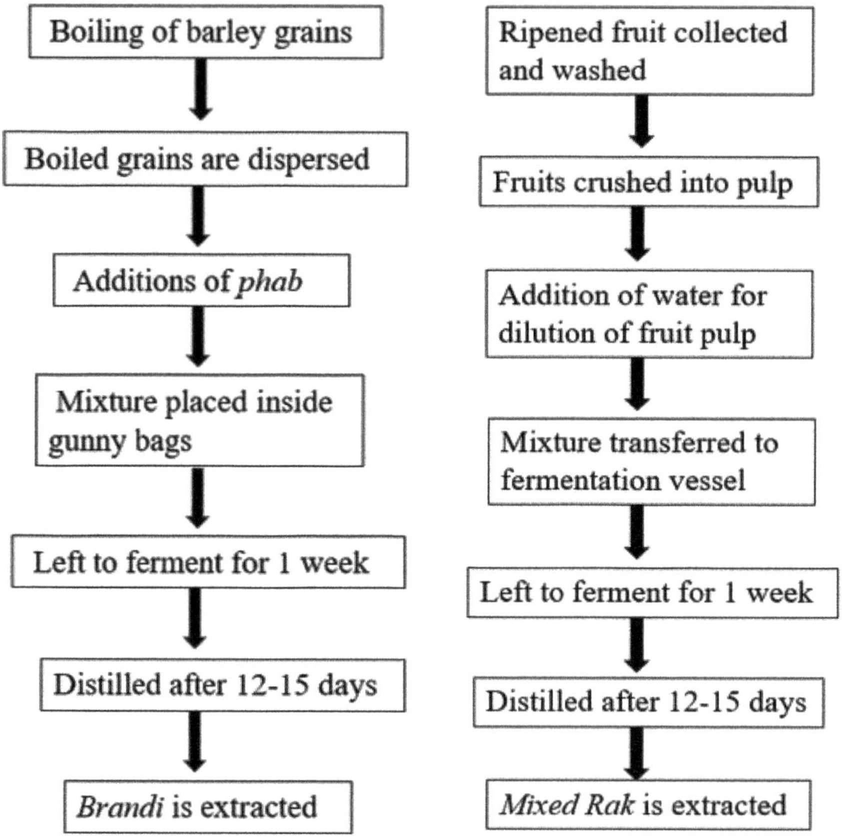

Fig. 14.5 Stepwise representation of the preparation processes for the indigenous alcoholic beverage *Nesang brandy* and *Mixed fruit rak*, respectively

Mixed Fruit Rak

Mixed fruit rak is a commonly available liquid liquor of Kinnaur, Himachal Pradesh. Fruits used in the preparation of mixed fruit rak include *Vitis vinifera* (grapes), *Prunus ceracifera* (cherry plum), *Prunus domestica* (plum), *Prunus armeniaca* (apricot), *Prunus persica* (peach), *Prunus mira* (smooth stone peach), and *Malus pumila* (apple) (Lata et al., 2021). The preparation of mixed fruit rak is predominantly associated with locally available seasonal fruits.

Ripened fruits are plucked, collected, and washed thoroughly for the preparation of mixed fruit rak. After washing, fruits are crushed properly into pulp inside a large container. Clean water is added to dilute the fruit pulp. For 1 part of fruit pulp, 1.5–2 parts of water is added for dilution. This mixture of pulp and water is transferred to a large fermentation vessel and allowed to undergo fermentation for

Fig. 14.6 **a** Ripened apples utilized in the preparation of mixed fruit rak. **b** Preparation of mixed fruit rak. **c** Steam distillation method used for the preparation of mixed fruit rak [Image credit: Ringchen chhomo, Kanam]. **d** Final distillation product of mixed fruit rak

7–15 days. The ambient temperature for fermentation lies between 12–32 °C. After fermentation, the fermented liquid is procced for distillation using a utensil, which is locally called *kun*. The final distillate product is clear, aromatic and ready to consume. Mixed fruit rak is sold locally at Rs 700/litre (Fig. 14.6).

Cultural Importance of Fermented Alcoholic Beverages in Kinnaura Tribe

Traditional or indigenous alcoholic beverages hold great cultural significance and are an integral part of customs and rituals among the Kinnaura tribe of Kinnaur district (Lata et al., 2021). In the tribal regions, social gatherings often involve the consumption of substantial amounts of alcoholic beverages (Tamang, 2001). In the Himalayas, distilled liquor is widely used by most ethnic communities both for social drinking and for religious rituals, including worshipping ancestors and deities (Tamang, 2022). Kinnaur is recognized for its distinctive and vibrant cultural heritage, which signifies a fusion of Hindu and Buddhist traditions (Kumari, 2025). The people of Kinnaur endure harsh conditions caused due to unfavourable weather and tough climatic conditions. As a result of environmental and cultural reasons, consumption of alcoholic beverages is widespread among people of Kinnaura tribe (Au - Singh, n.d.)

These indigenous beverages play a significant role during religious ceremonies in villages adding flavour to their rich cultural heritage. The local liquors have been a part of both Buddhist and Hindu ceremonies from ancient time including child birth, festivals, and marriages. In Buddhism, the locally made liquor is used as "Serkem" and in Hinduism, it is used as "Nimit" for performing tribal rituals. The locally available alcoholic beverages hold cultural importance in engagement ceremonies. The local beverage in the form of "Koring" is presented to the bride's family by the groom's family as a gesture of respect for marriage proposal. The engagement is finalize once the bride's family accepts the offering. As part of the tradition, the family worships the local deity in a ritual called 'Koringpajam' or 'Bottle Pooja' using the offered liquor (Lata et al., 2021). Apart from these rituals, the indigenous beverages are an integral part of various tribal festival like Losar, Mahang Sanga (Khawksa), Lamoth (Siskar), Shuparth (Faguli/Atharo), Beesh, Yakor, Fulech (Menthoko/Ukhayang/Dakhrane/Namgang).

Marriage rituals are believed to remain incomplete without the utilization of indigenous alcoholic beverages as they are the part of their cultural identity. These entire village of brides and groom family is served with alcoholic beverage as a gesture of respect and appreciation. A specially crafted copper vessel, locally called "Hurch" containing liquor inside it is decorated using butter and Chilgoza garlands. This vessel is used to ward off evil spirits.

Apart from marriages and festivals, alcoholic beverages are also associated with death ceremonies of community. A famous practice called 'Sheeza nu khu ran' is performed during the Fulaich festival, where indigenous beverages are offered to the dead soul as a symbol of gratitude and honour. All these examples shows that these traditional drinks are deeply embedded in the community's daily life, rituals, and seasonal practices, reflecting a rich heritage of indigenous knowledge and craftsmanship. The preparation and consumption of these alcoholic beverages at local level symbolizes hospitality along with communal bonding. These indigenous practices encourage social bonding and help in maintaining traditional knowledge systems (Fig. 14.7).

However, modernization has imposed significant effects on the culture of Kinnaura tribe (Kumari, 2025). Although, preparation of these traditional beverages does not require highly specialized skills, but proper handling of the ingredients and the storage techniques are crucial. If this indigenous knowledge is not passed down to future generations, it will be at the risk of erosion (Tomar et al., 2023).

Fig. 14.7 a–f Cultural significance of indigenous alcoholic beverages among the people of Kinnaura tribe in Kinnaur, Himachal Pradesh. **a–b** Significance of alcoholic beverages in marriage ceremonies, **c** Collection of fermented alcoholic beverages (Image credit: Chhatter Singh Shom, Nesang) **d–e** Significance of alcoholic beverages in Losar festival, **f** Significance of alcoholic beverages in tribal festivals

Conclusion

Traditional fermented alcoholic beverages hold a profound cultural, social, and religious significance among the Kinnaura tribes of Kinnaur, Himachal Pradesh. Rooted in centuries-old indigenous knowledge, these beverages are not only a vital part of their daily lives but also play an essential role in rituals, festivals, and social gatherings. The preparation methods, passed down through generations, reflect a deep understanding of local biodiversity and microbial fermentation processes, making these beverages unique to the region.

The diversity of ingredients and techniques used in brewing highlights the resourcefulness of the Kinnaura tribes in utilizing locally available fruits, grains, and other raw materials. Despite the socio-economic challenges posed by the region's remote geography, the preservation and continuation of these fermentation practices serve as a testament to their resilience and strong cultural identity. Moreover, the association of these beverages with religious ceremonies, from marriage rituals to death rites, underscores their intrinsic role in maintaining spiritual and communal harmony.

However, with modernization, changing consumption patterns, and the influence of commercial alcoholic beverages, there is a growing need to document and preserve these traditional practices. Scientific studies on the microbiology and health benefits of these beverages can further enhance their recognition and sustainability. Future research should focus on integrating traditional knowledge with modern brewing techniques to promote these indigenous beverages while ensuring their cultural authenticity.

The fermented alcoholic beverages of the Kinnaura tribes symbolize a rich heritage that intertwines tradition, spirituality, and biodiversity. Preserving and promoting these indigenous practices will not only safeguard an integral part of their cultural identity but also contribute to the sustainable utilization of local resources in the face of evolving socio-economic landscapes.

References

Au - Singh, Ty -. Jour. (n.d.) *SP–T1.*

Bhalla, T. 2007. *Traditional foods and beverages of Himachal Pradesh.*

Chawla, A., Kumar, A., Lal, B., Singh, R. D., & Thukral, A. K.(2012). Ecological characterization of high altitude Himalayan landscapes in the Upper Satluj River watershed, Kinnaur, Himachal Pradesh, India. *Journal of the Indian Society of Remote Sensing*, *40*(3), 519–539.

Kanwar, S. S., & Bhushan, K. (2020). Ethnic fermented foods and beverages of Himachal Pradesh. In *Ethnic fermented foods and beverages of India: Science history and culture* (pp. 189–208). Springer Singapore.

Kanwar, S. S., Gupta, M. K., Katoch, C., & Kanwar, P. 2011. *Cereal based traditional alcoholic beverages of Lahaul and Spiti Area of Himachal Pradesh.*

Kumar, R., Negi, N., & Changkum, V. (2023). Apple revolution in the tribal district of Kinnaur: From tradition to technology, horticulture redefined with quality assurance and sustainability. *International Journal For Multidisciplinary Research*, *5*(6).

Kumari, P. 2025. *Balancing tradition and modernity in Kinnaur, Himachal Pradesh, India.*

Lal, C., & Prasher, R. S. (2016). Traditional knowledge systems and conservation practices in tribal society of Western Himalayas: A case study of district Kinnaur of Himachal Pradesh. *International Journal of Farm Sciences*, *6*(3), 159–171.

Lata, S., Negi, P. S., Samant, S. S., Seth, M. K., & Shrama, S. (2021). *Documentation of Traditional Alcoholic Beverages and Their Indigenous Utilization Pattern by Kinnaura Tribes of Himachal Pradesh.*

Lata, S., Samant, S. S., Paul, S., & Negi, P. S. (2024). Festival of flowers (Fulaich): Unique traditional culture and custom of Kinnaura tribes in Himachal Pradesh, North Western Himalaya, India. *Indian Journal of Traditional Knowledge*, *23*(7).

Madilo, F., Kwashie, A., Kunadu, K. P.-H., Tano-Debrah, F., & Saalia, U. K. (2024). Diversity of production techniques and microbiology of African cereal-based traditional fermented beverages. *Journal of Food Quality*, *2024*(1).

Negi, S. 2017. *Religious beliefs, practices and change: A case study of Kinnaura tribe.*

Rawat, J. M., Pandey, S., Debbarma, P., & Rawat, B. 2021. Preparation of alcoholic beverages by tribal communities in the Indian Himalayan region: A review on traditional and ethnic consideration. *Frontiers in Sustainable Food Systems* 5 (July).

Roy, A. (2020). Traditional cereal-based alcoholic beverages of India: A rich source of unexplored microorganisms for potential health benefits. *IJBPAS*, *9*, 791–802.

Roy, B., Kala, C. P., Farooquee, N. A., & Majila, B. S. (2004). Indigenous fermented food and beverages: A potential for economic development of the high altitude societies in Uttaranchal. *Journal of Human Ecology (Delhi, India)*, *15*(1), 45–49.

Tamang, J. P., Okumiya, K., & Kosaka, Y. (2010). Cultural adaptation of the Himalayan ethnic foods with special reference to Sikkim. *Arunachal Pradesh and Ladakh.* ヒマラヤ学誌, *11*, 177–185.

Tamang, J. (2001). Food culture in the Eastern Himalayas. *Journal of Himalayan Research and Cultural Foundation*, *5*(3–4), 107–118.

Tamang, J. P. (2022). Dietary culture and antiquity of the Himalayan fermented foods and alcoholic fermented beverages. *Journal of Ethnic Foods*, *9*(1).

Tomar, S., Pant, K., Sharma, P., Sinha, S., & Mitra, D. (2023). Unravelling the hidden ethnic fermented treasure of the Himalayas - A review on the traditionally fermented beverages of the Northwest Indian Himalayan region. *Food Chemistry Advances*, *2*(100254), 100254.

Chapter 15
The Tribals and Rural Populations of India: Food and Livelihood Security

K. Murali Naik, M. P. Sukthija, U. V. Kiran, and Kamal Jaiswal

Abstract The plants known as "wild edible food plants" (WEFPs) are neither farmed nor cultivated, nonetheless during the times of food scarcity it is used as major vital food source for tribal people since they are close to nature. The major tribes in India, Kondh, Gond, Santal, Saora, Kolha, Shabar, Munda, Paroja, Bathudi, Bhuiyan, Oraon, Gadabas, Mirdhas, and Juang are the most numerous. The tribal people are dependent on the plants and animals for their food and other needs. They frequently gather and eat fruits, vegetables, tubers, flowers, mushrooms, and other stuff from nearby forests. They also know a great deal about edible wild plants. In this study, wild edible plants (WEPs) from various tribal dominated villages, Keonjhar, Mayurbhanj, Kalahandi, Bhitarkanika (Kendrapada), Rourkela (Sundargarh), Jeypore (Koraput), Rayagada, Ganjam, Gajapati, Nabarangapur, and Phulbani are identified, documented, and their food value is investigated. Despite their declining popularity, these wild food varieties are nutrient dense and should be used more often. It was discovered that a large variety of native foods, over half of which were leafy green vegetables, were excellent providers of micronutrients such as calcium, iron, beta-carotene, and folate. Maximizing the use of native foods can help this indigenous community's hidden hunger by providing a significant and long-lasting nutritional diversification strategy.

Keywords Wild edible food plants · Tribal's · Food security · Micronutrients · Hidden hunger

---

M. P. Sukthija
DSLD- College of Horticultural Engineering and Food Technology, Devihosur, University of Horticultural Science, Bagalkot, India

K. Murali Naik · U. V. Kiran (✉) · K. Jaiswal
Babasaheb Bhimrao Ambedkar University (A Central University), Satellite Centre, Amethi, Uttar Pradesh, India
e-mail: druvkiran@gmail.com

R. C. Sobti et al. (eds.), *Traditional Knowledge Systems in India: Sustainability, Ecology, and Resource Governance*, People, Cultures and Societies: Exploring and Documenting Diversities,
https://doi.org/10.1007/978-981-92-1346-7_15

Background

The safeguarding of the essential interaction between people and their surroundings is the fundamental survival paradigm. The growing disparity between the world's food supply and demand, coupled with the dual burden of diseases associated with both under- and over nutrition, point to the necessity of developing and putting into practice novel food-based methods that guarantee both nutritional quality and food sufficiency. Native American tribesmen have knowledge of their ancestral lands and food systems, which are based on regional historical continuity. Native foods are typically collected locally from the natural environment rather than being purchased, and they are then used in accordance with customary knowledge and wisdom.

Infant mortality and malnutrition among mothers and children are disproportionately higher in India's tribal tribes. The Indian tribal food system has a rich supply of natural foods, or traditional foods, which have been shown to improve the availability of food and nourishment. However, these neighbourhoods frequently suffer from nutritional deficiencies and poor health due to issues with geography, agricultural technology, cultural customs, lack of formal education, inadequate infrastructure, and poverty. In order to better understand the great diversity of their food systems and to take advantage of their agroforestry practices to enhance diversity in diets, state of nutrition, and availability of food, the indigenous people of Jharkhand offer an invaluable case study.

Aim:

The primary goal of this research was to assess how native foods can improve the nutritional status and nourishment of in danger tribal groups by increasing variety in diet and nutrient intake.

Specific Aims Include:

1. To describe the food environment and chart the accessibility, availability, and consumption of native foods by particular tribal groups.
    - Evaluate any changes in these tribes' eating habits as comparison to the native food intake.
    - Describe their farming practices in order to identify obstacles and chances to enhance the quantity of food produced locally.
2. Look at the obstacles and enablers to more Indian agriculture, food acquisition, and consumption.
3. To evaluate the nutritional content of commonly consumed native foods
4. To calculate the amount that native foods contribute to the populations' nutritional status and nutrient intake.

Introduction

Thirty states and union territories have designated Scheduled Tribes, although there are seven hundred and five separate ethnic groups have been designated as Scheduled Tribes. According to the 2011 census, the country's triple population is 10.43 core, or 8.6% of overall population. Of them, 10.03% reside in cities and 89.97% are rural residents. Compared to the overall population, which grew by 17.69%, the tribal population increased by 23.66% between the 2001 and 2011 censuses. In general, the Scheduled Tribes live in two different regions of India: Central India and the Northeast Region. The states of Madhya Pradesh (14.69%), Chhattisgarh (7.5%), Jharkhand (8.29%), Andhra Pradesh (5.7%), Maharashtra (10.08%), Orissa (9.2%), Gujarat (8.55%), and Rajasthan (8.86%) are home to the majority of Scheduled Tribe people in Dental India. Arunachal Pradesh, Assam, Nagaland, Mizoram, Manipur, Meghalaya, Tripura, Sikkim, and the Northeast constitute the other unique area. The percentage of Scheduled Tribes in each state is as follows: Mizoram has the highest percentage (94.43%), while Uttar Pradesh has the lowest [0.57%]. More people belong to Scheduled Tribes in around 17 states and 2 Union Territories than the 8.6% national average. Nearly 75% of the population of Scheduled Tribes resided in 49 districts of the nation as per the 2011 Census. According to the 2021 Census, there are 90 districts with a population of at least 50% Scheduled Tribes. In the 2001 Census, there were 75 districts in total. 48 of these 90 districts are located in eight states in the Northeast. More than 60% of people in Mizoram, Meghalaya, and Nagaland are members of Scheduled Tribes. Mizoram has the largest percentage of Scheduled Tribes among the states (94.43%), while Uttar Pradesh has the lowest percentage (0.57%). With 94.8% of its people belonging to Scheduled Tribes, Lakshadweep is the most populous Union Territory, while Daman & Diu has the lowest percentage (6.32%). The largest percentage of Scheduled Tribes (98.58%) is found in the districts of Korong Kamet in Arunachal Pradesh and Kanauji in Uttar Pradesh has the lowest population of Scheduled Tribes.

The perception of the livelihoods of tribal communities in India is a multifaceted, dynamic, and complex phenomenon that is influenced by various factors such as community type, age, gender, education, resource availability, infrastructure, and service quality, as well as social, economic, cultural, ecological, and political factors. For Native American tribes, agriculture is the main source of income. It is also crucial for the growth and survival of the nation, rural and urban development, employment and occupation, agro industries, food and nutrition security, growth and survival, and the reduction of poverty (Surayya et al., 2008). According to Chakraborty et al. (2009), almost 70% of the population is dependent on rain-fed agriculture, which is characterized by low productivity, erratic weather patterns and natural disasters, degraded soil with low fertility, ineffective irrigation, and depletion of natural resources. These factors made poverty, migration, unemployment, underemployment, food insecurity, and malnutrition worse for millions of tribal people living in India (Mourlin, 2007).

In spite of the fact that there are at least 3000 edible plant species known to science, just roughly 30 crops account for over 90% of the world's caloric intake, and only 120 crops have significant national economic importance (FAO, 1993). This demonstrates the multitude of species that different societies of humans continue to ignore or trash. Numerous plants that are edible are significant in terms of nutrition or other factors. One of the most abundant sources of protein in the world is Quinoa (Chenopodium quinoa), an Inca staple that is seldom known in the modern world (Reid & Miller, 1989). In the same way, many edible wild species and little-known crops are being unrecognized despite being essential to rural and tribal families' food security. For example, numerous families that live close to forests receive additional nourishment from the many wild species of Dioscorea, Colocasia, and Amaranthus, which are rich in vitamins and nutrients (Roy et al., 1998). Millions of communities that depend on forests also benefit financially from wild food. In India, people who gather goods for the market, such as honey, garcenia, and gooseberries, mostly rely on this as their source of income (Muralidharan et al., 1997). Many tribal and rural households in Kerala's Wayanad region still preserve knowledge about a variety of species for a variety of purposes, including dietary requirements (Anilkumar et al., 2008; Hema et al., 2006; Narayanan et al., 2011), medicinal purposes (Mathew & Unnithan, 1992; Mini & Sivadasan, 2007; Narayanan et al., 2011; Nisha & Sivadasan, 2007; Shanavaskhan et al., 2011), uses pertaining to spiritual and supernatural convictions (Pramod & Sivadasan, 2003), and various other purposes (Narayanan et al., 2011). Consuming wild food plants appears to be one of the key local survival strategies for some of these communities, and many of these species are not only regularly consumed as part of their regular diets but also during times of drought, food scarcity, and other adversity (Narayanan et al., 2011)

India is always known for its natural green cover and rich biodiversity. Variations in topography and climate have resulted into highly diverse forest ecosystems. Along with biotic and abiotic components, the 'forest people' have also been an integral part of these ecosystems. These people are completely amalgamated with the nature. Their livelihood, accommodation, food and many other requirements are satisfied by forests. Traditional crops and various parts of wild edible plants like fruits, flowers, leaves tubers, roots form an essential part of the tribal diet and play a major role in their nutrition. These wild plants in addition to meeting their food requirements satisfy the vitamin and mineral necessities. Most of them have medicinal properties. Thus, it can be said that though the term Nutraceutical has been coined recently, Indian tribals have been practicing them since the time immemorable.

Medicinal importance of most of these plants is known to the women of higher age group. As they go to collect vegetables, they have knowledge about their medicinal and nutritional values. But, it was observed that, this awareness has been decreasing as the generations pass. Several researchers have tried to find reasons for the loss of traditional knowledge.

One of the reasons could be, that Post British Era, India was in need of both food production as well as productions that provided raw material for industries, such as cotton, sugarcane etc. Green revolution brought in new ideas such as increased use of water, fertilizers, pesticides and high yielding varieties to increase production. Preference given to high yielding varieties resulted into decreased use, and finally disappearance of traditional crops, such as varieties of legumes, etc. (Gadgil & Guha, 2000) which contributed in the fulfilment of nutritional requirement, also started disappearing, resulting into situations such as nutritional deficiencies.

into situations such as nutritional deficiencies. Several researchers have documented the wild edible products used in the diet by tribes in various parts of India. (Sundriyal & Sundriyal, 2004; Jadhav et al., 2011; Rasingam, 2012; Rekha & Kumar, 2014). Additionally, these studies have emphasized the nutritious potential of these edible wild plants. They also emphasized how important it is that this expertise be revived. In addition to listing 50 wild edible plants from Maharashtra's Kolhapur area, Jadhav, Mahadkar, and Valvi also provided information on the food and medical uses of these plants. Deshmukh and Waghmode (2011) investigated 11 fruit plant species for their nutritional and therapeutic value and presented the traditional knowledge about the significance of wild edible fruits in the Western Ghats of Maharashtra as a food resource. The significance of certain fruits as dietary supplements was emphasized. Medicinal applications of these plants mentioned in classical Ayurvedic texts were already known to elderly women and are being practically implemented in their day to day lives. Deworming property of Shevga (Moringa oleifera), Takla (Cassia tora), Neem (Azadirachta indica), blood purifying property of Palash (Butea monosperma), antidiarrhoeal property of Pimpal (Ficus religiosa), Murudsheng (Helicterus isora), Kuda (Holarrhena antidysentrica), etc. were already known to them and they consciously use them in diet. Younger generation seems to be negligent about this and further, they are getting attracted towards modern junk foods like chips, burgers, etc especially in yeoor forest (Table 15.1).

Significance of ethnic foods of Northeast India:
1. Soybean based ethnic foods:

A. **Mizo tribe:** The Mizo tribe refers to the fermented dish made from soybeans as bekang. The soybean (bekang) seeds are ground, cooked for approximately thirty minutes, and then removed. To remove the excess water, boiled seeds are stored in bamboo plates. The belphuar tree's leaves eventually envelop the lukewarm seeds. Since they are said to be effective in regulating temperature during fermentation, only the leaves of the belphuar tree are used to wrap the bekang. The wrapped package is set on the bamboo shelf next to the kitchen hearth, and after three days the fermentation is finished. It can be eaten straight away or served after adding the appropriate amount of ginger and local chili. It is an essential component of all boiling dishes made with leafy vegetables. In some other areas in Mizoram, the tribal women use a different method of processing where they combine boiling seeds with ash and then wrap the mixture in belphuar tree leaves. Fermented soybean seeds are spread out in a sieve and dried in the sun to extend their shelf life. The dried seeds are then packaged into a bamboo container and sealed. It is removed, ground into

Table 15.1 List of wild edible plants used by tribals for their medicinal value (*Source* Gayatri Oak et al., 2015)

	Botanical name	Family	Part used	Uses
1	*Celosiaargentea*	Amaranthaceae		Treating kidney stones. Stems and leaves of celosia, bruised and applied as poultice, is used for treating of infected sores, wounds and skin eruptions Poultice of leaves, smeared with honey, used as cooling application to inflamed areas and painful affections such as buboes and abscesses Stools, hemorrhoidal bleeding, leucorrhoea and diarrhoea
2	*Cleome viscosa*	Capparidaceae	Leaf	Leaf paste on the wound
3	*Colocasia esculenta*	Araceae	Leaves, tuber	Leaves are cooked and Eaten as vegetable. Haematinic
4	*Cordia dichotoma*	Boraginaceae	Fruits, bark	Respiratory tract infections, Antihelminthic, Cough, Fruits consumed
5	*Curculigo orchioides*	Amaryllidaceae	Tubers	Urinary diseases, astonic
6	*Curcuma aromatica*	Zingiberaceae	Roots, stem	Antibiotic properties, helps indigestion
7	*Derris trifoliata*	Papilionaceae	Leaves	
8	*Dioscorea bulbifera*	Discoraceae		Anticancer, Weight gain
9	*Eleusine coracana*	Poaceae	Roots,	Used as staple food, root infusion for Abdominal distension
10	*Ficus racemosa*	Urticaceae	Leaves, fruits, bark, gum	Gum for treatment of acidity, stringent, antidiabetic, antiasthmatic, anti-inflammatory, antioxidant, antiulcer, anti-Pyretic and anti diarrheal in action
11	*Ficus religiosa*	Urticaceae	Bark, seeds, fruits, latex, leaves	Fruits consumed diarrhoea gastric problems, skin diseases
12	*Gloriosa superba*	Lliliaceae	Roots, bulb	Used in release of placenta, leaves used to Cure asthma in children
13	*Grewia tiliaefolia*	Tiliaceae	Fruits, bark	Fruits in digestive disorders, bark to Prevent hair fall
14	*Haldina cordifolia*	Rubiaceae	Fruits, leaves	The fruit treats dehydration, anti fungal use And acts as a coolant.
15	*Helicteris isora*	Sterculiaceae	Bark, fruit	Digestive infections in children
16	*Hibiscus cannabinus*	Malvaceae	Leaves, stem	Consumed as vegetable, fattening, purgative, and stomachic. Stem

(continued)

Table 15.1 (continued)

	Botanical name	Family	Part used	Uses
				peelings for anaemia, fatigue, lassitude In Gambia, the leaf infusion is used for coughs. leaves for dysentery and bilious, blood and throat disorders
17	*Holarrhena antidysenterica*	Apocynaceae	Fruits	Diarrhoea, Dysentery
18	*Holoptelea integrifolia*	Urticaceae	Leaves	Anti diarrhoeal property
19	*Hyoscyamus niger*	Solanaceae	Seeds	Oil is extracted from seeds
20	*Lagerstroemia nana*	Lythraceae	Bark, leaves	Purgative a stringent fruit
21	*Lannea coromandelica*	Anacardiaceae	Bark	Digestive infections, sour eyes, leprosy
22	*Macaranga peltata*	Euphorbiaceae	Leaves, bark	Decoction of leaves and bark used a svulnerary.
23	*Madhuca india*	Sapotaceae	Fruits	Urinary tract infections, Menstrual irregularities, Skin diseases
24	*Mangifera indica*	Anacardiaceae	Fruits, leaf	Urinary tract infections, Skin diseases, Anaemia, Bleeding
25	*Momordica dioica*	Cucurbitaceae		The roots are used in head trouble, treating urinary calculi. The leaves having aphrodisiac and anthelmintic properties
26	*Morinda citrifolia*	Rubiaceae		Analgesic, anti inflammatory, antioxidant, detoxifier
27	*Morinda tinctoria*	Rubiaceae	Leaf, roots, fruits	Diarrhoea, Dysentery
28	*Moringa oleifera*	Moringaceae		Wormicidal, Anti cancer
29	*Oroxylum indicum*	Bignoniaceae		Diarrhoea, fever, ulcer and jaundice
30	*Peucedanum grande*	Apiaceae		½ to 1 ounce like that of fennel seeds, as carminative, diuretic and stimulantinflatulency, gastric and Intestinal disorders etc
31	*Emblica officinalis*	Euphorbiaceae		The juice of amla fruit reduces burning sensation of skin. It strengthens the hair follicles and reduces the inflammation of scalp skin. Amla boosts immunity of skin and helps to prevent acne and pimples

(continued)

Table 15.1 (continued)

	Botanical name	Family	Part used	Uses
				Amla fruits help to reduce burning sensation in eyes and help to maintain health of eyes
32	*Phyllanthus niruri*	Phyllanthaceae	Roots, leaves, fruits, latex	Bronchitis, leprosy, anaemia, urinary discharge, anuria, biliousness, asthma
33	*Pongamia pinnata*	Papilionaceae	Twigs, seeds	Skin diseases, piles, worm infection
34	*Pterocarpus marsupium*	Papilionaceae	Leaves	Antidiabetic, leaves used to treat sores, boils and various Other skin diseases
35	*Rotheca serrata*	Verbenaceae	Leaves	Malays give a decoction for colic. Flowers eaten during childbirth. India, a decoction of root with ginger & coriander for nausea. Tender leaves eaten as a vegetable. Dutch Indies, roots eaten to clear the voice. Leaves. eaten during labour. Extract given for distended stomach. Ripe& unripe fruits chewed for coughs. Leaves. given to cattle for stomach diseases. Tender immature infl. eaten as a bitter seasoning, raw or toasted. Malays use it mostly externally. A poultice for ringworm, leprosy, headache, persistent fever
36	*Sterculia urens*	Sterculiaceae	Bark, seeds, gum	Trees exude gum karaya used in food stuffs as emulsifiers, stabilizers and thickeners. Seeds are eaten after roasting. Seeds and young tender roots are eaten in times of famine. The gum, leaf and bark of Karaya is used for wound and Throat infection
37	*Syzygium cumini*	Myrtaceae	Fruits	Wormicidal, Diabetes
38	*Tamerindus indica*	Fabaceae	Fruits	Rehydration, Improves digestion
39	*Tectona grandis*	Verbenaceae	Leaves, bark	Anaemia, inflammatory swellings, skin itching, Dysentery and Wood is good for headache, biliousness, burning sensation pain, and liver related troubles
40	*Terminalia bellerica*	Combretaceae	Fruit	Chronic diarrhoea and dysentery
41	*Terminalia chebula*	Combretaceae	Fruit	Purgative astringent fruit

(continued)

Table 15.1 (continued)

	Botanical name	Family	Part used	Uses
42	*Vernonia anthelmintica*	Asteraceae	Seeds	Skin diseases, fever, sore throat
43	*Wrightia tinctoria*	Apocynaceae	Fruit	Fever, stomach ache, skin diseases especially psoriasis and non-specific dermatitis
44	*Ziziphus jujuba*	Rhamnaceae	Fruit, bark	Cough, Mouth cleanser, Improves digestion
45	*Ziziphus rugosa*	Rhamnaceae	Bark, root, fruit	Bark, root and fruits are used for treatment of carbuncle, menorrhea, syphilis and ulcerated tongue

a paste, and kept inside the leaves of the belphuar tree after ten to twelve days. Through the official MHIP network (Mizo Hmeichhe Insuihkhawm Pawl), the women of the Mizo tribe sell this product.

B. **Galo tribe:** The Galo tribe women of Basar, Arunachal Pradesh, are skilled at making aagya, a fermented dish made from soybeans. The desired amount of soybean seeds are cooked for approximately thirty minutes in an aluminium container. To remove the water from boiled seeds, Oko leaves (Phrynium pubenerve Blume, family Maranphaceae) are used. Next, for 12 to 14 days, the seeds firmly enclosed in the Oko leaves are stored in the rapkho, a bamboo shelf, over the kitchen hearth. After that, it is removed and kept in a bamboo basket called an Udu. Now that the Aagya is prepared, it is combined with a variety of regional veggies, chutney, and cooked meals.

C. **Adi tribe:** The fermented food known as peron namsing is made by the Adi women of the East Siang district using soybean seeds. After being washed and cleared of chaff, soybean seeds are cleaned. The cleaned seeds are cooked for thirty to forty minutes, or until they are very tender. They are spread out across Epo to chill (bamboo mat). The seeds are placed into the leaves of the ekkam plant (Phrynium pubenerve, family Maranphaceae), and a bamboo or cane rope packet is created. It is then hung in the kitchen, on the bamboo-made peron shelf, to ferment. The seeds are removed after 10–12 days (in the summer) or 25–30 days (in the winter) of fermentation. Salt, local chili, fermented bamboo shoots, and a variety of boiling veggies are combined with these fermented seeds, known as peron namsing. After shaping it into a cake or ball and placing it close to the kitchen fire, some women proceed to process it further to create a paste and store it in ekkam leaves. Women from Adi community sell double-processed peron namsing in Pasighat market. It is rare to make the chutney from peron namsing. To make it, grind up fermented soybean seeds and combine them with local tomato, ginger, and chilli paste (preferably the Sibol variety), along with a splash of salt.

D. **Apatani tribe:** The Lower Subansiri district's Apatani plateau, home to the Apatani tribe, is renowned globally for its sustainable and environmentally friendly rice-fish farming techniques and associated food web. They also eat a broad variety of diverse ethnic dishes composed of forest and agricultural products. A fermented product made from soybeans known as yanni perung is highly well-liked. The seeds are prepared by cleaning, washing, and boiling in water for approximately half an hour. The kulu leaves (wild banana, Musa balbisiana Colla) are filled with lukewarm seeds, sealed tightly, and put over a fireplace to smoke. In summer, the seeds ferment in 3–4 days, but in winter, it could take up to 12 days. The fermented seeds are removed and combined with tomato, ginger, garlic, and chilli to make chutney and salt. Once dried and ground into a paste with dried ginger and red chili, it can be stored for several months.

E. **Sherdukpen tribe:** Arunachal Pradesh's Sherdukpen tribal women prepare chukchoro, a soybean-based dish, in the Shergaon (1981 m above MSL) area of the West Kameng district. To remove the husk and damaged seeds, soybean seeds are steeped in water. The seeds are then cooked in water for one to two hours, until they are quite tender. After the water is removed, lukewarm seeds are stored in banana leaves and covered with warm cotton fabric. During the winter, this package is stored in the bamboo-made shelf next to the fireplace for 15 to 30 days. The fermentation takes ten to twelve days to finish in the summer. Once the seeds start to release flavour, they are deemed fermented and removed from the basket. Once combined with various local veggies, it can be eaten straight away. It's formed into a paste and kept in banana leaves for extended preservation. In chutney, it's combined with salt, tomato, yak cheese, and the native chili, called solu. Once shaped as a ball or square and wrapped in banana leaves, women sell this product at the neighbourhood market.

2. Bamboo based ethnic foods:

A. Apatani tribe: Native bamboo shoots are used by Apatani women in several Ziro district villages to make a variety of fermented food products, including Hikhu, Hiring, and Hithyi.

Hikhu: Women gather the bamboo shoots from the jungle and give them a thorough wash before peeling and slicing them. Once the ekkam or banana leaves have been placed within the bamboo basket, they are cut into small pieces and moved within. Banana leaves are securely wrapped around the basket, and it is kept for fermentation for more than 6 to 8 days. Hikhu is the term for these fermented bamboo shoots. These fermented bamboo shoots are called Hikhu.

Hiring: The bamboo shoot sections are sliced thinly and then sealed airtight with ekkam leaves within a bamboo cylinder. After about a week of fermentation, the bamboo shoot-filled cylinder is ready for use. This approach produces fermented bamboo shoots with a greater flavour than the one using the earlier technique.

Hithyi: After being cut into slices, the bamboo shoots are sun-dried and kept in a bamboo basket. The name of this product is Hithyi.

B. Adi tribe: Since it was originally created by Adi women through casual experiment at different altitudes in the East Siang region, as well as in Rayang, Bolung, Jiya, and model villages of Roing in the Dibang valley area, the fermentation process of bamboo shoots demonstrates tremendous variety. The Adi women from numerous villages, including Balek, Napit, Renging, Mirbuk, Mirku, Yabgo, Jarku, Anini, and Roing, prepare the bamboo shoots into three major shapes. One is called ekung and is fermented in a variety of ways; the other is called eyup and consists of dried shoots; the third is called eting and is consumed fresh, boiled with other native green vegetables.

3. Ekung (Fermented bamboo shoot: Women from the forested areas harvest the required quantity of shoots of bamboo, which are then peeled and sliced into thin slices. After that, the slices are layered on top of the ekkam leaves and placed within the papur, an airtight bamboo basket. The bamboo shoots, which are subsequently termed ekung, are then allowed to fully ferment in the basket for 5–7 days in the summer and 8–10 days in the winter. The required quantity of bamboo shoots is harvested and added to the edung, a cylinder fashioned from the local variety of bamboo eye that is open on one side, in the second processing procedure.

The slices are well pressed and sealed using the ekkam leaves. The shoots are ready for use if the manure is left to ferment for more than four to five days. With this method, the fermentation process moves along faster than with the previous one, regardless of the season. Until the 1980s, the fermentation of bamboo was processed using a unique technique to reduce the cyanide content. Cyanide13 causes a variety of neurological system diseases, miscarriages, abnormal births, and goiter problems. For this reason, older ladies still tell expectant mothers not to eat anything made of bamboo unless it has been well processed.

When the bamboo shoots were being examined a few tiny holes were made in the edung to reduce the proportion of cyanide. In order to allow water to contact the dung's bottom during the fermentation process, these dungs are situated close to rivers or streams. As a result, the dangerous compounds drain out. It also helps to maintain a steady temperature in the edung. Over time, the younger generation has been rejecting this conventional wisdom at an alarming rate. This process yielded incredibly tasty and well-preserved bamboo shoots in comparison to earlier methods.

4. Processing using stones in water stream: Bamboo shoots are harvested from the forest by the Adi women (a Padam sub-tribe) of Roing, who then peel and thinly slice the shoots using a Dao, a tiny farming tool like a sword. In a bamboo basket, ekkam leaves are used to produce these slices. Encased in ekkam leaves, the small parcels of partially fermented shoots are sealed with cane or bamboo rope. Then, for three to four months, these packets are positioned beneath the stones near the stream that flows out of the hills' peak to reduce the bitterness. Additionally, this significantly helps in lowering the concentration of cyanogenic glycosides due to the action of bacteria and yeast such as Lactobacillus lactis, which causes serious issues with the nervous system, goiter, and miscarriage. mixing with a variety of local vegetables In ethnozoological recipes, fermented bamboo shoots are also used.

One major problem with small fish is that they often split into tiny pieces when being cooked. Adi women use ekung to improve and decrease its flavour. Ekung is a mainstay of the Adi tribe's cuisine and goes well with leafy vegetables, pig, and mithun meat (Bos frontalis Gaur). Ekung is used in a variety of traditional ceremonies and events, such as Solung, Etar, and Aran. Fresh raw bamboo shoots, chopped into small pieces, are also mixed with different leafy vegetables from the jungle, insects, and meals from ethnozoology. For example, dried bamboo shoots (tenga) coupled with a caterpillar that feeds on the green leaves of the Bengi tree (Cassia siamea Lam.) are boiled and eaten.

It is recommended for a quick recovery for the weak individual. The Galo tribe, who share this region, prepares this meal in a unique way. The caterpillar known as talo is cooked with coarse rice powder and eaten with some salt and chilli added after the rice grains have been broken up. This dish is called ammin. If there is an excess, the caterpillars are wrapped in ekkam leaves and cooked over a hardwood fire. It is then called as pamnam and is left over the hearth to dry. To produce the chutney, the dried caterpillars are crushed and mixed with local ginger paste, salt, and chili powder.

A. Manipur: Meitei community: The women take the newly emerging bamboo shoots, peel and cut them into small pieces, and then let them dry in the sun for ten to fifteen minutes. These women think that the inherent smell and taste of the bamboo shoots will be lost if they are cleaned with water. The dried bamboo shoot pieces are kept in an earthen container filled with black clay soil after a small amount of water and salt (thum) are added. This pot is stored for more than two or three months once it is sealed. After fermentation, bamboo shoots are taken out of the anaerobic environment and sun-dried until they have about 50% moisture content.

Then, it is mixed with a small amount of thum and thuntak (a bamboo or cane basket), and left to dry completely in the containers. Women used to also add turmeric powder in the past. This fermenting shot is ready to be served. Some of the ladies in the same area use somewhat different techniques for fermentation; for instance, they press the shoots correctly before putting them in the clay pot and discarding the extract. The goal of this process is to lessen the bitterness and diseases caused by cyanogenic glucosides. In hilly areas, some tribal women ferment bamboo shoots in enormous bamboo receptacles known as cots, which are nearly 2.13 meters high.

Because bamboo shoots require a specific amount of water to ferment, a large bamboo container is required. After combining with water, place the banana's green leaves inside the container before adding the bamboo slices. Once more, it is sealed tightly with a native type of banana leaf and allowed to ferment for two to three months. It is then prepared for serving. Like Nigol chakouka, the fermented bamboo stalk is more well-liked during festivals.

B. Meghalaya: Khasi tribe: the traditional dish fermented from Bambusa nutans. Lungseij, the way Khasi women call their former Murro. After being properly cleaned, peeled, and sliced into little pieces, bamboo shoots are harvested from the wild. After that, the parts are put inside a glass container or a conventional bamboo basket. Banana leaves are in the basket. The fermenting process is done in a bamboo basket if the fermented shoots will be consumed shortly; jars are used if the shoots will be preserved for a longer period of time. Water is applied to the shoots once they have been placed in the jar or basket. The shoots are allowed to ferment for more than 15–20 days in the winter and 7–10 days in the summer. It's then prepared for consumption. These fermented shoots are also used to make pickles. The Khasi tribe enjoys a number of regional delicacies prepared using bamboo shoots. The curry is made with fermented bamboo shoots and tungtep (little dry fish) and ktungmluh (large dry fish).

C. Tripura: Barman community: Compared to other tribes in the Northeastern region, the tribes of Tripura are not as fond of fermented food dishes made from bamboo. The women of the Barman community make Godhak, a traditional dish made from bamboo shoots. The borak variety of fresh bamboo shoots is harvested from the jungle, peeled, and chopped into little pieces. Banana pseudostems are gathered simultaneously, and little pieces are created when the epidermic layers are peeled off. These two ingredients are combined and heated to a soft boil in water. To enhance the flavour, add the dry fish, salt, chili, onion, and garlic pieces after ten to fifteen minutes. After this, Godhak is prepared to serve.

5. Leafy mustard (lai saag) based ethnic foods:
A. Nepali community of Mebo: Nepali women have developed the empirical knowledge to ferment several leafy vegetables for long-term use and preservation in the Adi community in Mebo, East Siang district, Arunachal Pradesh. In this regard, the most popular dish is gundruk, or Lai Saag's fermented meal. The required quantity of Lai saag (Tulang and Tuka varieties) must be taken, washed, and ground up properly in order to prepare it. Once inside, it is crushed with stones and placed inside a Doko, a bamboo container with holes on the bottom. Five to six days are allowed for this to ferment. Following that, the fermented leaves are taken out, crushed one more, and then spread out to dry for two to three days on a Dhari, or bamboo mat.

Once it is fully dried, it is called gundruk and is used as chutney or with curry. While chutney is made using gundruk, tomato, dried chile, ginger, garlic, and salt, curry is made with potatoes. Mixing gundruk with Dhedo, which is created by baking wheat or maize flour in a skillet over low heat and then gradually adding water, is the most popular way to make gundruk for breakfast. Stir constantly while baking and adding water for 30 to 40 minutes. Dhedo is the name of the last item.

B. Bhutia tribe of Sikkim: To produce Gundruk, an ethnic food, the women of the Bhutia tribe in Sikkim ferment lai saag using several methods. Before preparing Gundruk, the lai saag leaves are thoroughly cleansed, then crushed and squeezed to remove any leftover water. There are two ways to go about doing the fermentation.

(i) Fermentation in pit and soil: The size of the hole that is dug depends on how much lai saag needs to be fermented. The crushed green leaves are then put inside a tin box. Banana leaves are near the bottom of the box. Once the crushed green lai saag leaves are inserted, it is covered with banana leaves and squeezed to form an airtight seal. After inserting the tin container into the hole and adding dirt, it is left to ferment for a month. After being taken out of the tin container, the fermented lai saag is spread out onto a Nanglo (bamboo plate) and allowed to dry in the sun. After that, the banana leaves are removed and the tin container is removed. The dried gundruk keeps for two years in storage.

(ii) Fermentation in cow dung: This procedure has a faster rate of fermentation than the previous method. After being crushed, the green leaves of lai saag are tucked between two sheets of polythene and secured with thread. Within the cow dung, this package is fermented for longer than twenty days. After this period of time, the polythene sheet packet is taken out of the cow manure and carefully cleaned. Gundruk is taken out, spread out on the nanglo, and exposed to the sun to dry. In order to improve the efficacy of the nursing mother's milk, the Bhutia women use its soup. This fermented ethnic meal is suggested for senior citizens as a health tonic. In addition, it's utilized to prepare chutneys, curries, and other dishes.

6. **Rai (Brassica juncea) seeds based ethnic food:** The Bhutia tribal women in Sikkim also ferment the rai seeds. After being removed and cleaned, a sufficient amount of rai seeds are added to water and brought to a boil. The seeds are crushed with stones and kept in lukewarm containers to keep them impermeable. For two to three months, the pot is left to let the seeds to ferment internally. Many meals contain fermented seeds to aid with digestion, reduce gas, and ease stomach aches.

7. **Radish based ethnic food:** The Gurung tribe of Sikkim prepares sinki, a traditional dish made from fermented mula (radish). Radishes are cleaned and cut into little pieces, then sun-dried in a naaglo (a bamboo device used in the area to winnow grains) for three to four days. The desiccated portions are placed within a well and covered with plastic. The pit is covered with cow dung and earth paste and left to ferment for more than fifteen days. Soup is made using sinki, which are fermented root segments. Taken during the lean period, it is said to be especially helpful in alleviating diarrhoea and stomach pain.

8. Tree bean based ethnic foods:

A. **Meitei community:** The Meitei people, who inhabit the highland ecosystems of Manipur, have long been known for making a traditional dish made with tree beans, or Parkia roxburghii G. Don. The delicate pods of tree beans are used to make chutney, which has been shown to help with digestive problems. Iromba is one of the Meitei community's favourite delicious ethnic dishes. There are various types of Iromba depending on the materials that are used. The most popular is the seasonal Yongchak Iromba, available from winter to spring and made of tree beans. It is a necessary element of all feasts and festivities, such as Chhinaoba and Nugol Chakouba. The Christian community among the hill tribes also approves of this. The essential ingredients of Yongchak Iromba are Phutunis or Eromus dandricus, dry fish of the Puntus breed, red-skinned potatoes, Pullei (Alpini nigra (Gaertn.) Bentt), chilli, salt, and coriander. All the vegetables are prepared by boiling them after a container is filled to the full with water. For a vegetarian cuisine, fry Maroi Nakupi (Allium odorum Linn.) or soak nagari, and then mix a paste with salt and chile. Tree bean pods' green skin is removed with an iron tool called a yoingkhot. Peel the potatoes and tree beans, then thoroughly mix the contents with the prepared elements, coriander included. The elderly women of the Meitei community are familiar with this dish, which is regarded as an appetizer and represents their ethnicity.

Rural Foods Consumed in Different Parts of India:

Indian rural food ensures balanced diet and dietary diversity. It is usually plant based food rich in nutrition with immense health benefits. Rural food incorporates variety of food from different food groups like pulses, grains, cereals, vegetables etc. that can help provide essential nutrients and promote overall well-being. The food preparation methods differ from region to region, and play a significant role in determining the nutritional quality of rural Indian diets. The various components of rural Indian food is tabulated as follows (Table 15.2).

India is rich in culinary history and rural Indian food is traditional food shaped by centuries of history, regional influences, religious beliefs, and cultural practices. Each state in India has its own unique flavors, ingredients, and cooking techniques based on the climatic conditions & beliefs of that region. Coastal southern regions are rich in sea food varieties whereas northern regions are known for their spicy curries. States of Gujarat and Rajasthan are famous for vegetarian specialties and delicacies.

Vegetarianism has a long history in India, influenced by religious and cultural practices such as Hinduism, Buddhism, and Jainism. Many Indian households, particularly in the southern and western regions, have predominantly vegetarian diets. Vegetarian dishes are often celebrated for their rich flavours and variety, featuring ingredients like lentils, vegetables, dairy products, and spices. Following table depicts some traditional and very famous rural foods from across various states in India (Table 15.3).

Table 15.2 Different types of rural foods consumed in India

1	Milk Products	Milk, yogurt (dahi), and homemade cheese (cottage cheese/ paneer), buttermilk are commonly consumed in various rural areas of India. The consumption of these milk products is done either in raw form or as part of ingredient in recipes. They provide protein, calcium, vitamin D, and other nutrients essential for bone health and overall well-being
2	Vegetables	Rural food often include a variety of vegetables which are locally grown. Vegetables such as potatoes, onions, spinach, eggplant, okra, tomatoes, peas and various gourds. These vegetables contribute dietary fibre, vitamins (A, C, K), minerals (potassium, magnesium), and antioxidants to the diet
3	Fruits	Indian weather is suitable for growing various seasonal fruits which are consumed as part of diet. Fruits like mangoes, bananas, guavas, papayas, and citrus fruits are abundant in many rural regions. They are rich in vitamins, minerals, fibre, and antioxidants, promoting overall health and immunity
4	Grains &Cereals	Staple grains like rice, wheat, millets (such as bajra, jowar, and ragi), and pulses (lentils, chickpeas, beans) are significant sources of carbohydrates and plant-based protein. They also provide essential micronutrients like iron, zinc, and B-vitamins
5	Pulses (Dal)	Dals as they are called in India, are consumed on daily basis. lentils, chickpeas, and beans are staple sources of protein in rural Indian diets. They are also rich in fibre, iron, calcium, vitamins and minerals
6	Nuts & Dry Fruits	Nuts such as almonds, cashews, and peanuts, as well as seeds like sesame and pumpkin seeds, are often included in rural diets. They provide healthy fats, protein, fibre, vitamins (E, B-complex), and minerals (magnesium, zinc)
7	Oils	Cooking oils such as mustard oil, peanut oil, and coconut oil are commonly used in rural Indian cooking. While fats are essential for energy and absorption of fat-soluble vitamins, it's important to use them in moderation to maintain overall health
8	Sweetening Agents	Jaggery (unrefined cane sugar), honey, and palm sugar are traditional sweeteners used in rural Indian desserts and sweets. While they are high in calories and provide energy, they should be consumed in balanced way
9	Spices (Masala) And Herbs	Indian cuisine is renowned for its use of spices and herbs such as turmeric, cumin, coriander, ginger, garlic, and chili peppers. These ingredients not only enhance flavour but also offer various health benefits, including anti-inflammatory and antioxidant properties

Table 15.3 Table Traditional foods consumed in Villages foods of India

Sr. no.	Ethnic Food	Ingredients	Preferred location
1	LittiChokha	Roasted wheat dough ball stuffed with roasted gram powder and spices served with mash of roasted eggplant, tomatoes, potatoes with green spices	Uttar Pradesh, Bihar
2	Dal BatiChurma		Rajasthan
3	kafuli	Mix of spinach and fenugreek	Uttarakhand
4	Tehri (UP & Bihar) Kootanchoru (Tamil nadu) Masalebhaat (Maharashtra) Khichadi (Gujarat, Rajasthan) BisiBele Bath (Karnataka)	Rice, lentils and vegetables mixed with local spices	
5	Sarso da saag	Green leafy vegetable	(Punjab, Haryana)
6	Millets (jowar, Bajra, makka—corn)		
7	Ulundhu kali laddu	Black urad dal, rice, jiggery, sesame oil	Tamil Nadu
8	Akki roti (Karnataka)	rice flour, vegetables	
9	Upsaaru (Karnataka) Letha (Orissa)	lentils with vegetables and spices	
10	GoanKhatkhate (Goa)	coconut curry with lentils and vegetables	
11	KambuKoozh (Tamil Nadu)	Pearl millet kheer	
12	GongaruPachadi (Andra Pradesh)	Sorrel Leaves Pickle	
13	AamPoraSharbat (Bengal) AamPanna (Maharashtra, Gujarat)	Raw mango drink	
14	Yakhiyn (J&K)	Lamb cooked with curd and spices	(J&K)
15	Hakh (J&K)	Green leafy vegetable	(J&K)
16	Lotus Stem curry		(J&K)
17	Appam& Stew	Rice pancakes with vegetable or mutton stew	(Kerala)
18	Banana Kinathappam	sweet steam dish made from banana, coconut and jaggery	
19	bamboo shoot fry	vegetable	(Assam)
20	ChhenaPoda	semolina, cottage cheese and sugar	Orissa

References

Anilkumar, N., Narayanan, M. K. R., & Satheesh, K. (2008). Traditional knowledge of three 'mycophilic' communities on wild edible mushrooms of Wayanad district, Kerala. *Ethnobot, 20*(1), 41–47.

Chakraborty, P., Tewari, H. R., & Jha, M. K. (2009). Sustainable rural livelihoods through participatory natural resource management A case study. *Journal of Rural Development, 28* (1), 85–100.

Deshmukh, B. S., & Waghmode, A. (2011). Role of wild edible fruits as a food resource: Traditional knowledge. *International Journal of Pharmacy & Life Sciences, 2*(7).

FAO. 1993. *The State of Food and Agriculture 1993: Water policies and agriculture.*

Gadgil, M., & Guha, R. (2000). *The Use and Abuse of Nature.* Oxford University Press.

Hema, E. S., Sivadasan, M., & Anilkumar, N. (2006). Studies on edible species of Amaranthaceae and Araceae used by Kuruma and Paniya tribes in Wayanad district, Kerala, India. *Ethnobot, 18*(1), 122–126.

Jadhav, V. D., Mahadkar, S. D., & Valvi, S. R. (2011). Documentation and ethnobotanical survey of wild edible plants from Kolhapur District. *Recent Research in Science and Technology, 3* (12), 58–63,2076–5061.

Mathew, P. J., & Unnithan, C. M. (1992). Search for plants having anti-cancer properties used by the tribals of Wayanad, Malappuram and Palakkad Districts of Kerala, India. *Aryavaidyan, 6* (1), 61–67.

Mini, V., & Sivadasan, M. (2007). Plants used in ethnoveterinary medicine by the Kurichya tribe of Wayanad district in Kerala, India. *Ethnobot, 19*(1-2), 94–99.

Mourlin, K. (2007). Nrega – A Key to Sustainable Rural Development an empirical evidence from Betul district. *Vikas Vani Journal, 1*(4), 14–23.

Muralidharan, P. K., Sasidharan, N., & Seethalakshmi, K. K. (1997). Biodiversity in Tropical Moist Forest: Study of Sustainable Use of Non-Wood Forest Products in the Western Ghats. KFRI Research Report No. 133. KFRI, Thrissur, Kerala, India.

Narayanan, M. K. R., Anilkumar, N., Balakrishnan, V., Sivadasan, M., Alfarhan, H. A., & Alatar, A. A. (2011). Wild edible plants used by the Kattunaikka, Paniya and Kuruma tribes of Wayanad District, Kerala, India. *Journal of Medicinal Plants Research, 5*(15), 3520–3529. https://doi.org/10.5897/jmpr.9001295

Narayanan, M. K. R., Mithunlal, S., Sujanapal, P., Anilkumar, N., Sivadasan, M., Alfarhan, A. H., & Alatar, A. A. (2011). Ethnobotanically important trees and their uses by Kattunaikka tribe in Wayanad Wildlife Sanctuary, Kerala. *Indian Journal of Medicinal Plants Research, 5*(4), 604–612.

Nisha, V. M., & Sivadasan, M. (2007). Ethnodermatologically significant plants used by the traditional healers of Wayanad district, Kerala, India. *Ethnobot, 19*(1–2), 55–61.

Oak, G., Kurve, P., Kurve, S., & Pejaver, M. (2015). Ethno-botanical studies of edible plants used by tribal women of Thane District. *Journal of Medicinal Plants Studies, 3*(2), 90–94.

Pramod, C., & Sivadasan, M. (2003). Ethnobotany of religious and supernatural beliefs of Kurichya of Wayanad District, Kerala, India. *Ethnobot, 15*, 11–18.

Rasingam, L. (2012). Ethnobotanical studies on the wild edible plants of Irula tribes of Pillur Valley, Coimbatore district, Tamil Nadu, India. *Asian Pacific Journal of Tropical Biomedicine*, S1493–S1497.

Reid, W. V., & Miller, K. R. (1989). *Keeping the Options Alive: The Scientific Basis for Conserving Biodiversity.* World Resource Institute.

Rekka, R., & Senthil Kumar, S. (2014). Ethnobotanical notes on wild edible plants used by Malayali tribals of Yercaud Hills, Eastern Ghats, Salem District, Tamil Nadu. *International Journal Of Herbal Medicine, 2*(1), 39–42.

Roy, B., Halder, A. C., & Pal, D. C. (1998). *Plants for Human Consumption in India.* Botanical Survey of India.

Shanavaskhan, A. E., Sivadasan, M., Alfarhan, A. H., & Thomas, J. (2011). Ethnomedicinal aspects of angiospermic epiphytes and parasites of Kerala, India. Indian J. Tradit. Know. (In press).

Sundriyal, M., & Sundriyal, R. (2004). Wild edible plants of the Sikkim Himalaya: Nutritive values of selected species. *Economic Botany*, *58*(2).

Surayya, T., Krishna Kumar, K. N., Sharma, R., Karla, S., Kujur, S. S., Bala, S., & Basnayat, B. (2008). Sericulture Based Micro Enterprise as a Source of Rural Livelihood and Poverty Alleviation: A case study of Anantapur district (Andhra Pradesh). *Journal of Rural Development*, *27*(1), 149–176.

Chapter 16
Food Security Through Indigenous Agricultural Practices – a Case Study of Birbhum District, West Bengal, India

Vibhash C. Jha and Sajal Ghosh

Abstract Traditional farm practices enhance climate resilience, conserve indigenous species, enhance soil fertility, manage crop failure, harness local resources, minimize carbon footprint, conserve balanced ecosystems, and develop resilient farm infrastructure. They minimize environmental degradation, conserve indigenous species, and employ organic methods to enhance productivity. They also contribute to the conservation of genetic diversity and have a low carbon footprint. In poor countries and rural areas, 795 million individuals, or one out of every nine individuals, are hungry; however, technological advances like genetic modification, enhanced soil productivity, and irrigation are capable of boosting food production. Indigenous women and food sovereignty play a key role in combating world food poverty, climate change, and to assist poor farmers to combat world food poverty and climate change. Birbhum, the most northern district of the Burdwan division, is a region of 4545 km^2 the district consists of a fractured plateau of hard, impermeable Archean crystalline rocks overlain by laterite and alluvium Gondwana deposits. The district is famous for its diversified landscape and tropical climatic conditions, with agriculture as a large industry, albeit in adversity, through traditional methods like mixed cropping and organic manures. The research will define indigenous knowledge strategies, measure sustainability in terms of agriculture, determine strengths and weaknesses, evaluate opportunities and threats, and determine diversification of crops in Birbhum district. This research employs a literature review, field survey, interview, group discussion, priority calculation, AHP, SWOC analysis, data analysis, suggestions,

V. C. Jha (✉)
Professor and Former Head, Department of Geography, Visva-Bharati University, Santiniketan, West Bengal, India
e-mail: vcjha@asia.com

S. Ghosh
Research Scholar, University Department of Geography, Ranchi University, Ranchi, Jharkhand, India
e-mail: sajalghoshsuri@gmail.com

R. C. Sobti et al. (eds.), *Traditional Knowledge Systems in India: Sustainability, Ecology, and Resource Governance*, People, Cultures and Societies: Exploring and Documenting Diversities,
https://doi.org/10.1007/978-981-92-1346-7_16

recommendations, and opportunities Indigenious knowledge and sustainable agricultural practices such as Birbhum district traditional land management ensure food security and sustainability. Food security is a global goal on the basis of availability, utilization, and access but still persists in high-income nations, which leads to undernourishment and obesity. The sustainable agriculture practices involving maximum utilization of land, risk minimization, and enhancement of biodiversity are intercropping, crop rotation, agroforestry, organic manure, mulching, jhum cultivation, intercropping, vermicomposting, bee farming, biodynamic farming, conventional pest management, and water harvesting methods.

Keywords Food security · Birbhum district · Vermicomposting · Sustainable farming practices

Introduction

The information and abilities gathered over many generations that direct human civilizations in their interactions with their surroundings are referred to as indigenous, traditional, and local knowledge (Ihenacho et al., 2019). In order to produce food that is secure, wholesome, and sustainable, indigenous farming is a sustainable production method that restores and strengthens the natural functions of the farm environment (Kanagasabapathi & Sakthivel, 2012). Indigenous farming systems, often overlooked, are productive, adaptive, and ecologically principled practices that shape high-value natural and cultural landscapes and provide numerous benefits to people's quality of life (Guo et al., 2021). Indigenous knowledge, which has been gathered over many generations, unites local people to protect their cultural values and forest resources (Camacho et al., 2016). Indigenous knowledge is represented in language, social structures, institutions, values, and laws. It covers spiritual, environmental, resource-use, and people-to-people ties (Kazmi et al., 2014). Indigenous knowledge (IK) is essential to the social capital and means of subsistence of the local population, especially in developing nations where farmers strategize their agricultural output to ensure food security and sustainability (Nnadi & Ezudike, 2013). Conventional agriculture disregards the long-term ecological effects on agro-ecosystems and biodiversity in favour of production and profit maximization (Mohamed et al., 2016). Indigenous people preserve their cultural expression via stories, music, dances, and paintings, and they see the world as an interconnected whole (Koocheki, 2004). For the sustainable management of agricultural resources, indigenous knowledge including traditional, local, communal, and rural people's knowledge is essential (Chaudhry & Chaudhry, 2011). The study investigates the function of the Indigenous Agricultural System in a Punjabi hamlet, emphasizing the methods that have developed locally (Chaudhry & Chaudhry, 2011). Indigenous knowledge enhances the design of sustainable agricultural systems, encouraging rural communities to adopt and maintain innovations and provide a basis for decision-making in the face of both known and unknown obstacles (E, 2019). An essential component of tribal culture in Arunachal Pradesh is traditional land management, which respects their way of life and the natural environment while fostering sustainability and better livelihoods (Wangpan et al., 2014). Crucial planting and harvesting are part of traditional agriculture, which is frequently accompanied by ceremonies and significant holidays that strengthen ties among the community (Wangpan et al., 2017). By providing marginal farmers and pastoralists with financial stability, traditional agriculture enhances the nutritional security and diet of the world's people (Patel et al., 2020).Community values are frequently represented in goods, customs, or historical practices that are handed down to succeeding generations (Handoyo et al., 2022). Recent advancements have gradually revitalized the acceptance of indigenous knowledge for sustainable development in agricultural and rural regions (Ponge, 2013). Genetic variety must be preserved through indigenous knowledge and traditions in order to improve food security and prepare for climate change (Innovation & Goal, n.d.). To lessen the consequences of climate change, indigenous groups are employing a variety of agricultural diversification strategies (Cheplogoi et al., 2021). With 70% of their work in shifting agriculture case studies, indigenous women make a substantial contribution to food security and sustainable resource management (Erni, 2015). The link between undernourishment scale and agricultural sector characteristics in developing nations, considering variations in agricultural production potential (Pawlak & Kołodziejczak, 2020). Sustainable agriculture has the potential to address the immediate and long-term needs of populations susceptible to food insecurity (Tonui et al., 2024). Throughout history, Indigenous Peoples' food systems (IPFS) have offered wholesome meals in a variety of settings (IFAD, 2022). Indigenous farming practices offer both advantages and challenges for enhancing the resilience of smallholder farmers in the face of climate change (Nyashilu & Nyomora, 2023). Traditional indigenous agriculture is a knowledge repository and adaptation of farming practices to local environments, requiring alignment with modern innovations and practices (Pradesh & Security, n.d.). Food sovereignty refers to the right of people to produce healthy, culturally appropriate food using sustainable methods and to define their own food and agriculture systems (Buxton et al., 2016). Food security is a complex and dynamic issue that requires attention in both the short and long term (Jeder et al., 2020). Food security refers to the provision of adequate, high-quality food from domestic sources or imports to ensure nutritional well-being and meet all physiological needs (Sekaran et al., 2021). Indigenous knowledge refers to a distinct understanding of a particular community's world, unique to their culture or society (Masango & Mbarika, 2022). Over half a billion people worldwide suffer from chronic food insecurity, largely due to sub-standard living conditions in developing countries (Galhena et al., 2013). Population growth surpasses linear food production increment, requiring a 70% increase in crop productivity in the next decades to meet human needs (Temesgen, 2021). Climate change directly impacts food systems and food security, potentially increasing competition for resources needed for agriculture through the need for mitigation measures (Khan, 2020). Local and indigenous foods have the potential to improve food and nutrition security, promote economic empowerment, and benefit marginalized farmers and consumers in Sub Saharan Africa. The Sustainable Development Goals (SDGs) for food and nutrition security require the active participation of female and youth-headed farming households in their production (Shelembe et al., 2024). Improved agriculture enhances the MDGs by boosting food production, diversity, and environmental solutions, while also fostering partnerships for long-term agricultural benefits (IAEA, 2007). Sustainable development goals aim to eradicate hunger, ensure food security, and enhance nutrition, but climate change is already impacting agriculture and food security, exacerbating the challenge (Gitz et al., 2016). The 1996 World Food Summit aimed to reduce malnutrition by half by 2015, but the reduction in undernourished individuals has been minimal, making it unlikely to be achieved (de Graaff et al., 2011). Indigenous people have preserved their agrifood systems, which can be utilized in areas facing food poverty (Lugo, 2020). For centuries, resource-poor farmers in developing countries have been utilizing indigenous technologies through crop-livestock-forest integration (Upreti, 1999). Food security is crucial for countries with varying economic development levels, with the agricultural sector playing a strategic role in enhancing food availability (Burchi et al., 2011). In the current century, meeting the growing global population's sustainable food and dietary needs presents a significant challenge (Singh & Singh, 2017). Hunger remains a significant issue in developing and emerging countries, with nearly 1 billion people still hungry. Progress towards addressing this issue is slow. (Burchi et al., 2011). The essay emphasizes the connection between agriculture and climate change and the rising interest in traditional agriculture throughout the world as a climate-smart, sustainable food production approach. (Singh & Singh, 2017). Survey reports reveal that despite the significant GDP growth in the past two decades, hunger and malnutrition persist among the majority of the Indian population (Brahmanand et al., 2013). The promotion and utilization of indigenous knowledge in food handling, processing, preservation, and storage can significantly reduce postharvest loss and enhance food security (Kuyu & Bereka, 2020). Indigenous knowledge is crucial for sustainable food system transformations, yet it is frequently marginalized in policy and practice (Vijayan et al., 2022). Since the 1960s, the green revolution has significantly improved global food security through conventional input-intensive and agro-chemical-based agriculture (Reddy et al., 2022). Traditional vegetables from a single species provide local farmers and consumers with a consistent, nutritious supply of fresh vegetables throughout their life cycle, enhancing food security (Hart & Mouton, 2005). To guarantee water availability for agriculture and consumption, Potato Park combines conventional and innovative water collecting techniques with family and community micro-reservoirs (Handoyo et al., 2022). According to the study, in order to survive, households at the study location upgraded their agricultural practices, varied their food sources, and developed storage and preservation techniques (Cheplogoi et al., 2021) (Fig. 16.1).

Food is a basic human need that comes in three forms: affordable, easily accessible, and readily available. All people should have access to it, even those who receive government assistance, and it should be reasonably priced to provide

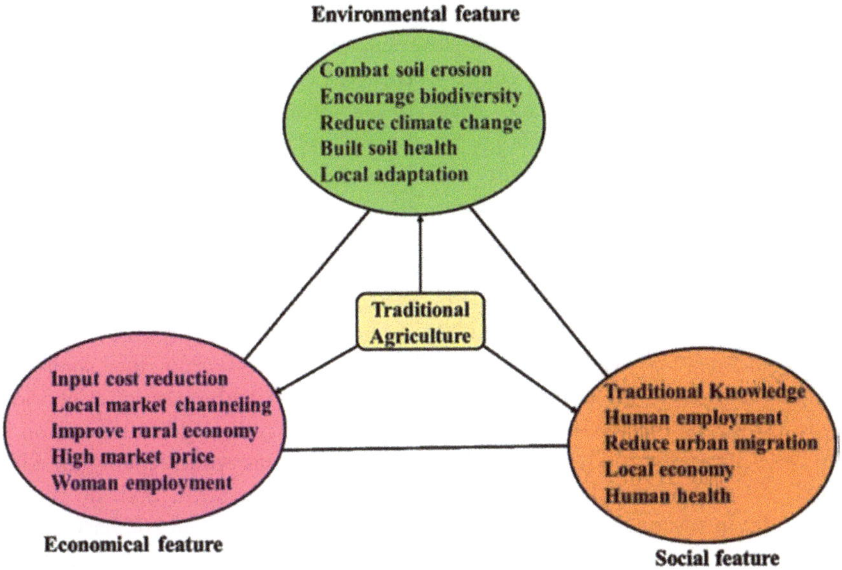

Fig. 16.1 Traditional agriculture practices (Patel et al., 2020)

basic necessities (Jain, 2016). Three pillars are typically thought of as supporting food security: availability, access, and use (Camire, 2021). Food security, which is defined as having adequate access to nourishing food for healthy growth, is still a global objective (Zhu et al., 2022). Despite agricultural and economic policies guaranteeing food supply, food insecurity persists in high-income nations, as seen by reports of rising need of emergency food aid (Gallegos et al., 2023). Due to increased expenses, stress, and physiological adjustments to dietary limitation, food insecurity contributes to high rates of overweight, obesity, and undernutrition in many nations (FAO, 2018). Not just production but also sustainability, availability, access, and usage are required for food security. To provide a fair and effective global food supply, a "new agriculture" that takes into account ecosystem services and biodiversity is required (Devereux, 2001). In developing nations and rural regions, 795 million people, or one in nine people, suffer from undernourishment; nevertheless, advances in technology such as genetic modification, improved soil fertility, and irrigation can increase the availability of food (United Nations, 2017). 1.9 billion people globally experienced moderate to severe food insecurity in 2017, with South Asia and SSA having the greatest percentages. 9.2% of people worldwide experienced acute insecurity in 2018 (Beyene, 2023). The uneven distribution of agricultural advancement is the root cause of the world's food difficulties, since it prevents price reduction even in cases where technological advancements enable higher yields in some nations (Beyene, 2023). In order to combat hunger and malnutrition and ensure global food security, a number of agencies are collaborating to increase agricultural output, especially in smallholder farms (Bhoite & Athawale, 2018). The COVID-19 epidemic has hastened the steady collapse in global food security that began in 2015, resulting in acute hunger and long-term aggravation of inequality (Teeuwen et al., 2022). The food production systems at the international, national, and local levels are predicted to be greatly impacted by climate change, which will have an effect on the availability, accessibility, use, and sustainability of food (Singh, 2016). Despite the ambitious nature of the initial aim, the World Food Summit of 1996 approved a revised target that enabled 150 million additional people to remain hungry by the year 2015 (Butler, 2017). In order to ensure a sustainable future, it is critical to connect the aims of food safety and security, which calls for careful management of evidence-based trade-offs (Vågsholm et al., 2020). In order to examine the worldwide food supply and demand with an emphasis on global food security, this study use scenario analysis (van Dijk & Meijerink, 2014). With the use of high-yield seeds, fertilizers, land reforms, and irrigation projects, India's green revolution in the 1960s increased food grain output by 15 million tons while feeding an additional 150 million people (Narkhede, 2014). Definitions, ideas, and the role of agriculture in sustainability are covered in this overview, with an emphasis on the potential effects of population expansion and economic development on global food security (Stefanis, 2014). In order to characterize the food security situation of Canadian households, the government now employs a measure of food insecurity rather than a metric to evaluate the FPC's food security target (Mahmood et al., 2023). In order to shed light on the persistent problem of food insecurity in developing nations, this study looks at the socioeconomic traits of rural families in South-East Nigeria as well as their level of food security during the COVID-19 epidemic and its determinants (Deaton & Scholz, 2022). Food insecurity is a serious public health concern that has been connected to depression and ill health in adults, as well as delayed cognitive and emotional development in children (Chilton & Rose, 2009). The goal of food security indicators and measures is to capture food availability, access, and usage. These indicators and measures are developed from family income and spending surveys to estimate per capita caloric availability (Dr. Parvathamma, 2015). Food crises are greatly worsened by violent wars because they destroy infrastructure, irrigation systems, and agricultural land, this results in widespread famine, relocation, and long-term negative repercussions on the impacted communities (Kemmerling et al., 2022). Traditionally, food security has been assessed using an ordinal scale that goes from minimal food security to extreme hunger in children (Derrickson et al., 2000). By 2015, the World Food Summit of 1996 sought to cut the number of undernourished people in half and reduce hunger by 20 million each year (Fawole et al., 2016). Since 1970, studies on food security have concentrated on global, regional, and domestic concerns; originally, they examined changes in the cost, availability, and accessibility of staple foods (Bobe et al., 2019). The degree to which families have access to sufficient amounts and qualities of food is referred to as food security or food insecurity (Cheber Bezu, 2018). 805 million people worldwide are affected by the problem of food security every day, and one in nine of them suffer from hunger (Roy et al., 2019). Since the 1974 World Food Conference, the concept of food security has evolved, expanded, and diversified, traditionally focusing on the availability and access to adequate food (Bishwajit et al., 2014). Microfinance is a comprehensive financial solution aimed at disadvantaged individuals, particularly women and marginalized farmers, who often lack access to formal financial services (Rahman et al., 2019). High yields, climatic adaptation, genetic advancement, pest control, and innovative farming techniques are all necessary to feed nine billion people (Stefanis, 2014). Despite an annual demand of 20 million tonnes, India faces a paradoxical position with its food-grain reserves at an all-time high of 62 million tonnes (Kumar et al., 2012). Goal 2 of the Sustainable Development Agenda is to advance sustainable agriculture, improve nutrition, and increase food security (Mahmood et al., 2023). The occurrence of food insecurity in families is a noteworthy issue as it indicates which households are most susceptible to fluctuations in food prices (Deaton & Scholz, 2022). Feeding one's family is the responsibility of household heads, and food security is a matter of national security, a country must also raise its citizens' level of food security (EGWUE et al., 2021) (Fig. 16.2).

An estimated 200 million people in India suffer from malnutrition, indicating a high level of food insecurity there (McKay et al., 2023). Food security ensures individuals can consistently obtain sufficient, safe, and healthy food to meet their daily needs and maintain a healthy, happy, and effective lifestyle (Bala, 2020). In sub-Saharan Africa, smallholder farmers' livelihoods and food security are seriously threatened by climate change (Wakweya, 2023). Climate change is posing a significant threat to agriculture, increasing its vulnerability to climate risks and posing challenges to sustainable agrifood production (Vatsa et al., 2023). Two-thirds of India's people work in agriculture, which accounts for 16% of the nation's GDP and

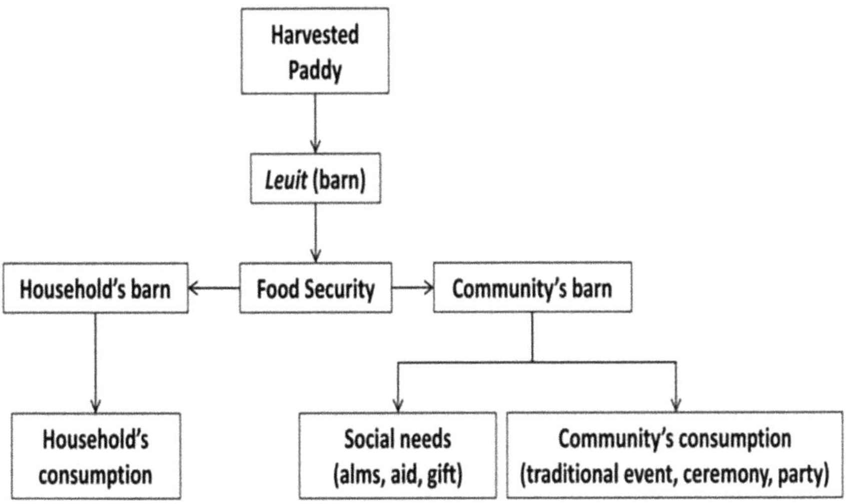

Fig. 16.2 Food security (Handoyo et al., 2022)

greatly enhances both the economic and social well-being of the nation (Gurditta & Singh, 2016). Due to migration, fatalities, and damage to property, unfavourable weather events can have a substantial negative influence on the welfare of urban impoverished households, which mostly depend on food (Martins, 2020). In addition to severe famine and droughts on the flip side, half of India's population struggles to make ends meet (Bala, 2020). In light of the fact that 690 million people, or 8.9% of the world's population, are predicted to be hungry in 2020, the agricultural sector is essential to maintaining food security (Ajatasatru et al., 2024). A number of variables impact food security on a global, regional, national, and individual level, including income, spending, household size, gender, and age of the head of home (Saad et al., 2023). Food availability is crucial for food security, and India is mostly self-sufficient in cereals but deficient in pulses and oilseeds (Bala, 2020). India's culture is centred around agriculture, with 60% of the population living in its villages, largely dependent on monsoons for their livelihood (Rao, 2020). The general public is experiencing significant dread and anxiety around their ability to eat due to the COVID-19 lockdown (Mph & Kannan, 2021). India's government has implemented various food security schemes to provide food to rural communities, poor families, schools, labourers, and pregnant women, including the Essential Commodities Act, 1997, and 2007 (Zhou, 2018).

Agroforestry is a sustainable agricultural practice that combines trees, crops, cattle, and other land uses to provide long-lasting livelihood options. It enhances food and nutritional security by providing tree foods, boosting staple crop production, raising farmer incomes, and promoting ecosystem services like pollination. Agroforestry systems, such as aqua forestry, agri-silviculture, and agrihorticulture, are used in India to generate income from farms' tree use. They also promote sustainable forest management, environmental conservation, and mitigate environmental consequences. Agroforestry systems, such as aqua forestry, agri-silviculture, and agrihorticulture, support livelihoods, alleviate climate change effects, and generate food, fuel, and shrubs. However, challenges such as bureaucracy and lack of substitute market facilities exist. Agroforestry combines science and traditional wisdom to increase agricultural resilience and forest cover, tackles climate change, and yields greater economic rewards for farmers. Wild edible plants (WEPs) are essential for rural communities to maintain revenue, encourage nutritional variety, and ensure food security.(Sarveswaran S et al., Sarveswaran et al., 2023), (Gassner. & Dobie., 2022), (Miccolis et al., 2016), (Ramni et al., 2013), (Yadav, 2018). (Kraft et al., 2021). (VERMA et al., 2016). (Sobola & Amadi, 2015), (Leakey, 2017), (Handa et al., 2020). (Jain & Verma, 2021). (Getachew, 2014). (Sarvade et al., 2014), (Doshi et al., 2015). (Akter et al., 2022). (Johar, 2021). (Asfaw & Tadesse, 2001), (Kloos, 2023), (Kidane & Kejela, 2021), (Berihun & Molla, 2017), (Bharucha & Pretty, 2010).

Indigenous knowledge is used by tribal communities in agricultural activities, making efficient use of lands because of ecological constraints, eating small millets, analysing seeds, preserving soil fertility, and preserving landraces (The et al., n.d.). Despite the many obstacles to food security, the growing understanding of the significance of attaining more variety in agricultural systems is a fundamental component of sustainable development (Kahane et al., 2013). The study looks at how farming techniques affect biological variety, including genetics and geography, and how it affects rural communities, economies, and livelihoods in addition to rural society (Brookfield & Stocking, 1999). For vast communities across the world, unofficial networks of food production, processing, distribution, retailing, and transportation provide a safe and wholesome food source (Zimmerer & de Haan, 2020).

Objectives

1. To pinpoint the precise indigenous knowledge tactics employed by the Birbhum district people.
2. To evaluate sustainability maintain through indigenous agricultural practices.
3. To assess and identify strengths and weaknesses of indigenous agricultural practices.
4. To analyse opportunities and challenges influences impacting indigenous agriculture.
5. To identify crop diversification of Birbhum district.

Study Area

The Birbhum district is located in the latitude range of 23°32′30″ to 24°35′00" N and the longitude range of 87°5′25" E to 88°1′40" E. Birbhum, the northernmost district of the Burdwan division, spans 4545 km^2. Birbhum is bounded to the east by the Murshidabad district, to the south by the Burdwan district, which is divided by the Ajoy, and to the north and west by the Santal Parganas region of Jharkhand state. Administratively, it is split up into three subdivisions (Suri, Bolpur, and Rampurhat), twenty-seven villages, and nineteen community development blocks. The western section of Jharkhand district features a divided plateau, with hard, impermeable Archean crystalline rocks and laterite and alluvium Gondwana deposits. Numerous rivers and plateau streams drain the territory. Riverine and lateritic features are what define the research region. The Ajay, Mayurakshi, Dwarka, Brahmini, and Bansloi river basin are all included in the study region. This area is geographically a part of the 'Rarh' area; the western section of this district consists of the south-south projecting, shattered plateau of Jharkhand. West Bengal, India's Birbhum district is a rural region with a varied landscape and a tropical climate. Rice, wheat, legumes, fruits, vegetables, horticulture, oilseeds and pulses are among the major crops. To preserve soil health and encourage self-sufficiency, indigenous methods such as sustainable farming, water conservation, and seed preservation are employed. The district of Birbhum, located in West Bengal, India, is renowned for its varied terrain and warm temperature. Agriculture is an important industry. Despite difficulties, traditional methods like mixed crops and organic fertilizers are still used (Fig 16.3 and 16.4).

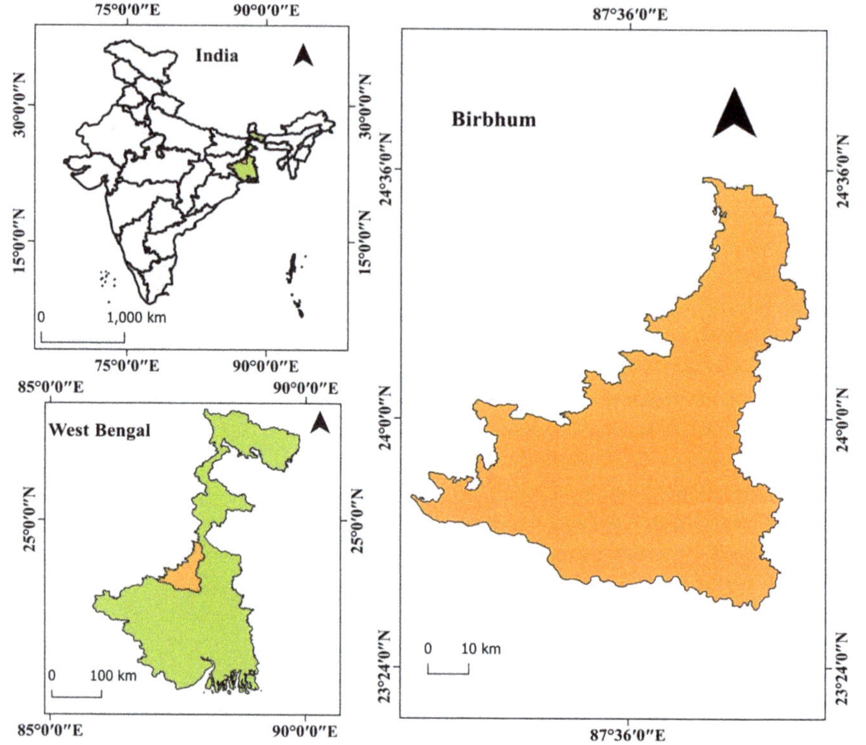

Fig. 16.3 Study area

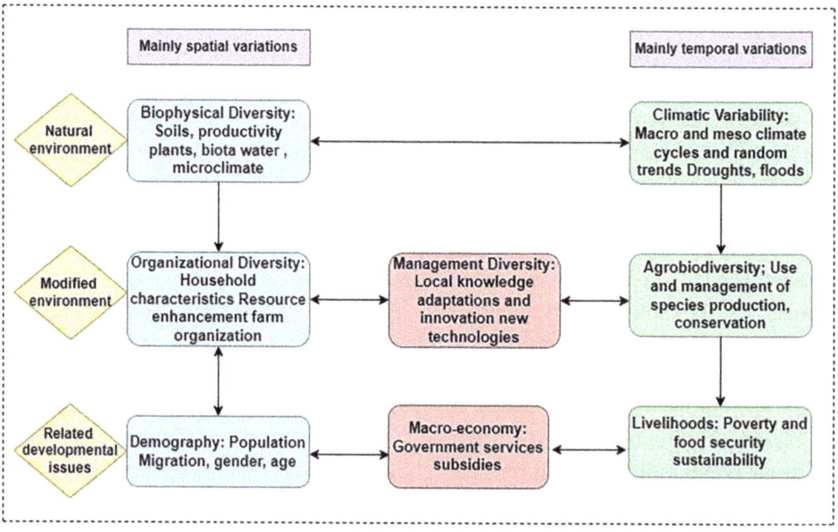

Fig. 16.4 Agroforestry (Brookfield & Stocking, 1999)

Methodology

The findings are expected to guide policy interventions and community initiatives promoting sustainable agriculture and local food security. The purpose of the project is to determine whether using traditional farming methods in the Birbhum district can increase food security. To assess these practices, a Likert scale and SWOT-AHP research methods are used. Through a thorough assessment of the literature and the author's own practical observations made in several Birbhum district tourism destinations, these strengths, flaws, possibilities, and dangers were determined. In the end, each question was given a five-point Likert scale score, with a five being the highest possible. The respondents were asked to indicate how much they agreed or disagreed with each of the questionnaire's questions. Each question's response was assigned a numerical score that accurately reflected the sentiments of the respondents (Fig. 16.5)

Likert scales are psychometric response scales that are used in surveys to gauge a single characteristic by asking participants to rate their preferences or level of agreement with a statement or series of statements (Aasa, 2019). The Likert scale analysis approach determines additional statistical treatment based on study objectives and frameworks by assigning item response variables into ordinal or interval scales (Joshi et al., 2015). Two quantitative studies can be conducted on Likert type data: scoring, which measures common properties of items, and ranking the relative significance of a group of objects based on individual qualities, without averaging or factor analysis (Mumu et al., 2022). In social science and attitude research initiatives, Likert scales are helpful (James & Lee, 2011). One part of a person's affect or cognition that may be operationalized and quantified is psychological constructs, which are most commonly measured using Likert-scale instrument (Beglar & Nemoto, 2014). The kind of Likert scale is appropriate for research in the social and behavioural sciences concerning people's perceptions, attitudes, feelings, views, personalities, and descriptions of their surroundings (Olaniy, 2019).

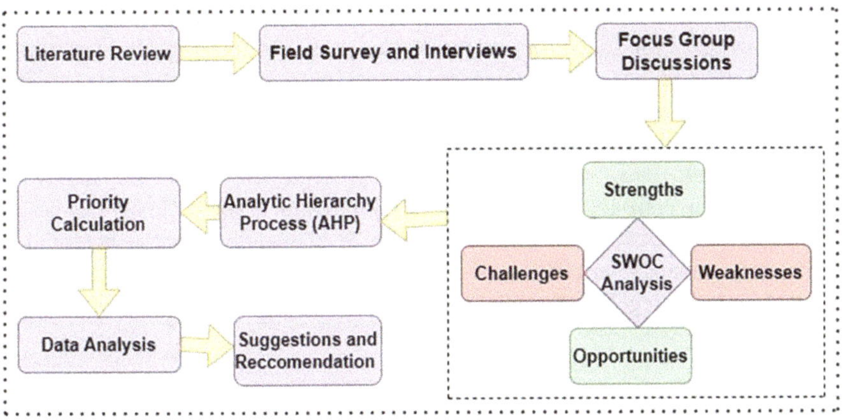

Fig. 16.5 Methodology of this study

Crop Diversification

Crop diversification is a tactic used by smallholder farmers to improve agroecosystem functioning and reduce the effects of climate change on the planet. It has difficulties in semi-arid agricultural settings, nevertheless (McCord et al., 2015). Through sustainable intensification, higher productivity, and lower input costs made possible by technical advancements, crop diversification enables farmers to raise their income levels and introduce new crops that are well-suited to their ecoregions (Paroda, 2022). The capacity of smallholder-dominated Indian agriculture to diversify into high-value crops and perhaps boost farm incomes as a result of growing consumer demand for high-value food goods is under question (Birthal et al., 2011). Agro-ecosystem resilience to stresses, agricultural productivity, and biodiversity are all improved by introducing a variety of crop types. Plant production, quality, health, and nutritional value are all improved by this technique. Crop diversification, as opposed to specialized farming, is an endeavour to increase crop variety by crop rotation, intercropping, or multiple cropping in order to improve the availability, sustainability, and productivity of ecological systems (Rahman et al., 2024). Because of its closeness to the Chhotonagpur plateau, the western portion of Birbhum district lacks the agricultural prosperity and diversity of the northern and eastern parts, however there is a somewhat good correlation between irrigation intensity and diversification there (Sahal & Rudra, 2020).

The Gibbs-Martin approach, which requires continuous data while maintaining homogeneity of diversification, was used to compute diversification on the areal extent of crops. is computed on discrete data using the Simpson's Index.

$$\text{Gibbs and Martin's Technique} = 1 - \left(\sum X^2\right) / \left(\sum X\right)^2$$

In this case, X represents the portion of land that a certain crop occupies at any given moment. His index value falls between 0 and 1. Greater diversity would follow higher index values, and vice versa. In fact, this method aids in our comprehension of the extent of variety.(Rahaman & Singh, 2020). In agriculture, crop diversification has a big influence on social and economic situations, it gives the impoverished more resources and gives young people jobs, which benefits farmers all year long (Dalal & Shankar, 2022). A key tactic to increase farm income, generate jobs, reduce poverty, and preserve important soil and water resources is diversifying agriculture into high-value, competitive businesses (Othieno & Shinyekwa, 2011). In India, crop diversification has the potential to significantly enhance farmers' well-being, nutritional security, and agricultural sustainability, turning farming into a profitable and sustainable business venture

Table 16.1 Crop diversification index of Birbhum district

Sl no.	Name of the block	Crop diversification index
1.	Murarai I	0.670087
2.	Murarai II	0.648411
3.	Nalhati I	0.662328
4.	Nalhati II	0.654821
5.	Rampurhat I	0.578288
6.	Rampurhat II	0.627269
7.	Mayureswar I	0.579207
8.	Mayureswar II	0.684762
9.	Md. Bazar	0.404479
10.	Sainthia	0.50529
11.	Dubrajpur	0.374704
12.	Rajnagar	0.131724
13.	Suri I	0.40013
14.	Suri II	0.553293
15.	Khayrasol	0.238453
16.	Bolpur-Sriniketan	0.548657
17.	Labhpur	0.556397
18.	Nanoor	0.623644
19.	Illambazar	0.480931

(Anuja & Ray, 2021). A first indicator of the breadth and pattern of agricultural diversification is the temporal variations in cropping patterns and the growth of cultivated areas for various crop groupings (Basantaray et al., 2022). Crop diversification may maintain productivity, whereas agricultural intensification increases crop output but simplifies production, diminishing genetic uniformity and variety and raising worries about the environment and biodiversity (Hufnagel et al., 2020). The trend, degree, and effects of food crop diversity on nutritional metrics in India using district-level data from the most current time frame (Anuja et al., 2022) (Tables 16.1 and 16.2).

Table 16.2 Blockwise crop diversification index

Sl. No	Class	No. of blocks	Name of blocks	Reasons
1	0.131720–0.238450 (Very Low)	1	Rajnagar	Rajnagar, and are as follows: the local climate; economic issues; lack of expertise; paucity of water; traditional farming methods; hard rock area low agricultural productivity
2	0.238451–0.404480 (Low)	4	Khayrasol, Md.Bazar, Dubrajpur, Suri-I	Khoyrasol, and Md. Bazar blocks are as follows: the local climate; economic issues; lack of expertise; paucity of water; traditional farming methods; lateritic prone area facing gully erosion
3	0.404481–0.556400 (Medium)	5	Sainthia,Suri-II, Bolpur-Sriniketan, Labhpur, Illambazar	Due to a large amount of waste land and a limited utilization of contemporary technologies, the region's soil has modest levels of diversity and is deficient in fertility elements including potash, lime, nitrogen, and phosphoric acid.
4	0.556401–0.627270 (High)	4	Rampurhat-I, Rampurhat-II, Mayureswar-I, Nanoor	Rich soil types, such as clay-loam and sandy-loam, that are favourable for development of agriculture.
5	0.627271–0.684760 (Very High)	5	Murarai-I, Murarai-II, Nalhati-I, Nalhati-II, Mayureswar-II	In terms of agriculture, these areas are distinguished by comparatively rich soil types, such as clay-loam and sandy-loam, that are ideal for double- or multiple-cropping

Result and Discussion

Gender Contrasts in Birbhum's Indigenous Literacy

In Birbhum district, male-female literacy disparity amongst native groups is pronounced, with men leading women in all respects. While men would gain the most from greater exposure to new farm information and technology, women particularly illiterate women are exempted. According to studies, Adivasi men in villages in Birbhum are 60% literate, then plummet drastically to about 38% among women Overall, more than 56% of inhabitants of such villages are illiterate and cannot read or write, and about 58% of the women fall in this bracket (Table 16.3).

Table 16.3 Literacy rate among tribal population of Birbhum district

Total population	Total male population	Total female population	Total literate	Total male literate	Total female literate	Total illiterate	Total male illiterate	Total female illiterate
15,087	7565	7522	6605	3990	2615	8482	3575	4907
(%)	50.14	49.86	43.78	60.40	39.60	56.22	42.15	57.85

Impacts on Agriculture & Food Security

Traditional agriculture is based on tradition: proven methods passed on through stories, observation, and practice. While literacy advantages men, however, they learn and put into action new techniques at the cost of female farmers, who often carry much of the farm burden, and so cannot participate equally in rewards. This gap can limit the evolution and broader application of traditional methods vital for food security. While these age-old practices remain crucial to maintaining resilient food systems, literacy imbalances hinder full participation and adaptation by women the very people most involved in farming. Educational empowerment especially for women can transform this dynamic: Enhanced innovation: Educated women can more easily learn, adapt, and improve traditional planting, seed preservation, pest control, and harvesting techniques. Stronger household food security: Literacy and learning enable women to contribute meaningfully to strategies that secure stable food supplies. Preserving traditional knowledge: Women literate in both modern and indigenous practices can document, strengthen, and pass on valuable cultural and agricultural heritage (Table 16.3)

Traditional Pest Management: Rather of using chemical pesticides, natural predators, plant extracts, and cultural methods are used to manage pests. Throughout the nation, several traditional tribal pest management methods have been modified and put to new uses (Firake et al., 2013).

Water Harvesting Techniques: Gathering and preserving rainfall for use in farming during arid times. It is encouraged that communities adopt water collecting strategies to deal with water scarcity and flooding-related disasters (Subagyono & Pawitan, 2008) (Fig. 16.6 and Table 16.4).

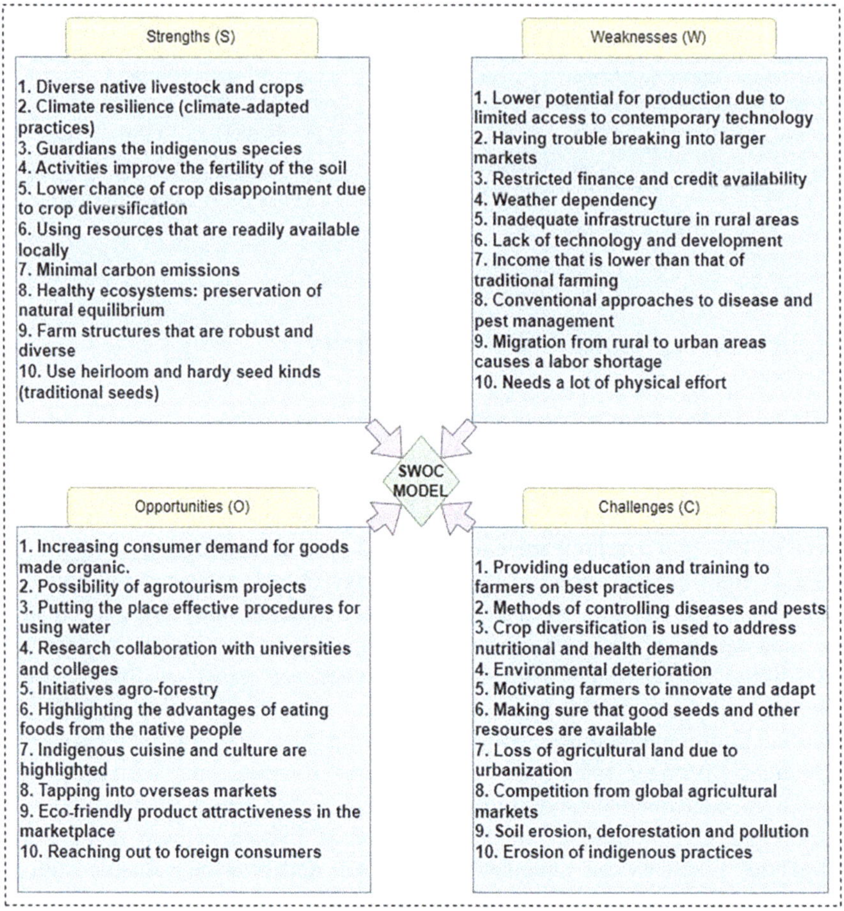

Fig. 16.6 SWOC model

Strength factors analysis

Climate Resilience Practices (16.4%): Indigenous knowledge that adapt to regional climates decline the impact of environmental changes, encouraging sustainable agriculture.3. Protectors of Native Species (9.7%): Conserving native species aids in sustaining genetic diversity and overall agricultural sustainability.4. Improvement of Soil Fertility (11.1%): Indigenous agricultural practices often employ organic methods to boost soil quality, crucial for ongoing agricultural productivity. Maintaining Crop Failures (5.1%): By cultivating multiple crops, farmers reduce the risk of losing everything to pests, weather, or disease. Using Local Resources (16.5%): Dependence on materials such as farm-saved seeds, compost, and local biomass lessens reliance on external inputs and supports

Table 16.4 Strengths factors

	S1	S2	S3	S4	S5	S6	S7	S8	S9	S10	Percentage (%)
S1	1	2	3	2	8	2	2	4	7	5	22.9
S2	1/2	1	2	3	5	1/2	2	5	6	4	16.4
S3	1/3	1/2	1	1	2	1/2	1	3	5	4	9.7
S4	1/2	1/3	1	1	5	1/3	2	2	4	5	11.1
S5	1/8	1/5	1/2	1/5	1	1/2	1	1	2	3	5.1
S6	1/2	2	2	3	2	1	3	4	4	2	16.5
S7	1/2	1/2	1	1/2	1	1/3	1	1	4	1	6.8
S8	1/4	1/5	1/3	1/2	1	1/4	1	1	3	2	5.1
S9	1/7	1/6	1/5	1/4	1/2	1/4	1/4	1/3	1	2	2.9
S10	1/5	1/4	1/4	1/5	1/3	1/2	1	1/2	1/2	1	3.6

Table 16.5 Weaknesses factors

	W1	W2	W3	W4	W5	W6	W7	W8	W9	W10	Percentage (%)
W1	1	3	4	6	6	2	5	4	6	7	28.0
W2	1/3	1	2	2	4	1/3	3	3	3	6	13.0
W3	1/4	1/2	1	3	1	1/3	2	2	2	3	8.5
W4	1/6	1/2	1/3	1	2	1/2	4	2	3	2	8.3
W5	1/6	1/4	1	1/2	1	1/2	2	1/3	2	4	6.1
W6	1/2	3	3	2	2	1	2	5	5	8	18.3
W7	1/5	1/3	1/2	1/4	1/2	1/2	1	2	4	2	6.0
W8	1/4	1/3	1/2	1/2	3	1/5	1/2	1	2	4	6.2
W9	1/6	1/3	1/2	1/3	1/2	1/5	1/4	1/2	1	1	3.0
W10	1/7	1/6	1/3	1/2	1/4	1/8	1/2	1/4	1	1	2.5

sustainability. Lower Carbon Footprint (6.8%): These minimum input methods often need minimal energy and decline greenhouse gas emissions compared to industrial farming. Balanced Ecosystems (5.1%): Traditional agricultural practices control ecological harmony, enhancing biodiversity like the "dongor" systems where multi-crop plantings naturally control pests. Resilient Farm Structures (2.9%): Diverse agricultural setups, including varied farm infrastructure, increase the stability and robustness of production systems (Table 16.5).

Weakness factors analysis

Weather Dependency (8.3%): Indigenous customs are vulnerable to climate changes as they often depend significantly on local weather patterns. Insufficient Infrastructure in Rural Regions (6.1%): The efficiency of agricultural activities is influenced by insufficient infrastructure in rural regions concerning storage, transportation, and overall operations. 8. Indigenous knowledge to Disease and Pest Management (6.2%): Compared to modern techniques, traditional methods for measurement of pests and diseases might be less efficient, possibly leading to agricultural losses.9. Labor deficit Resulting through Migration from Rural to Urban Areas (3.0%): Migration from rural regions to urban centres influence productivity by creating a deficit of labour for agricultural sectors. Decline Yield Due to Outdated Tools (28.0%): Dependence on conventional farming practices usually results in lesser harvests compared to contemporary machinery or superior seed types, consequently diminishing overall productivity. Challenges in Accessing Broader Markets (13.0%): Small indigenous farmers often face barrier in market entry, limiting their earning capacity and potential for business expansion. Minimum Access to Financing and Capital (8.5%): In the absence of adequate financial resources such as loans, grants, or microfinance farmers are unable to invest in expanding their operations or improving their farms. Deficiency in Technological Assistance (18.3%): Lack of modern resources like soil analysis tools, irrigation technology, and expert advice undermines the effectiveness and success of sustainable farming methods. Lower Profit Margins Compared to Conventional Agriculture (6.0%): Although sustainable and traditional, indigenous agricultural practices may at times result in lower financial gains than conventional or intensive farming models (Table 16.6).

Table 16.6 Opportunities factors

	O1	O2	O3	O4	O5	O6	O7	O8	O9	O10	Percentage
O1	1	5	2	6	4	2	5	4	9	9	28.2
O2	1/5	1	4	2	5	1/2	3	3	9	5	15.7
O3	1/2	1/4	1	1	2	1/3	3	3	3	3	9.1
O4	1/6	1/2	1	1	3	1/3	4	2	3	5	9.4
O5	1/4	1/5	1/2	1/3	1	1/5	2	1/2	2	2	4.6
O6	1/2	2	3	3	5	1	2	5	8	3	18.0
O7	1/5	1/3	1/3	1/4	1/2	1/2	1	2	2	1	4.7
O8	1/4	1/3	1/3	1/2	2	1/5	1/2	1	3	5	5.0
O9	1/9	1/9	1/3	1/3	1/2	1/8	1/2	1/3	1	2	2.3
O10	1/9	1/5	1/3	1/5	1/2	1/3	1	1/5	1/2	1	3.0

Opportunity factors analysis

Increasing Demand for Organic Produce (28.2%): Consumers worldwide are leaning towards organic foods due to health, flavour, and environmental factors. The organic food sector achieved approximately $230 billion in 2024 and is projected to expand at an annual rate exceeding 10% in the next ten years. Birbhum's indigenous agricultural practices by tribal people, emphasizing native crops and less chemical fertilizer use and basically organic fertilizer uses, are suitable positioned to capitalize on this rising consumer interest. Potentiality in Agrotourism (15.7%): Agrotourism appeals to those wanting to relationship with local agriculture, merging indigenous agricultural practices with cultural practices. Development of Water Efficiency (9.1%): Adopting suitable water conservation strategies like drip irrigation, rainwater harvesting can significantly boost productivity while conserving a limited resource. These methods enable farmers to attain higher production with reduced water usage. Partnerships with Educational Institutions (9.4%): Collaborating with universities and research institutions can improve innovation in regional agricultural methods. Agroforestry: An Expanding Chance (4.6%): Incorporating trees and shrubs into agricultural land is not only environmentally friendly it's also financially savvy. Agroforestry enhances soil nutrients, minimizes erosion, and improves water retention. Additionally, it provides farmers with consistent yields and extra income sources such as firewood, timber, fruit, and fodder. It also boosts farm resilience by establishing microclimates that mitigate extreme conditions and aid wildlife all while improving biodiversity (Table 16.7 and Fig. 16.7).

Table 16.7 Challenges factors

	C1	C2	C3	C4	C5	C6	C7	C8	C9	C10	Percentage (%)
C1	1	3	4	5	7	2	4	6	8	5	27.7
C2	1/3	1	3	4	5	1/2	2	4	6	6	15.9
C3	1/4	1/3	1	2	5	1/2	1	3	3	3	9.4
C4	1/5	1/4	1/2	1	3	1/3	3	2	4	5	8.8
C5	1/7	1/5	1/5	1/3	1	1/4	1	1	2	2	4.0
C6	1/2	2	2	3	4	1	2	5	7	4	17.0
C7	1/4	1/2	1	1/3	1	1/2	1	2	4	3	7.0
C8	1/6	1/4	1/3	1/2	1	1/5	1/2	1	3	4	4.7
C9	1/8	1/6	1/3	1/4	1/2	1/7	1/4	1/3	1	3	2.8
C10	1/5	1/6	1/3	1/5	1/2	1/4	1/3	1/4	1/3	1	2.5

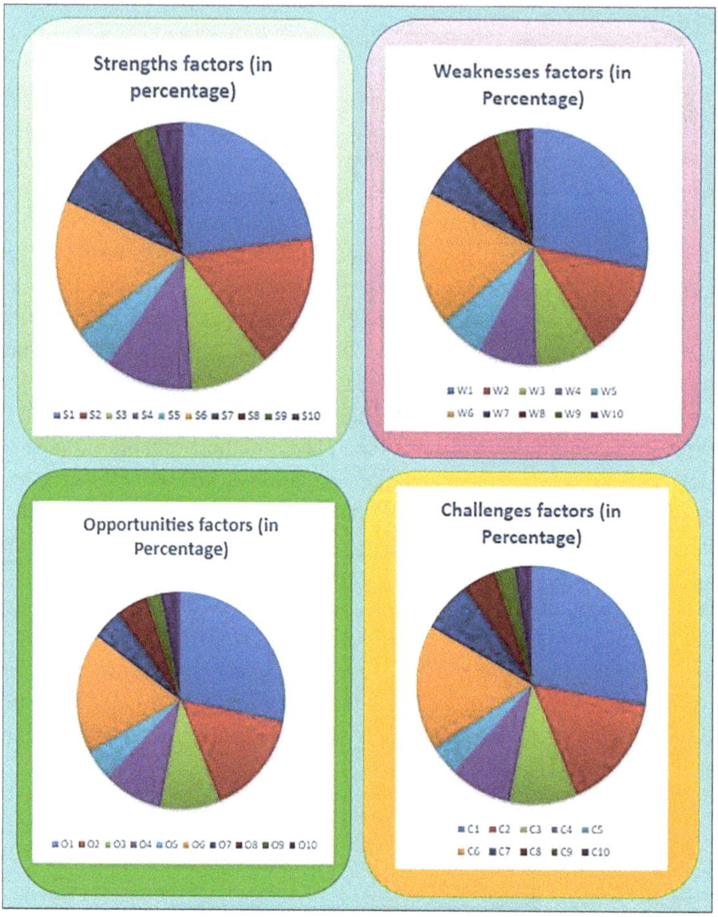

Fig. 16.7 Strengths, weaknesses, opportunities and challenges factors

Challenge factors analysis

Food security in Birbhum District is impact by various challenges within local agricultural practices, such as: 1. Providing Farmers with proper Training on Optimal Practices (27.7%): Instructing farmers on innovative, effective agricultural techniques is necessary but challenging, impact the overall productivity and sustainability of conventional methods. Proper Management of Pests and Diseases (15.9%): management of pests and diseases remains a major concern that impact agricultural production and well-being. Many kind of crops for health and nutritional growth (9.4%): Diversified crops helps different dietary needs, but require meticulous planning and implementation. Environmental Degradation (8.8%): Risks to the long-term sustainability of agricultural methods include soil pollution and erosion. Certainly, share the text you want me to rephrase. Inspiring Farmers to Adopt Innovation and Adjust (4.0%): A continual issue is urging farmers to embrace new ideas and adjust to changing conditions.6. Guaranteeing Access to Quality Seeds and Resources (17.0%): While inconsistent, obtaining premium seeds and resources is crucial for maintaining productivity. Loss of Agricultural Land Due to Urban Expansion (7.0%): The expansion of urban areas invades vital agricultural land, reducing the area available for farming. Pressure from International Agricultural Markets (4.7%): The presence of cheaper alternatives in global markets can undermine local practices and threaten the economic stability of local farmers. Soil erosion, deforestation, and pollution (2.8%): Environmental issues affecting ecosystem well-being and farming efficiency comprise deforestation and soil erosion.

Plate: A-Indigenous agricultural practices often employ organic methods to boost soil quality, crucial for ongoing agricultural productivity, B-Adopting suitable water conservation strategies by santal tribe like drip irrigation, rainwater harvesting can significantly boost productivity while conserving a limited resource, C-Indigenous knowledge by santal tribe that adapt to regional climates decline the impact of environmental changes, encouraging sustainable agriculture, D-Indigenous customs are vulnerable to climate changes as they often depend significantly on local weather patterns, E-Agroforestry enhances soil nutrients, minimizes erosion, and improves water retention, F-Traditional agricultural practices control ecological harmony, G-Indigenous seed collection and store for future, H-Dependence on materials such as farm-saved seeds, compost, and local biomass lessens reliance on external inputs and supports sustainability, I-Diverse agricultural setups, including varied farm infrastructure, increase the stability and robustness of production systems.

Conclusion

The study shows that traditional farming methods are essential for food security but encounter major obstacles. Indigenous practices such as climate-smart agriculture, water collection, and crop variety (index values spanning 0.13–0.68) support soil vitality and ecological diversity. Major weaknesses consist of obsolete tools (28%), a technological gap (18.3%), and obstacles to market access. Opportunities like increasing organic demand (28.2%) and agrotourism continue to be underexploited. Tackling educational inequalities, ensuring financial accessibility, and incorporating contemporary pest control methods are crucial. In the end, supporting and enhancing indigenous knowledge with focused policy measures can convert traditional agriculture into a sustainable, food-secure future for the communities in Birbhum.

Thus, it is concluded that such type of research has described the indigenous knowledge strategies, measure sustainability in terms of agriculture, determine strengths and weaknesses, evaluate opportunities and threats, and determine diversification of crops in Birbhum district following a process of literature review, field survey, interview, group discussion, priority calculation, AHP, SWOC analysis, data analysis, suggestions, recommendations, and opportunities for Indigenous knowledge and sustainable agricultural practices where traditional land management ensure food security and sustainability.

References

Aasa, O. (2019). Analyses and methods. *Likert Scales Analysis and Method, July*, 1–8. https://doi.org/10.13140/RG.2.2.32784.02569/1

Gassner, A., & Dobie, P. (2022). *Agroforestry: A primer*. https://doi.org/10.5716/cifor-icraf/bk.25114

Ajatasatru, A., Prabhu, V., Pal, B. D., & Mukhopadhyay, K. (2024). Economy-wide impact of climate smart agriculture in India: A SAM framework. *Journal of Economic Structures, 13*(1). https://doi.org/10.1186/s40008-023-00320-z

Akter, R., Hasan, M. K., Kabir, K. H., Darr, D., & Roshni, N. A. (2022, April). Agroforestry systems and their impact on livelihood improvement of tribal farmers in a tropical moist deciduous forest in Bangladesh. *Trees, Forests and People*, *9*, 100315. https://doi.org/10.1016/j.tfp.2022.100315

Anuja, A. R., & Ray, M. (2021). *Project report crop diversification: Pattern, determinants and its impact on nutritional security in India Rajesh T.* June 2022.

Anuja, A. R., Shivaswamy, G. P., Ray, M., & Singh, K. N. (2022). Pattern of crop diversification and its implications on undernutrition in India. *Current Science*, *122*(10), 1154–1160. https://doi.org/10.18520/cs/v122/i10/1154-1160

Asfaw, Z., & Tadesse, M. (2001). Prospects for sustainable use and development of wild food plants in Ethiopia. *Economic Botany*, *55*(1), 47–62. https://doi.org/10.1007/BF02864545

Bala, M. (2020). Food security in India–policy, issues and challenges. *SSRN Electronic Journal, March*. doi: https://doi.org/10.2139/ssrn.3555193

Basantaray, A. K., Paltasingh, K. R., & Birthal, P. S. (2022). Crop diversification, agricultural transition and farm income growth: Evidence from Eastern India. *Italian Review of Agricultural Economics*, *77*(3), 55–65. https://doi.org/10.36253/rea-13796

Beglar, D., & Nemoto, T. (2014). Developing likert-scale questionnaires. *JALT2013 Conference Proceedings*, 1–8.

Berihun, T., & Molla, E. (2017). Study on the diversity and use of wild edible plants in bullen district Northwest Ethiopia. *Journal of Botany*, *2017*, https://doi.org/10.1155/2017/8383468

Beyene, S. D. (2023). The impact of food insecurity on health outcomes: Empirical evidence from sub-Saharan African countries. *BMC Public Health*, *23*(1), 1–22. https://doi.org/10.1186/s12889-023-15244-3

Bharucha, Z., & Pretty, J. (2010). The roles and values of wild foods in agricultural systems. *Philosophical Transactions of the Royal Society B: Biological Sciences*, *365*(1554), 2913–2926. https://doi.org/10.1098/rstb.2010.0123

Bhoite, A. A., & Athawale, G. H. (2018). *Food insecurity to food security: An overview at global level. 3*(4), 73–77.

Birthal, P., Joshi, P., Roy, D., & Thorat, A.. (2011). Diversification in Indian agriculture towards high-value crops: The role of smallholders. *IFPRI Discussion Paper*, *00727*, 40.

Bishwajit, G., Barmon, R., & Ghosh, S. (2014). Reviewing the status of agricultural production in Bangladesh from a food security perspective. *Russian Journal of Agricultural and Socio-Economic Sciences*, *25*(1), 19–27. https://doi.org/10.18551/rjoas.2014-01.04

Bobe, M., Procopie, R., & Bucur, M. (2019). Exploring the role of individual food security in the assessment of population's food safety. *Amfiteatru Economic*, *21*(51), 347–360. https://doi.org/10.24818/EA/2019/51/347

Brahmanand, P. S., Kumar, A., Ghosh, S., Roy Chowdhury, S., Singandhupe, R. B., Singh, R., Nanda, P., Chakraborthy, H., Srivastava, S. K., & Behera, M. S. (2013). Challenges to food security in India. *Current Science*, *104*(7), 841–846. https://doi.org/10.9790/0837-194293100

Brookfield, H., & Stocking, M. (1999). Agrodiversity: Definition, description and design. *Global Environmental Change*, *9*(2), 77–80. https://doi.org/10.1016/S0959-3780(99)00004-7

Burchi, F., Fanzo, J., & Frison, E. (2011). The role of food and nutrition system approaches in tackling hidden hunger. *International Journal of Environmental Research and Public Health*, *8*(2), 358–373. https://doi.org/10.3390/ijerph8020358

Butler, C. D. (2017, July). Global food security, population and limits to growth. *Health of People, Places and Planet: Reflections Based on Tony McMichael's Four Decades of Contribution to Epidemiological Understanding*. https://doi.org/10.22459/hppp.07.2015.14.

Buxton, M., Carey, R., & Phelan, K. (2016). Balanced urban development: Options and strategies for liveable cities. *Balanced Urban Development: Options and Strategies for Balanced Development for Liveable Cities*, *72*, 153–170. https://doi.org/10.1007/978-3-319-28112-4

Camacho, L. D., Gevaña, D. T., Carandang, A. P., & Camacho, S. C. (2016). Indigenous knowledge and practices for the sustainable management of Ifugao forests in Cordillera, Philippines. *International Journal of Biodiversity Science, Ecosystem Services and Management*, *12*(1–2), 5–13. https://doi.org/10.1080/21513732.2015.1124453

Camire, M. E. (2021, *February*). Food Insecurity. *Handbook of Rural Aging*, 170–175. https://doi.org/10.4324/9781003128267-31

Chaudhry, A. G., & Chaudhry, H.-R. (2011). Indigenous farming practices and sustainable rural development: A case of Indigenous agricultural practices in a Punjabi Village of Sheikhupura district #. *FWU Journal of Social Sciences*, *5*(2), 98–128.

Cheber Bezu, D. (2018). A review of factors affecting food security situation of Ethiopia: From the perspectives of FAD, economic and political economy theories. *International Journal of Agriculture Innovations and Research*, *6*(6), 2319–1473.
Cheplogoi, S. K., Ombati, J. M., & Udoto, M. O. (2021). Diversification of indigenous agricultural practices and implications on household food security: Practices and lessons from local communities of Baringo County, Kenya. *International Journal of Advanced Research*, *9*(10), 1232–1240. https://doi.org/10.21474/ijar01/13672
Chilton, M., & Rose, D. (2009). A rights-based approach to food insecurity in the United States. *American Journal of Public Health*, *99*(7), 1203–1211. https://doi.org/10.2105/AJPH.2007.130229
Dalal, S., & Shankar, T. (2022). Crop diversification and its importance in agriculture: A review. *Indian Journal of Natural Science*, *13*(72), 44540–44548.
de Graaff, J., Kessler, A., & Nibbering, J. W. (2011). Agriculture and food security in selected countries in sub-Saharan Africa: Diversity in trends and opportunities. *Food Security*, *3*(2), 195–213. https://doi.org/10.1007/s12571-011-0125-4
Deaton, B. J., & Scholz, A. (2022). Food security, food insecurity, and Canada's national food policy: Meaning, measures, and assessment. *Outlook on Agriculture*, *51*(3), 303–312. https://doi.org/10.1177/00307270221113601
Derrickson, J. P., Fisher, A. G., & Anderson, J. E. L. (2000). The core food security module scale measure is valid and reliable when used with Asians and Pacific Islanders. *Journal of Nutrition*, *130*(11), 2666–2674. https://doi.org/10.1093/jn/130.11.2666
Devereux, S. (2001). Food security and nutrition: The global challenge. *Food Policy*, *26*(1), 97–99. https://doi.org/10.1016/s0306-9192(00)00026-9
Doshi, P., Brockington, J., & Brook, R. (2015). Assessing agroforestry adoption in tribal areas of Maharashtra, India. *Agriculture for Development*, *25*, 12–16.
Parvathamma, G. L. (2015). An assessment on food security in developing economiesproblems \nand policy initiatives\n. *IOSR Journal of Economics and Finance (IOSR-JEF*, *6*(1), 01–10. https://doi.org/10.9790/5933-06110110
E, E. N. (2019, July). *Indigenous knowledge system: The need for reform and the way forward indigenous knowledge system: The need for reform and the way forward.*
EGWUE, O. L., Agbugba, I. K., & Mukaila, R. (2021). Assessment of rural households food insecurity during covid-19 Pandemic in South-East Nigeria. *International Journal of Research -GRANTHAALAYAH*, *8*(12), 182–194. https://doi.org/10.29121/granthaalayah.v8.i12.2020.2713
Erni, C. (2015). Shifting cultivation, livelihood and food security: New and old challenges for indigenous peoples in Asia. *Shifting Cultivation, Livelihood and Food Security*, *3*, 12.
FAO. (2018). The state of food security and nutrition in the world 2018: Building climate resilience for food security and nutrition—policy support and governance. In *Food and Agriculture Organization of the United Nations*, 7(7). http://www.fao.org/publications%0A, http://www.fao.org/policy-support/tools-and-publications/resources-details/en/c/1152267/
Fawole, W. O., Ozkan, B., & Ayanrinde, F. A. (2016). Measuring food security status among households in Osun State, Nigeria. *British Food Journal*, *118*(7), 1554–1567. https://doi.org/10.1108/BFJ-11-2015-0425
Firake, D., Lytan, D., Thubru, D., Behere, G., Firake, P., & Thakur, N. (2013). Traditional pest management practices and beliefs of different ethnic tribes of Meghalaya, North Eastern Himalaya. *Indian Journal of Hill Farming*, *26*(1), 58–61.
Galhena, D. H., Freed, R., & Maredia, K. M. (2013). Promising aproach. *BioMed Central*, 1–13.
Gallegos, D., Booth, S., Pollard, C. M., Chilton, M., & Kleve, S. (2023). Food security definition, measures and advocacy priorities in high-income countries: A Delphi consensus study. *Public Health Nutrition*, *26*(10), 1986–1996. https://doi.org/10.1017/S1368980023000915
Getachew, M. (2014). Evergreen agriculture: Agroforestry for food security and climate change resilience. *Journal of Natural Sciences Research*, *4*(11), 80–90.

Gitz, V., Meybeck, A., Lipper, L., Young, C., & Braatz, S. (2016). Climate change and food security: Risks and responses. In *Food and agriculture organization of the United Nations*. https://doi.org/10.1080/14767058.2017.1347921

Guo, T., García-Martín, M., & Plieninger, T. (2021). Recognizing indigenous farming practices for sustainability: A narrative analysis of key elements and drivers in a Chinese dryland terrace system. *Ecosystems and People*, *17*(1), 279–291. https://doi.org/10.1080/26395916.2021.1930169

Gurditta, H., & Singh, G. (2016). Climate change, food and nutritional security: Issues and concerns in India. *Journal of Climate Change*, *2*(1), 79–89. https://doi.org/10.3233/jcc-160009

Handa, A. K., Sirohi, C., Chavan, S. B., Dhillon, R. S., Ahlawat, K. S., & Rizvi, R. H. (2020). Agroforestry in Haryana: Status and way forward agroforestry in Haryana: Status and way forward. *Indian Journal of Agroforestry*, *22*(1), 1–+10.

Handoyo, F. W., Hidayatina, A., Nadjib, M., & Purwanto,. (2022). Perspectives of indigenous communities on food security: Understanding local values in agricultural practice. *IOP Conference Series: Earth and Environmental Science*, *1105*(1). https://doi.org/10.1088/1755-1315/1105/1/012007

Hart, T., & Mouton, J. (2005). Indigenous knowledge and its relevance for agriculture: A case study in Uganda. *Indilinga*, *4*(1), 249–263.

Hufnagel, J., Reckling, M., Ewert, F., Hufnagel, J., Reckling, M., & Ewert, F., 2020. Diverse approaches to crop diversification in.pdf. *Agronomy for Sustainable Development.*

IAEA, T. C. P. (2007). Agriculture and food security the IAEA contribution TC projects help member states enhance agricultural productivity and improve food security. *Iaea.*

IFAD (2022). *Sustainable and resilient indigenous peoples' Food Systems about digital toolbox.* 1–37.

Ihenacho, R. A., Orusha, J. O., & Onogu, B. (2019). rural farmers use of indigenous knowledge systems in agriculture for climate change adaptation and mitigation in Southeast Nigeria. *Annals of Ecology and Environmental Science*, *3*(1), 1–11. https://doi.org/10.22259/2637-5338.0301001

Innovation, S., & Goal, S. D. (n.d.). *Brie fing.*

Jain, S. (2016). Food security in India: Problems and prospects. *OIDA International Journal of Sustainable Development*, *9*(1), 11–20. http://www.ssrn.com/link/OIDA-Intl-Journal-Sustainable-Dev.html

Jain, S., & Verma, N. (2021, *June*). Recent research in Agriculture for doubling of farmer's income. *Empyrealpublishinghouse.Com.* https://www.empyrealpublishinghouse.com/pdf/edited-book-recent-research-in-agriculture-for-doubling-of-farmer-income.pdf

James, T. C., & Lee, O. (2011). Using Likert-type scales in the social sciences. *Journal of Adult Education*, *40*(1), 19–22.

Jamnadass, R., Place, F., Torquebiau, E., Malézieux, E., Iiyama, M., Sileshi, G. W., Kehlenbeck, K., Masters, E., McMullin, S., Weber, J. C., & D, I. K. (2013). Agroforestry, food and nutritional security: Background paper n° 4. *ICRAF Working Paper No. 170. Nairobi, World Agroforestry Centre*, 20 p.

Jeder, H., Hattab, S., & Frija, I. (2020). An econometric analysis for food security in Tunisia. *New Medit*, *19*(4), 1–14. https://doi.org/10.30682/nm2004a

Johar, V. (2021). Socio-economic and environmental assets sustainability by agroforestry systems: A review. *International Journal of Agriculture Environment and Biotechnology*, *14*(4). https://doi.org/10.30954/0974-1712.04.2021.6

Joshi, A., Kale, S., Chandel, S., & Pal, D. (2015). Likert scale: Explored and explained. *British Journal of Applied Science & Technology*, *7*(4), 396–403. https://doi.org/10.9734/bjast/2015/14975

Kahane, R., Hodgkin, T., Jaenicke, H., Hoogendoorn, C., Hermann, M., Dyno Keatinge, J. D. H., D'Arros Hughes, J., Padulosi, S., & Looney, N. (2013). Agrobiodiversity for food security, health and income. *Agronomy for Sustainable Development*, *33*(4), 671–693. https://doi.org/10.1007/s13593-013-0147-8

Kanagasabapathi, K., & Sakthivel, V. (2012). Constraints in the adoption of indigenous farming practices. *Journal of Extension Education*, *28*(3), 5723. https://doi.org/10.26725/jee.2016.3.28.5723-5726

Kazmi, T., Chaudhry, A. G., Ahmed, A., & Khan, S. E. (2014). Farmers beliefs about indigenous farming practices and sustainable agricultural development. *Pakistan Journal of Agricultural Research*, *27*(1), 51–57.

Kemmerling, B., Schetter, C., & Wirkus, L. (2022, June). The logics of war and food (in)security. *Global Food Security*, *33*, 100634. https://doi.org/10.1016/j.gfs.2022.100634

Khan, A. (2020). City food security. In *Resilient Sustainable Cities*. https://doi.org/10.4324/9780203593066-20

Kidane, L., & Kejela, A. (2021). Food security and environment conservation through sustainable use of wild and semi-wild edible plants: A case study in berek natural forest, Oromia special zone, Ethiopia. *Agriculture and Food Security*, *10*(1), 1–16. https://doi.org/10.1186/s40066-021-00308-7

Kloos, H. (2023). Challenges and prospects of medicinal plant sustainability in Ethiopia. *Journal of Pharmacy and Pharmacology Research*, *07*(04), 233–242. https://doi.org/10.26502/fjppr.088

Koocheki, A. A. (2004). Indigenous knowledge in agriculture with particular reference to saffron production in Iran. *Acta Horticulturae*, *650*, 175–182. https://doi.org/10.17660/ActaHortic.2004.650.17

Kraft, P., Rezaei, E. E., Breuer, L., Ewert, F., Große-stoltenberg, A., Kleinebecker, T., Seserman, D. M., & Nendel, C. (2021). Modelling agroforestry's contributions to people—a review of available models. *Agronomy*, *11*(11), 1–25. https://doi.org/10.3390/agronomy11112106

Kumar, A., Bantilan, M. C. S., Kumar, P., Kumar, S., & Jee, S. (2012). Food security in India: Trends, patterns and determinants. *Indian Journal of Agricultural Economics*, *67*(3), 445–463.

Kuyu, C. G., & Bereka, T. Y. (2020). Review on contribution of indigenous food preparation and preservation techniques to attainment of food security in Ethiopian. *Food Science and Nutrition*, *8*(1), 3–15. https://doi.org/10.1002/fsn3.1274

Leakey, R. R. B. (2017). Definition of agroforestry revisited. *Multifunctional Agriculture, February*, 5–6. https://doi.org/10.1016/b978-0-12-805356-0.00001-5

Lugo, D. (2020). Las comunidades indígenas y sus sistemas alimentarios: Una contribuci6n al debate actual. *Journal of Ethnic Foods*, *7*(1), 1–10.

Mahmood, T., Kumar, R., Ali, T. M., Naeem, N., & Pongpanich, S. (2023, October). Determinants of the food insecurity at household level in Pakistan: A multilevel model approach. *PLoS ONE*, *18*(10), 1–12. https://doi.org/10.1371/journal.pone.0291343.

Martins, B. O. (2020). The impact of climate change on food security. *International Journal of Multidisciplinary Education and Research*, *5*(4), 26–32. https://doi.org/10.13140/RG.2.2.24853.35042

Masango, C. A., & Mbarika, V. W. (2022). Indigenous knowledge application in increasing food security: A measure to consider? *Inkanyiso*, *14*(1), 1–10. https://doi.org/10.4102/ink.v14i1.20

McCord, P. F., Cox, M., Schmitt-Harsh, M., & Evans, T. (2015). Crop diversification as a smallholder livelihood strategy within semi-arid agricultural systems near Mount Kenya. *Land Use Policy*, *42*(2015), 738–750. https://doi.org/10.1016/j.landusepol.2014.10.012

McKay, F. H., Sims, A., & van der Pligt, P. (2023). Measuring food insecurity in India: A systematic review of the current evidence. *Current Nutrition Reports*, *12*(2), 358–367. https://doi.org/10.1007/s13668-023-00470-3

Miccolis, A., Mongeli, F., Henrique, P., Marques, R., Luis, D., Vieira, M., Francia, M., Maurício, A.-V., Hoffmann, R., Rehder, T., Vinicius, A., & Pereira, B. (2016). *Agroforestry Systemsfor Ecological Restoration*.

Mohamed, Z., Terano, R., Sharifuddin, J., & Rezai, G. (2016). Determinants of Paddy Farmer's unsustainability farm practices. *Agriculture and Agricultural Science Procedia*, *9*, 191–196. https://doi.org/10.1016/j.aaspro.2016.02.120

Mph, R. J., & Kannan, S. (2021). *Maintaining food security during dicult times: An experience of COVID-19 in Kerala*. 1–8. https://doi.org/10.21203/rs.3.rs-864547/v1

Mumu, J., Tanujaya, B., Charitas, R., & Prahmana, I. (2022). Likert scale in social sciences research: Problems and difficulties. *FWU Journal of Social Sciences*, *16*(4), 89–101. https://doi.org/10.51709/19951272/Winter2022/7

Narkhede, P. A. (2014). *Title of the paper: "Food security in India: Concepts, realities & innovations" author (s): Mailing address*: *June*.

Nnadi, F. N., & Ezudike, J. (2013). Assessment of indigenous knowledge practices for sustainable agriculture and food security in idemili south local government area of anambra state, Nigeria. *Journal of Resources Development and Management-An Open Access International Journal*, *1*, 14.

Nyashilu, I. M., & Nyomora, S. M. A. (2023). *Research Challenges and opportunities for up-scaling indigenous farming practices for adaptation to climate change induced food insecurity in Busega District, Simiyu Region, Tanzania*. *19*(4), 419–427. https://doi.org/10.5897/AJAR2020.14759

Olaniy, A. A. (2019). Application of likert scale's type and Cronbach's alpha analysis in an airport perception study. *Scholar Journal of Applied Sciences and Research*, *2*(4), 1–5. http://innovationinfo.org/articles/SJASR/SJASR-4-223.pdf

Othieno, L., & Shinyekwa, I. (2011). This document is discoverable and free to researchers across the globe due to the work of AgEcon search. Help ensure our sustainability.. *Trade, Revenue and Welfare Effects of the East African Community Customs Union Principle of Asymmetry on Uganda*.

Paroda, R. (2022). Crop Diversification for Sustainable Agriculture. *Ecology, Economy and Society*, *5*(1), 15–21. https://doi.org/10.37773/ees.v5i1.611

Patel, S. K., Sharma, A., & Singh, G. S. (2020). Traditional agricultural practices in India: An approach for environmental sustainability and food security. *Energy, Ecology and Environment*, *5*(4), 253–271. https://doi.org/10.1007/s40974-020-00158-2

Pawlak, K., & Kołodziejczak, M. (2020). The role of agriculture in ensuring food security in developing countries: Considerations in the context of the problem of sustainable food production. *Sustainability (Switzerland*, *12*(13). https://doi.org/10.3390/su12135488

Ponge, A. (2013). Integrating indigenous knowledge for food security: Perspectives from the millennium village project at Bar-Sauri in Nyanza Province in Kenya. *Nternational Conference on Enhancing Food Security in the Eastern and Horn of Africa Regions. A Conference Held at the Imperial Royale Hotel in Kampala, Uganda on 16 – 17 November 2011.*, 43.

Pradesh, A., & Security, F. (n.d.). *the Tribals of Andhra Pradesh and Telangana The Trajectory of Transition and Impacts on Livelihoods and Food Security*.

Rahaman, H., & Singh, S. (2020). De-stunting growth of crop diversification Index: A case study from Malda District, India. *Copyright@ EM International*, *26*(September).

Rahman, M. M., Islam, A., & Ferdousee, S. (2024). Crop diversification, sustainable production, and consumption (SDG-12) in rural Bangladesh: Insights from the northern region of the country. *Circular Agricultural Systems*, *4*(1), 0–0. https://doi.org/10.48130/cas-0024-0005

Rahman, M., Paul, S., & Saha, J. (2019). Food security and livelihood of tribal people in Bangladesh: The role of microfinance. *Fundamental and Applied Agriculture*, *0*, 1. https://doi.org/10.5455/faa.37825

Rao, P. S. (2020). *Food security bill—issues and implications food security bil—Issues and Implications 2018 IJLERA—all right reserved. April 2018*.

Reddy, A. A., Melts, I., Mohan, G., Rani, C. R., Pawar, V., Singh, V., Choubey, M., Vashishtha, T., Suresh, A., & Bhattarai, M. (2022). Economic impact of organic agriculture: Evidence from a pan-India survey. *Sustainability (Switzerland*, *14*(22). https://doi.org/10.3390/su142215057

Roy, D., Dev, D. S., & Sheheli, S. (2019). Food security in Bangladesh: Insight from available literature. *Journal of Nutrition and Food Security*, *4*(1), 66–75. https://doi.org/10.18502/JNFS.V4I1.401

Saad, A. Z., Radi, T., Saida, S., & Al-Najjar, K. (2023). Estimating some indicators of food security for rural households. *Asian Journal of Agriculture and Food Sciences*, *11*(2). https://doi.org/10.24203/ajafs.v11i2.7171

Sahal, P., & Rudra, S. (2020). *Analysis of crop diversification and irrigation intensity: A study of Birbhum analysis of crop diversification and irrigation intensity:' A study of Birbhum district. December 2017.*

Sarvade, S., Singh, R., Vikas, G., Kachawaya, D. S., & Khachi, B. (2014). Agroforestry: An approach for food security. *Indian Journal of Ecology*, *41*(1), 95–98.

Sarveswaran, S., Johar, V., Singh, V., & Ragunanthan, C. (2023). Agroforestry: A way forward for sustainable development. *Ecology, Environment and Conservation*, *29*(Suppl). https://doi.org/10.53550/eec.2023.v29i04s.046

Sekaran, U., Lai, L., Ussiri, D. A. N., Kumar, S., & Clay, S. (2021). Role of integrated crop-livestock systems in improving agriculture production and addressing food security – A review. *Journal of Agriculture and Food Research*, *5*, 100190. https://doi.org/10.1016/j.jafr.2021.100190

Shelembe, N., Hlatshwayo, S. I., Modi, A., Mabhaudhi, T., & Ngidi, M. S. C. (2024). The association of socio-economic factors and indigenous crops on the food security status of farming households in KwaZulu-Natal Province. *Agriculture (Switzerland*, *14*(3), 1–18. https://doi.org/10.3390/agriculture14030415

Singh, G. (2016). Climate change and food security in india: Challenges and opportunities. *Irrigation and Drainage*, *65*(2), 5–10. https://doi.org/10.1002/ird.2038

Singh, R., & Singh, G. S. (2017). Traditional agriculture: A climate-smart approach for sustainable food production. *Energy, Ecology and Environment*, *2*(5), 296–316. https://doi.org/10.1007/s40974-017-0074-7

Sobola, O. O., & Amadi, D. C. (2015). *The role of agroforestry in environmental sustainability*. *8* (5), 20–25. https://doi.org/10.9790/2380-08512025

Stefanis, C. (2014). Global food security: An agricultural perspective. *Journal of Agriculture and Sustainability*, *6*(1), 69–87.

Subagyono, K., & Pawitan, H. (2008). Water harvesting techniques for sustainable water resources management in catchments area. *International Workshop on Integrated Wastershed Management for Sustainable Water Use in Humid Tropical Region*, *8*, 18–30.

Teeuwen, A. S., Meyer, M. A., Dou, Y., & Nelson, A. (2022). A systematic review of the impact of food security governance measures as simulated in modelling studies. *Nature Food*, *3*(8), 619–630. https://doi.org/10.1038/s43016-022-00571-2

Temesgen, B. (2021). Global role of plant breeding in tackling climate change. *International Journal of Agricultural Science and Food Technology*, *July*, 223–229. https://doi.org/10.17352/2455-815x.000111

The, F., Agricultural, G. I., & Systems, H. (n.d.). Format for proposals of candidate systems for the globally-important ingenious agricultural heritage systems (GIAHS) programme. *Most*, 1–24.

Tonui, K. K., Onyango, A. C., & Ouma, C. (2024). Seasonality of associations between production of indigenous foods and food security status of mother-child dyads in Kisumu County, Kenya. *BMC Nutrition*, *10*(1), 1–15. https://doi.org/10.1186/s40795-024-00820-6

United Nations. (2017). The role of science, technology and innovation in ensuring food security by 2030. *United Nations Conference on Trade and Development*, 55. https://unctad.org/system/files/official-document/dtlstict2017d5_en.pdf

Upreti, Y. G. (1999). *Indigenous knowledge, agricultural practices and food security in developing countries: Opportunities and challenges*. 25.

Vågsholm, I., Arzoomand, N. S., & Boqvist, S. (2020). Food security, safety, and sustainability—getting the trade-offs right. *Frontiers in Sustainable Food Systems*, *4*(February), 1–14. https://doi.org/10.3389/fsufs.2020.00016

van Dijk, M., & Meijerink, G. W. (2014). A review of global food security scenario and assessment studies: Results, gaps and research priorities. *Global Food Security*, *3*(3–4), 227–238. https://doi.org/10.1016/j.gfs.2014.09.004

Vatsa, P., Ma, W., Zheng, H., & Li, J. (2023). Climate-smart agricultural practices for promoting sustainable agrifood production: Yield impacts and implications for food security. *Food Policy*, *121*(April), 102551. https://doi.org/10.1016/j.foodpol.2023.102551

VERMA, S., SINGH, V., VERMA, D. K., & Giri, S. P. (2016). Agroforestry practices and concepts in sustainable land use systems in India. *International Journal of Forestry and Crop Improvement*, *7*(1), 126–131. https://doi.org/10.15740/has/ijfci/7.1/126-131

Vijayan, D., Ludwig, D., Rybak, C., Kaechele, H., Hoffmann, H., Schönfeldt, H. C., Mbwana, H. A., Rivero, C. V., & Löhr, K. (2022). Indigenous knowledge in food system transformations. *Communications Earth and Environment*, *3*(1), 9–11. https://doi.org/10.1038/s43247-022-00543-1

Wakweya, R. B. (2023). Challenges and prospects of adopting climate-smart agricultural practices and technologies: Implications for food security. *Journal of Agriculture and Food Research*, *14*(June), 100698. https://doi.org/10.1016/j.jafr.2023.100698

Wangpan, T., Taka, T., Tangjang, S., & Arunachalam, A. (2014). Sustainability of Indigenous farming practices: A case study among apatani, Nocte and Wancho tribes in Arunachal Pradesh, Northeast India. *International Journal on Agricultural Sciences*, *6*(1), 103–108.

Wangpan, T., Tangjang, S., & Arunachalam, A. (2017). Tribal agriculture: Tradition in transition in the Indian Eastern Himalaya. *Current Science*, *112*(7), 1327–1329.

Yadav, R. P. (2018). Sustainable agroforestry systems for livelihood security and their economic appraisal in Indian Himalayas. *Economic Affairs*, *63*(3), 633–639. https://doi.org/10.30954/0424-2513.3.2018.7

Zhou, Z.-Y. (2018). Current status of food security in China. *Achieving Food Security in China, January*, 88–109. https://doi.org/10.4324/9781315751092-4

Zhu, P., Hietala, R., Motsholapheko, M., Yusriadi, Y., & Cahaya, A. (2022). *Open access edited by food security systems in rural communities: A qualitative study.*

Zimmerer, K. S., & de Haan, S. (2020). Informal food chains and agrobiodiversity need strengthening—not weakening—to address food security amidst the COVID-19 crisis in South America. *Food Security*, *12*(4), 891–894. https://doi.org/10.1007/s12571-020-01088-x

Chapter 17
Traditional Healing Practices: A Study Among the Western Himalayan Gaddi Tribe

Shivanshi Nehria

Abstract The nomadic Gaddi people have long lived in the western Himalayas. Gaddis predominate in Kangra and Chamba districts in Himachal Pradesh. This study identifies the customary healthcare practices of Gaddis in the Bharmour tehsil of Chamba district and the Dharamshala *tehsil* of Kangra district. They raise sheep and goats and divide their time between winters in the Himachal plains and summers in alpine regions on mountains. They are familiar with wild species and their varied uses because they roam and live close to forests. Elderly residents, women, and traditional specialists were questioned to gain insight into the community's traditional healing practices. The respondent was chosen via purposive sampling. Because the Gaddis lived in the highlands without access to medical services, they created their own special methods for treating illnesses. They had little trouble recognising plants that were harmful to humans and animals alike. Gaddis today is split into two groups: pastoralists and those who live permanently in one location. Traditional knowledge of wild species and belief in spiritual healing are prevalent in both groups. According to the exploration, both humans and animals are treated with conventional healing methods. The tribe is permeated with this wisdom. The wild edible plants are eaten seasonally and have a wealth of medicinal properties. Gaddis has no difficulty identifying medicinal plants that are suitable for humans and animals. The community's traditional healers utilise medicinal plants to treat their patients, but they are reluctant to impart their plant expertise and mantras used in curing to others outside the family. The study describes the Gaddi tribe's customary healthcare practices in detail and explains how younger people come to believe in these practices.

Keywords Gaddi tribe · Pastoralist · Medicinal plants · Spiritual healing

---

S. Nehria (✉)
Department of Anthropology, Eliezer Joldan Memorial College, leh, Ladakh, India
e-mail: nehriashivanshi@gmail.com

R. C. Sobti et al. (eds.), *Traditional Knowledge Systems in India: Sustainability, Ecology, and Resource Governance*, People, Cultures and Societies: Exploring and Documenting Diversities,
https://doi.org/10.1007/978-981-92-1346-7_17

Introduction

Traditional, indigenous, or non-western forms of treatment are still practiced in many parts of the world. The term "traditional medicine" or "traditional system of health care" describes indigenous healthcare systems in developing countries that serve indigenous populations. Traditional medical systems see people as closely tied to nature (Bhasin, 2002). Communities that prioritise religion as a legislative framework develop traditions that are passed down through generations. Religious rituals shape our understanding of the cosmos, creation, human psychology, well-being, ill-health, and ultimately mortality (Farooqi, 2006). Indigenous peoples define community as people in their immediate surroundings, as well as connections to the environment or the entire planet, relationships with ancestors who have gone before, and relationships with those who will come after them. Therefore, "community means the living, the unborn, the dead, and nature as a whole" (Wangoola, 2000, p. 271, cited in Rowkith & Bhagwan, 2020).

The World Health Organisation defines holistic systems as those that consider man as a whole within a wider ecological spectrum. They emphasize that illness or disease is caused by an imbalance in man's overall ecological system rather than just by a causal factor or pathological evolution (Bhasin, 2002). A traditional medical practitioner or traditional healer is someone who is recognized by the community in which he lives as competent to provide health care using vegetable, animal, and mineral substances, as well as certain other methods, based on the community's social, cultural, and religious backgrounds, as well as the prevailing knowledge, attitudes, and beliefs about physical, mental, and social well-being and the causes of disease and disability. Traditional healers employed therapeutic formulas derived from various natural substances (animal, mineral, and vegetable). They have a considerable understanding of how to use plants and herbs for medical and nutritional purposes (Shankar et al., 2012).

Jakovljevic (2017) described spirituality as a force of integration that includes the awareness of spiritual concepts and principles that create an understanding of greater significance and connection between oneself, others, the natural world, and the cosmos itself. It has also been interpreted as a sense of togetherness among all other individuals, incorporating close links with kin, communities, their nature, and the cosmos (McGoldrick and Hardy 2019). Spirituality plays a vital role in tribal people's culture and acts as an indicator for how they spend their daily lives (Chatterjee & Sharma, 2018). Tribes contend that an immortal or supernatural power gifted creation with abundant resources while also serving as an advisor for their safety and progress. As a result, the use of omens, dreams, and divination is an important component of tribal spirituality.

The Gaddis are a people who reside in the Dhauladhar range of the Western Himalayas, primarily in the districts of Chamba and Kangra. Variously described as semi-pastoral, semi- agriculturists, semi-nomadic, semi-sedentary, transhumant, migratory, and largely agriculturists in earlier studies, In the present scenario, Gaddis have been divided into two groups: one is a pastoral group, and the other is a settled one. Gaddis are primarily agricultural. Gaddis, who may be a transhuman tribe, follow well-defined routes and exploit pastures on a seasonal basis. The harsh winters and snow push them to shift to new biological zones and switch pastures. Pastoralism is a historic profession with spiritual significance among the Gaddis. Actually, they consider themselves Shaivic or practice Shivaism, and everything that happens in their lives is somehow related to Lord Shiva. The Shiva story can be viewed as an underlying mythology in terms of how identities are formed and how places are perceived.

Singh & Kaushal (2000) researched the enormous plant variety of the Gaddis of Himachal Pradesh's Kangra Valley, discovering that the Gaddis travelled long distances in search of food for their animals, collected wild plants for medication, and used herbal medicine to cure human and animal ailments. Gaddi women were mindful of curative herbs grown nearby and used to heal the illnesses of their children (Sharma & Maheshewari, 2004). According to Kishor et al. (2018), Gaddis consume a wide variety of wild edible plants, including leaves, fruits, roots, stems, and bark. They employed wild creatures to treat ailments such as arthritis, diabetes, jaundice, nephritis, colds, coughing, stomach problems, dental issues, and more. Gaddis are Hindus by religion, but their belief in nature, spirits, and supernatural creatures confuses their religion. They seek guidance from oracles and shamans to cure their ailments and believe in medicinal plurality (Bhasin, 2008). Thakur et al. (2016) discovered that the Gaddis younger generation showed a lack of curiosity and belief in the traditional medical system, but the elderly had good recognition abilities. The Gaddi tribal group of the Bharmour in Chamba district and Dharamshala in Kangra district of Himachal Pradesh is the focus of the current paper, which examines their traditional healing practices to cure their ailments. The paper will look at the views of the Gaddis on the conventional healing method.

Research Methodology

First-hand information on traditional knowledge regarding medicinal plant utilisation by the inhabitant was gathered through semi-structured and in-depth interviews with two tehsils: Bhramour in Chamba district and Dharamshala in Kangra district. The participant observation technique was used. Qualitative data was collected from the traditional healers, elder citizens of the community through interviews who had the experience of living in *dhars* (seasonal settlements in mountains), and other knowledgeable informants of the area. Quantitative datawas collected from randomly selected 400 Gaddi households in Bharmour and Dharamshala *tehsil*. The photographs were clicked of wild medicinal plants that were collected by the informant or that I found on my way with the guidance of the informant. Narratives are used for the analysis of qualitative data. Quantitative data is assessed using statistical tools.

Result and Observation

Gaddis have a propensity to adhere to traditional herbal and spiritual therapies. They have a co- existing belief in both natural causes and supernatural causes. The use of biomedicine was hampered by the absence of health services and communication, along with historical migration routes. The Gaddis are renowned for their knowledge of the identification of wild plants, and they begin to look for valuable medicinal and edible plants as time passes for harvest. Herbs from the forest are used by *vaids* (medicine men) and *dali bane walas* (traditional healers) to cure ailments in villages. *Dali bane wala* performs *jhad phuk*, which is the act of blowing away evil spirits, on the afflicted person by using *bana* (chinaberry) leaves or Cannabis sativa with *mantras*. He treats issues pertaining to the skin. Traditional healers never reveal the healing plants and *mantras* they use. They believe that by disclosing their knowledge, their healing powers will wane and their capacity to heal others will be diminished. Of all the ancient healers, only the v*aid* was able to prescribe medicinal plants and charge for his services. Within the family, the expertise of traditional healers is verbally transmitted from one generation to the next. For their services, the *dali bane wala* does not charge. Doctors are becoming more and more popular, although *dai* (midwives), used to deliver babies at home earlier. Pregnant women continue to visit midwives in the last trimester to receive massages and to find out the baby's position.

Malkauta village resident Sant Ram (*vaid*/85 years old) stated that,"

> *Initially, I was not very knowledgeable about plant identification. All I knew came from my twelve years as a young pastoralist who lived with his father in an era of grazing for animals. In Bharmour and the lower hills, people trusted and flocked to my father for treatment of any illness because he was a highly esteemed vaid. During my travels, I became knowledgeable about various types of therapeutic herbs. I leave the village today and go to the dhars (mountain dwellings), which are far away, to collect medicinal plants.* (Table 17.1)

Ram Singh from Dharamshala (*dali bane wale,56* years old) said,

> *I treat five to seven illnesses, such as lunddhe (chicken pox), breast fibrosis, punjia zehar (hand and foot infection), janeu, patka, and darasal (skin diseases). In the process, bekal leaves, senspa, vulture feathers, and bana (chinaberry) leaves are employed. Additional*

Table 17.1 Table represents the medicine used by *vaid* for ailments

Medicine	Ailment
Jeevani	It is made of *kalkutiyazeher* (wild medicinal plants), and used for boils
Nepali	Given at the last stage of life
Jalab	For constipation
Hajmoli	For indigestion
Saraj	For headache and dizzyness
Gaudandiya hadtal	For warm effects in the body and dizzyness

Table 17.2 Table represents the cure used by *dali bani wali* for ailments

Ailment	Cure
Helliorcharmrog (skin problem)	Medicinal plant oil and *bana* leaves are used in *dali bani* therapy. Only on Fridays is oil distributed. *March* (black pepper) seeds are added to the oil, which is then completely infused with mantras during preparation. Restrictions are imposed on the affected person. The restrictions include avoiding particular foods, staying away from ladies wearing red clothes or who are menstruating, and combing. Depending on the nature of the issue, the affected person will need to take the *Dali* at least three times, depending on the type of problem
Kalunda (germs in teeth)	It gives me terrible oral pain. Dali bana is then conducted following the giving of ash (*vibhuti*), which is made using mantras. There is relaxation for the patient. *Mantras* are repeated for total release if problems continue, and an iron nail is fastened to the tree. It is called the *kilna* process.
Kangugu (ear pain)	To solve the issue, *dali* is carried out and provided *kaar* (restricted)
Adhharipedh (headache on half side of head)	Many people used to endure pain, even if it is rare these days. Sun down marks the conclusion of the suffering, which begins at sunrise. Dali is performed as a remedy
Chickenpox	Dali is offered. The last day is when the *lual* (last day ritual) is held. Depending on how bad the illness is, the *lual* can take place five to seven days later, or even later on unusual days

illnesses include Adhhari pedh (headache on half of the head), Kangugu, Kalunda (tooth-germs),hellior charmrog (body itching), and Dandiya zeher (Tables 17.2 and 17.3).

Deshraj (59 years old), a bone setter from Dharamshala, said,

I treat both humans and animals. I practice without any formal training. With my touch and outer look, I identify the problem. I prepare paste and oil with wild medicinal plants for sprains and body aches.

Satni Devi (74 years old) was a *dai* from Bhagsunag village, Dharamshala. She shared her experience with Dai and stated,

I've made about 70 deliveries total. Home deliveries are typically carried out in the kitchen in Gaddis culture since it has a hearth for warmth and quick availability of hot water. For a healthy foetus, the head should be down and the feet should be up. In our customary method of birthing, massage begins in the seventh month and lasts until the eighth month. More intense massage techniques may result in the infant developing abnormalities. By the ninth month, the growing baby had matured completely, and it was possible to feel where it was located inside the womb. In the ninth and tenth months of pregnancy, a woman is given a til(sasame seed) kada. I advise the mother to calm down and avoid becoming anxious when the labour pains start. Home births are undoubtedly 100 times better than hospital births if women have patience and trust in their doctors. The sole distinction is that incisions are administered and heartbeats are monitored in hospitals; otherwise, the procedures are identical.

Table 17.3 List of wild medicinal herbs used by Gaddis to treat illness

Wild medicinal plants	Illness
Chaura (Angelica glauca)	For paralysis, constipation and body ache
Kuth (Saussurea lappa)	Bodyache and Massage
Ghughiban(Pleurosperumcanddlii)	Bodypain and asthma, *duni* (smoke) is given when there is effect of evileye
Kansu (castor)	For body ache
Pajja(Prunuscerasoides)	For bodyache, cold and internal injury
Nelkanthi(Ajugabracteosa)	Diabetes and cold
Bnaksha (Viola pilosa)	Cold, cough
Nhani	For *pooja and hawan*
Ain (Urtica diocia)	Blood purification and kidney
Chukri(Oxyria digyna)	For warmness in body and jaundice
Tulsi(holybassil)	For cold
Gandhla (Murraya koenigii)	Sprain and joint pain
Kainth(Pyrus pashia)	Boils and blisters in mouth
kiyayi(Solanumnigrum)	Burn and headache
Tirmira(Zanthoxylumarmatum)	Oral hygiene and blood purifier
Kakru (Hedera nepalensis)	For stomachache
Khor(Juglansregias)	For oral hygiene
Pudina (Mentha congifolia)	Treating vomiting and diarrhoea, headache
Banajwain(Micromeria biflora)	Stomach pain, cough, cold and fever

The Gaddis tend to perform shamanic practices, and the traditional specialist is known as *chela*. *Chela* are divided into two types: oracles and shamans. Oracles act as intermediaries between gods and humans, while shamans heal people of evil spirits. Men are more likely than women to participate in oracle practice. Oracles and shamans could be from any caste, whether upper- or lower-caste Gaddi. Under this effect, the oracle revealed God's will. There is a separate oracle for each temple, and everything he says during a procession is considered a divine utterance. When people question the Oracle about the basis for their suffering during their interactions with God, he replies as God has instructed. Young individuals commonly express their belief in black magic, spirits, and other supernatural beings, as do the old. In the event of a disease, whether it affects humans or animals, they first consult a shaman. The *chela* recites mantras and blows the negative spirit out of the ailing individual's head three times with a *mutha* (peacock feather). *Chela* wisdom is passed down from one generation to the next among households. Both men and women can be shamans. Practitioners are more likely to be guys. A shaman is the first to be informed of any ailment, whether it affects humans or animals.

When someone visits the *chela*, he asks that the patient does *dhyan* (meditation). After a short period of *dhyan*, the patient begins to shake and enters a trance while describing what has occurred. The *chela* frequently questions him about the images he receives in his unconscious. Ash was enclosed in a paper pallet and put into a little silver box that was meant to be worn as an amulet by the *chela*. They also offered "charmed water" (*pyasa*), which the victim was required to drink every day till they recovered (Bhasin, 2008; Sharma, 2010). These amulets were used to regulate or treat many forms of possession, such as ghosts, the evil eye, depression, infertility, and sorcery (Bhasin, 2008; Sharma, 2010). The amulet ritual focused on violating taboos, especially those surrounding contact and food, as well as combining the purity-pollution conflict. Sharma (2010) discovered that the sufferer was labelled unclean and healed since it was believed that possessing spirits did not transcend the caste structure and would not stay in ritually dirty beings. The physical contact between the possessed spirit and the patient, as well as the poisoned water they prepared, rendered both impure. The healing process began when the spirit, which desired liberty and the chance to come together with the pure, abandoned the customarily polluted body. Many women in Gaddi communities went through spirit possession (Bhasin, 2008).

Maya Devi (79 years old) from Dharamshala stated that,

> *The amulets were presented by Chela or Purohit. Amulets are used to protect from black magic, the evil eye, natural spirits, and other diseases. Silver amulets with front openings are utilised. Whatever causes the illness, multiple substances are stored within the amulets. While reciting mantras, Chela ties many knots in the amulet strand. You can use red, black, white, or multicoloured thread. People of all ages wear amulets across their necks, arms, and waist, and they cannot be swapped. It is produced specifically for the concerned individual. Chela employs paper or bhuj patras (birch leaves) to write mantras into amulets and sometimes plants.*

Autur is an ancestral demon that may be the spirits of childless spouses, the first spouse, or a deceased kid. The ghost of a childless spouse will seek out any member of the descent group, whereas the departed soul of the first wife will target the second wife, and the spirit of the deceased kid will trouble the younger sibling. Spirits emerge in the afflicted individual's dreams, and the *chela* can be employed to determine the cause. The afflicted person worships the spirit and wears silver amulets with *Autur* carvings. *Dhoop* (incense), *prashad* (cooked offerings), vermillion, and animals are offered to assist the *Autur* in recovering from disease. *Auturs* are engraved on stones and stored near water sources or beyond the hamlet for the advantage of the descendant group.

Maya (79 years old) reported significant leg pain that did not respond to medication, despite seeing many doctors. After much hustle, she went to meet the *chela*, who told her that she was attacked by her husband's uncle (who died without having children). She then placed an amulet with his image around her neck and revered him. She immediately felt relieved from pain after that (Fig. 17.1).

In Bharmour, 200 families (100%) were picked, with 125 (62.5%) believing in black magic and the evil eye and the other 75 (37.5%) not. In Dharamshala, 132

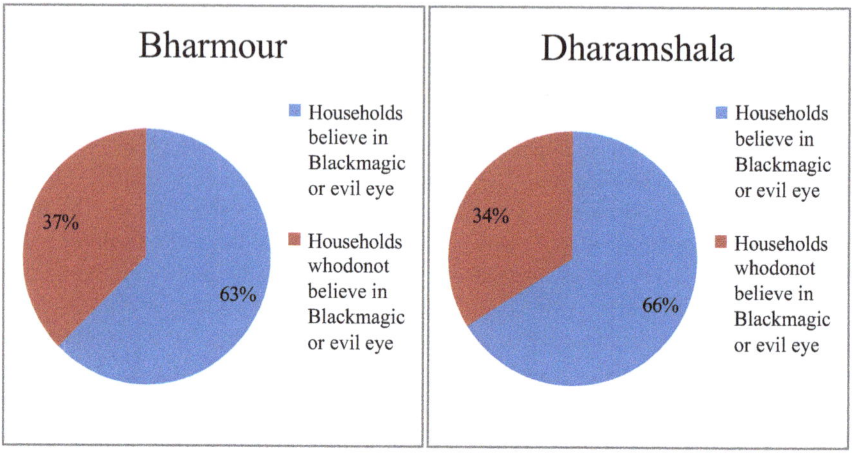

Fig. 17.1 Illustration of Gaddis' belief in evil eye and black magic

(66%) families held such ideas, whereas 68 (34%) did not. According to the investigation, the vast majority of them hold supernatural beliefs in things like sorcery. They are constantly seeking assistance from both contemporary healthcare and spiritual healers' conventional therapeutic approaches. The well-educated Gaddis also believe in black magic and the evil eye, which is why they consult shamans and utilize home remedies to combat the evil eye. Even within the field, many people were hesitant to admit to believing in supernatural events for fear of being labelled as backward. As stated by Simpson (2023), the vast majority of Gaddi people dispute witchcraft's continued importance and danger. On a more fundamental level, however, there are noises of witchcraft and disasters caused by witches, as well as black magic, or curses that somebody can place, that are spread among specific localities, caste neighbourhoods, and households.

Along with the aforementioned spiritual healing, Gaddis practice home treatments for the evil eye. *Satrana* is a healing process that requires seven different materials: rice, *mah ki dal* (black lentils), *channe ki dal* (gramme lentils), *hari moong* (green lentils), *sarson* (mustard seeds), salt, and an iron nail. All items are placed on a single piece of paper or fabric, revolved in a clockwise circle over the person who is affected, and then kept far from the home. *Chad chadna* is a procedure where *chela* often performs the treatment on adults; only a few households employ it on affected youngsters. They cook a goat called *bakri* with wheat flour and paint it with vermillion. The affected child was instructed to sit, and *Chad* was placed away from the dwelling and rotated clockwise. *Mirch alekna* is the process where three dried red chilies with ends are used, and they are turned clockwise and anticlockwise around the sick child's body. Chilies are then added to the flames. If the fumes from the fires melt like burned chilly, everything was well; otherwise, the child is believed to be the target of an evil eye. Almost all Gaddi households with young children practice this (Table 17.4 and Fig. 17.2).

Table 17.4 Percentage of households used medicinal plants to cure ailment in selected villages of Bharmour and Dharamshala *tehsils*

S.No.		Bharmour	Dharamshala
1.	Households who use medicinal plants for curing illness	103(51.5%)	54(27%)
2.	Households who use medicinal Plants for curing illness	97(48.5%)	146(73%)

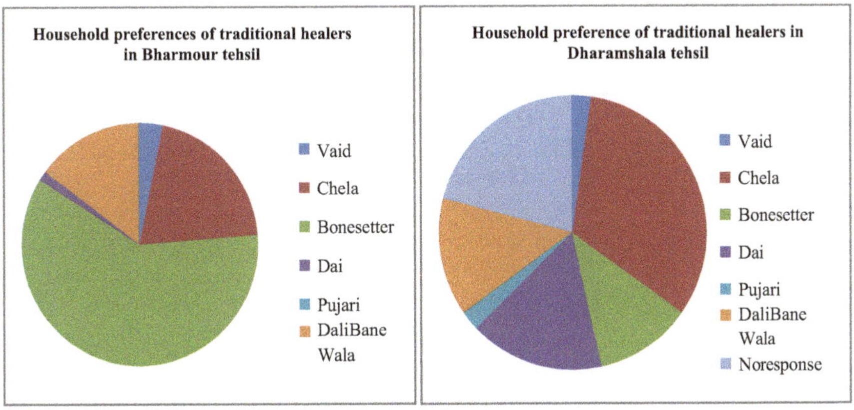

Fig. 17.2 The choices of traditional healers in Bharmour and Dharamshala households

In selected villages of the Bharmour *tehsil*, research indicated that 103 (51.5%) of 200 households use medicinal herbs to treat illness, while 97 (48.5%) do not. While 146 (73%) of the selected villages in the Dharamshala *tehsil* did not use medicinal herbs, 54 (27%) of the 200 families did. It is clear that Gaddis from Bharmour *tehsil* use medicinal herbs more frequently than Dharamshala Gaddis. The Gaddis of Bharmour depended more on traditional healers than the Gaddis of Dharamshala. Bonesetters (83.18%) are most frequently consulted by households in the selected Bharmour villages, followed by *chelas* (28.31%) and *dali bane walas* (20.35%). *Vaid* (4.42%) and *dai* (1.76%) receive fewer visits. No one visits *pujaris*; instead, they go to *chela* for home-related matters. However, in some Dharamshala *tehsil* villages, *chelas* (32.55%) were more commonly consulted than *dai* (16.27%) and *dali bane walas* (13.95%). *Vaids* (2.3%) and *pujaris* (2.3%) were rarely consulted. Gaddis employed home treatments for a variety of diseases. They employed household goods to treat ailments, believing that they were more beneficial than any other contemporary treatment (Table 17.5).

Table 17.5 List of illness and home treatments used by Gaddis

S. no	Illness	Home treatment
1.	Constipation(children)/ cold and cough	*Mulethi* (Licorice), *badi elaichi* (large cardamom), *dalchini* (cinnamon), *kuth* (costus), *magha* (melon seeds), and *march* (blackpepper) decoctions are given, or *ghee* (clarified butter) is poured in the nose for cold
2.	Cold and cough	Ginger tea
3.	Body heat	Mint tea, lassi (buttermilk)
4.	Beesting	Iron substance is rubbed over affected area
5.	Mouth ulcers	Blacksheep milk
6.	Skin rashes	*Thalchjoda* (goat'shairrope) and *pula* (lady shoes made up of cannabis sativa fibre and barley straw)
7.	Skin disease	Cow urine is used (preferred local cows), maximum of medicines given traditional healers for skin disease are mixed with cow's urine and then applied
8.	Ear pain	A few drops of lukewarm mustard oil is poured in ear and plugged with garlic clove
9.	Toothache	Salt and mustard oil is rubbed over the affected area or paste of clove with mustard is applied in affected area
10.	Stomach ache	Carom seeds and sugar or kala zira kada
11.	Wounds	Paste of turmeric is applied
12.	Vomiting and diarrhea	Salt and sugar water is given to prevent dehydration or mint tea
13.	Low platelets counts	Goatmilk
14.	Piles	Buttermilk from buffalo's milk or radish or lemon in milk
15.	Bodyache	Oil is prepared that includes mustard oil, garlic, and fenugreek seeds
16.	Boils	Scale of onion along with mustard oil kept on effected area
17.	Chickenpox	The affected person is asked to sleep over *neem* leaves and coconut oil is applied

Discussion

The government's desire to modernise the isolated hills (Bharmour) has resulted in significantchanges. Traditional values, religious beliefs, and practices have weakened as a result of shifting social and demographic patterns (Bhasin, 1997). Despite the establishment of a health service in Bharmour's outlying villages, no doctors are present and only visit the centres on rare occasions. In case of an emergency, they must travel to Bharmour Hospital. The hamlet relies on traditional therapy for minor illnesses. According to Bhasin (2008), tribal people are forced to rely on private medical practitioners even for their most basic medical needs, given that the government-run healthcare system does not perform well. In contrast to Dharamshala, Gaddis provide both public and private medical services. Malik et al. (2015) discovered that people living at higher altitudes were strongly reliant on wild species since higher altitudes had more therapeutic plants than lower altitudes. It was due to two factors: first, a lack of medical facilities; and second, their belief in traditional herbal remedies. There is less reliance on traditional medical techniques. Gaddis fall into three distinct groups: home remedy experts, herbal specialists, and spiritual healers. Home remedies specialists are older persons who do not consider themselves healers but do recommend and give herbal remedies in cases of illness. Sharma & Maheshwari (2004) examined the Gaddi tribes' traditional medical practices in Kangra.

The Gaddi women are aware of the medicinal benefits of every ingredient in the mixture used for in-home treatments. They were aware of surrounding farms that grew therapeutic herbs. Herbal specialists are those who treat patients with locally grown herbs. Ritual specialists and spiritual healers use divination and therapeutic cult rituals to treat patients. They learn the trade secrets from their family or another expert in the relevant sector (Bhasin, 2008). Gaddis use natural remedies to cure a wide range of diseases. They treat ailments with readily available at-home items that they believe are more successful than any contemporary treatment. They are conversant with the surrounding environment and utilise wild herbs to both consume and heal minor ailments. Despite the existence of modern medicine, both static and migrant Gaddis prefer traditional methods of treating illness. Gaddis, from both tehsils, believed in spiritual healing. In situations of disease, people seek basic care from traditional healers such as *vaid* (medicine man), *dali bane wala* (traditional healer), *haddi chadane wala* (bonesetter), *dai* (midwife), *chela*, and *pujari or purohit*. In villages, *vaids* and *dali bane walas* use therapeutic plants. *Dali bane wala* administers *jhad phuk* (bad spirits removal) on the affected individual while reciting *mantras* and utilising *bana* (chinaberry) or cannabis sativa leaves. He makes a paste out of forest-derived natural components. Hemangioendothelioma-related disorders. Yogiraja (1972, p. 110–50) described the native healer as "*dareh*" (*dali-banewala*).He handled animals for foot rot, colic, and food poisoning, as well as basic human ailments including boils, measles, mumps, and smallpox. Making an infusion of local herbs and breathing the right *mantra* (incantation) over it worked as a remedy. The event comprised an agriculturalist's rite, in which the earth, plants, and sickle(implements) were used as symbols. According to Sharma (2010), the illness can also arise as a result of polluting factors, such as a breach of purity rituals, the use of banned foods, a polluting contact, or an inability to offer sacrifice to the corresponding deity, such as *Sitla mata* in the case of smallpox. Using wild medicinal plants and a sickle, the ancient healer swore to combat pollution.

Traditional healers never mention the plants and mantras they use. They feel that exposing their abilities will weaken their healing powers and make it difficult for them to treat people properly. Singh & Lal (2008) discovered that the indigenous healers, known as "*amchis*" in Spiti and "*larje*" in Lahaul, had an in-depth knowledge of the natural flora; hence, the people of Lahaula and Bodh trusted them. They believed that if they shared their traditional expertise with others, the medicinal properties of herbs would be diminished. They never asked for their assistance. Every society's cultural healers integrated philosophy, assessment, and therapy. In addition, an understanding of the medicinal properties of plants, herbs, and other natural materials was acquired. They were taught about creation myths, cultural projections and illustrations, the precise pace of sustaining equilibrium between humans, the natural world, and the universe, and the symbolic and transcendent connection with the sacred. They were proficient at gathering and processing plants and other natural assets with the highest curative properties (Bhasin, 2008).

According to all of the traditional healers interviewed, no one in their family has shown an interest in learning traditional healing skills. Although there was no income and labour is simply for the well-being of others, presently everyone desires to earn money, and nobody likes to be bound by ancient norms. One of the *dai* (midwives) interviewed by the researcher died, and no other person in her family learned how to function as a midwife, indicating that the method has been eliminated. The situation is the same for the other healers. The medicine men might only be accessible to the present generation of Gaddis, but there is a good probability that no one else in the village will be familiar with traditional medicine in the future. The eldest members of the family have knowledge of wild medicinal herbs, but they rarely venture out to harvest them.

They rely increasingly on market-based medications. Gaddis previously depended on forestry goods due to their fragile economy (Thakur et al., 2020). Their conventional way of living has changed. Thakur et al. (2016) discovered that older Gaddi generations had strong identification skills, whereas the younger generations had a lower interest and faith in recognition.

Kumar et al. (2000) state that modernization and market trends had lessened Gaddis' dependence on conventional understanding of plant therapeutic effects. When Kandari et al. (2012) investigated the medicinal and aromatic plants utilised by the Bhotia tribe in the Chamoli district, they discovered that roadside households collected fewer wild species than houses in isolated areas. With the onset of tourism, the profession and knowledge in that region declined, and older, more knowledgeable people did not pass on their knowledge of how to identify wild medicinal plants to future generations. According to Appadurai (1995), a locale is a cultural territory that has evolved over time and is socialised through ritual activities. This space is linked to the larger cultural and political construct by referencing and connecting it to the larger order of both ethical and divine, embedded in the social system and economic agency—for example, the organisation of kinship and social interaction or stratification, as well as the agricultural or shepherding cycles. The village's sacred space allowed for the simultaneous coexistence of the dominant pan- Indian classical religious tradition and local beliefs based on a variety of region-specific myths, mythologies, gods, and godlings.

The community was made up of numerous professional castes that engaged in agricultural operations (Sharma, 2010). Gaddis believe strongly in the capacity of faith healing to bring healing, as seen by the growing number of both religious and non-medical spiritual healers. In today's world, Gaddis thought that diseases have supernatural causes, and it was critical for them to determine whether a patient's ailment was caused by a deity's wrath, an evil spirit's action, sorcery, witchcraft, or a transgression of a taboo. Once this is established, all you have to do to get healthy is seek the advice of a faith healer. Those who believe a condition has been triggered by a god's wrath, or because of an evil spirit, sorcery, or a transgression of a taboo, take appropriate precautions to avoid and cure the illness. According to Malinowski (1923), magic primarily meets psychological needs, providing man with the confidence he requires to perform critical duties, to remain composed and mentally intact during outbreaks of fury, and to avoid needless love, sadness, and stress (Upadhyay and Pandey,Upadhyay & Pandey, 1997,p.184). He believes that trust in magical power stems from inherited tales that depict multiple magical feats. Myth provides the foundation for the magical system, and magic connects the mythological legacy to the contemporary (Upadhyay & Pandey, 1997, p. 211).

Sorcerers, witches, and other supernatural beings, as well as the wrath of deities and spirits, can cause devastation in human lives (Bhasin, 2008). In the field, it was observed that the Gaddis believe that spirits have distinct names and exist inside societal limits around forests, rivers, lakes, mountains, and desolate areas. Nature spirits, or ancestral beings, are supposed to be responsible for a wide range of discomforts and illnesses. Many people in India believe that furious deities, malevolent spirits, and deceased humans possess victims and cause them to suffer (Bhasin, 2008). According to, who are cited in Bhasin (2008), spirit possession is akin to anxiety, originating from the individual's intra-psychic strain and an impulsive condition provoked by an incident or setting causing unusual pressure or grief. Getting possessed by a spirit is mostly an emotional state. Nabokov (2000) describes the practice of worshipping several Hindu deities for their alleged healing qualities in Tamil Nadu, aprovincein southern India. The Tamil people are a mixed-race group that includes both demon-possessed victims and communicators who communicate with the deities. According to Bhasin (Bhasin, 2008),both men and women seek treatment from spirit possession, which she refers to as a Gaddi's Cultural Bound Disorder. Spirit possession is a type of therapy employed by those who suffer from illnesses that cannot be treated using modern medical methods, psychology, or the legal system.

The inquiry showed that medical pluralism was common amongst the Gaddis, and in addition to modern medicine, they sought help from traditional spiritual healers. According to Bhasin's (2008) research, the majority of India is home to a variety of treatment systems and health behaviour patterns. Indian healthcare is distinguished by scientific diversity, home cures, consultations with indigenous specialists, and/or primary healthcare services (PHC). These medical methods are considered alternative, non-traditional, and supplemental. According to her, the diversified medical system of doctors and gods in tribal societies allows for adaptability and meets a wide range of societal goals. Each Gaddi home includes elders who know how to use locally accessible plants to cure basic concerns. According to Pederson et al. (1989), information from traditional healthcare can be used to develop creative healthcare paradigms and widen the scope of the classic epidemiologic frame work. Anelderly woman in the field remarked that modern culture rejects spiritual healing, black magic, and other indigenous spirits that can bring illness. Supernatural powers have existed since the beginning and are going to continue to do so. People's minds have been corrupted by learning; people are looking for medical care in circumstances of supernatural sickness.

Unfortunately, the *chela* is no longer as potent as he once was. Traditional medical practices and biomedicine coexist in rural and tribal India, according to studies. However, knowledge of and belief in traditional healing is limited to the older generation. The younger generation has more faith in allopathy than in traditional medicine and wild medicinal herbs, as evidenced by the example of the Gaddis, particularly in Kangra. Researchers felt that these two were distinct and that biomedicine would eventually replace conventional medicine (Foster & Anderson, 1978). Traditional healing procedures are on the decline among the Gaddi, as the new generation is unwilling to absorb this skill.

Conclusion

Gaddis are quite knowledgeable about the plants that thrive in their area, and they use them in their daily lives as remedies and veggies. They introduced numerous researchers to different types of wild food plants. Modernization has altered their perspectives on traditional knowledge. Only the elderly population believes this, and relatively few people in this new age recognize therapeutic species. Traditional healing traditions are vanishing because no one in the current generation wants to learn about healing. The establishment of biomedical facilities, and education in Gaddi regions also reduced reliance. Traditional knowledge should be passed on verbally or shown by the elders of every household to the present generation to ensure that the knowledge remains in the tribe and future generations also take advantage of resources in their surroundings.

References

Appadurai, A. (1995). The production of locality. In R. Fardon (Ed.), *Counterworks: Managing the diversity of knowledge*. Routledge.

Bhasin, V. (1997). Problem in educating the gaddis: A transhumant population. In K. C. Mahanta (Ed.), *People of the Himalayas ecology, culture, development and change* (58–73). Kamla Raj Enterprises.

Bhasin, V. (2002). Traditional medicine among tribals of Rajasthan. *Journal of Social Sciences*, *6*(3), 153–172.

Bhasin, V. (2008). Gaddi's folk medicines: A source of healing. *Studies on Ethno-Medicine*, *2*(1), 1–27. https://doi.org/10.1080/09735070.2008.11886310

Farooqi, Y. N. (2006). Traditional healing practices sought by Muslim psychiatric patients in Lahore, Pakistan. *International Journal of Disability, Development and Education*, *53*(4), 401–415.

Foster, G. M., & Anderson, B. &. B. (1978). *Medical anthropology*. Wiley.

Jakovljević, M. (2017). Psychiatry and religion: Opponents or collaborators? The power of spirituality in contemporary psychiatry. *Psychiatria Danubina*, *29*(suppl. 1), 82–88.

Kandari, L. S., Phondani, P. C., Payal, K. C., Rao, K. S., & Maikhuri, R. K. (2012). Ethnobotanical study towards conservation of medicinal and aromatic plants in upper catchments of Dhauli Ganga in the Central Himalayas. *Journal of Mountain Science*, *9*(2), 286–296. https://doi.org/10.1007/s11629-012-2049-7

Kishor, A., Kumar, A., Tomer, V., Kumar, V., & Gupta, K. (2018). Wild food plants of Himachal Pradesh: A review. *Plant Archives*, *18*(2).

Kumar, K., Singh, K. K., Asthana, A. K., & Nath, V. (2000). Ethnotherapeutics of bryophyte Plagiochasma appendiculatum among the Gaddi tribes of Kangra valley, Himachal Pradesh, India. *Pharmaceutical Biology*, *38*(5), 353–356. https://doi.org/10.1076/phbi.38.5.353.5963

Malik, Z. A., Bhat, J. A., Ballabha, R., Bussmann, R. W., & Bhatt, A. B. (2015). Ethnomedicinal plants traditionally used in health care practices by inhabitants of western Himalaya. *Journal of Ethnopharmacology*, *172*, 2015),133–144.

Malinowski, B. (1923). *Argonauts of the western Pacific*. Routledge.

Nabokov, I. (2000). *Religion against the self.* Oxford University Press.

Pedersen, D., & Baruffati, V. (1989). Healers, saints and doctors: Elements for the analysis of medical systems. *Social Science and Medicine*, *29*(4), 487–496. https://doi.org/10.1016/0277-9536(89)90194-9

Rowkith, S., & Bhagwan, R. (2020). Honoring tribal spirituality in India: An exploratory study of their beliefs, rituals and healing practices. *Religions*, *11*(11), 549.

Shankar, R., Lavekar, G. S., Deb, S., & Sharma, B. K. (2012). Traditional healing practice and folk medicines used by Mishing community of North East India. *Journal of Ayurveda and Integrative Medicine*, *3*(3), 124.

Sharma, B., & Maheshewari, S. (2004). Traditional medical practices of Gaddi tribes in Kangra district, Himachal Pradesh. *Indian Journal of Traditional Knowledge*, *4*(2).

Sharma, M. (2010). Symbols of empowerment: Possession, ritual and healers in Himachal Himalaya (North India). *Journal of Asian and African Studies*, *45*(2), 196–208. https://doi.org/10.1177/0021909609357413

Sharma, M. R. (2018). Belief of tribal's in supernatural power and its relation with religious life (with special refernce to Indian tribal society). *IJRAR-International Journal of Research and Analytical Reviews*, *5*(3), 48–55.

Simpson, N. (2023). Encountering the dain. *Himalaya The Journal of the Association for Nepal and Himalayan Studies*, *42*(2), 70–85. https://doi.org/10.2218/himalaya.2023.7809

Singh, K. K., & Kaushal, K. (2000). Observations on ethnoveterinary medicine among the Gaddi tribe of Kangra Valley, Himachal Pradesh. *Ethnobotany*, 12,42–44.

Singh, K. N., & Lal, B. (2008). Ethnomedicines used against four common ailments by the tribal communities of Lahaul-Spiti in Western Himalaya. *Journal of Ethnopharmacology*, *115*(1), 147–159. https://doi.org/10.1016/j.jep.2007.09.017

Thakur, A., Singh, S., & Puri, S. (2020). Exploration of wild edible plants used as food by Gaddis-A tribal community of the Western Himalaya. *The Scientific World Journal*, 6280153. https://doi.org/10.1155/2020/6280153

Thakur, M., Asrani, R. K., Thakur, S., Sharma, P. K., Patil, R. D., Lal, B., & Parkash, O. (2016). Observations on traditional usage of ethnomedicinal plants in humans and animals of Kangra and Chamba districts of Himachal Pradesh in north-western Himalaya, India. *Journal of Ethnopharmacology*, *191*, 280–300. https://doi.org/10.1016/j.jep.2016.06.033

Upadhyay, V. S., & Pandey, G. (1997). *History of anthropological thought*. Concept Publishing Company.

Wangoola, P. (2000). Mpambo, the African multiversity: A philosophy to rekindle the Africanspirit. *Indigenous Knowledges in Global Contexts*, 265–277.

Yogiraja, Y. (1972). *Sidha Shabarmantra*. Randhir Books.

Chapter 18
The Therapeutic Influence of Yoga on Well-Being and Happiness

Suman Rawat and Saamdu Chetri

Abstract This short chapter delves into the intricate link between well-being and happiness, stressing the pursuit of happiness as a core human goal. It explores correlations between happiness and longevity, immunity, productivity, and social engagement, drawing insights from the Dalai Lama, Aristotle, and William James. Definitions of happiness, mental balance, and health are examined from various perspectives, including Webster's Dictionary, Seligman, and Ayurveda. The impact of modern lifestyles, technology, and psychosomatic diseases on mental well-being is highlighted, suggesting that health, harmony, and bliss are innate but often disrupted by external factors. The pursuit of happiness involves aligning with our natural state of well-being. Yoga's role in fostering health and happiness is underscored, emphasising its holistic approach covering physical, mental, cultural, and spiritual dimensions. The Chitta Prasadanam practice promotes positive attitudes towards diverse individuals for mental clarity and happiness. The eight limbs of yoga, as advocated by Maharshi Patanjali, are explored, illustrating how ethical disciplines, postures, breathing, and meditation contribute to holistic health. Yoga principles are relevant for disease prevention, mental harmony, and sustainability in modern life. Ultimately, the chapter reinforces that health and happiness are intertwined and achievable through embracing yoga principles, encouraging prioritisation of physical, mental, and emotional well-being for enduring happiness.

Keywords Happiness · Health · Yoga · Ayurveda · Chitta prasadanam · Ashtanga yoga

---

S. Rawat (✉) · S. Chetri
Dr. Vishwanath Karad MIT World Peace University, Pune, India
e-mail: sumanrajput2014@gmail.com
e-mail: dr.saamdu.chetri@gmail.com

R. C. Sobti et al. (eds.), *Traditional Knowledge Systems in India: Sustainability, Ecology, and Resource Governance*, People, Cultures and Societies: Exploring and Documenting Diversities,
https://doi.org/10.1007/978-981-92-1346-7_18

Introduction

Countless individuals come into existence, some destined for fleeting lives marked by illness or adversity. Conversely, others endure, reaching a hundred or more age, embracing life's spectrum from triumph to despair, joy to hatred, and love. Regardless of the brevity or longevity of our time on Earth, a fundamental question persists: What lends purpose to life? What infuses significance into our existence? (Dalai Lama, p. 6).

The purpose of our existence is the pursuit of happiness, a concept endorsed by thinkers from Aristotle to William James in the Western tradition. A contented life yields numerous benefits, including longevity, boosted immune function, higher earnings, and enhanced likability. Happy individuals display heightened optimism, resilience, political and civic engagement, and increased productivity, creativity, and adeptness in handling life's challenges compared to their unhappy counterparts. Numerous surveys underscore that unhappy individuals tend to be more self-absorbed, socially isolated, brooding, and sometimes antagonistic (Dalai Lama, p.6–7). Joyful individuals cultivate stronger marriages, foster closer friendships, and exhibit more philanthropy and altruism (Chase, 2016). Happy individuals demonstrate a propensity for love, assistance, and forgiveness. While establishing causality between happiness and these traits is challenging, these correlations are significant. Happiness pervades every facet of our lives, whether we are reaching for material goals or spiritual enlightenment. Understanding our pursuit of happiness can enlighten our quest for fulfilment (Chase, 2016).

Webster's dictionary, dating back to 1993, defines happiness as "good fortune" or "prosperity." Seligman (2002) outlines conditions for happiness, stressing positive emotions (like contentment, happiness, and hope), character traits (such as love, courage, compassion, curiosity, integrity, and moderation), and supportive institutions (including justice, responsibility, and effective parenting). According to the same dictionary, mental balance refers to "the healthy psychological state of someone with good judgment." In Buddhism, mental suffering, like anxiety and depression, is attributed to an unbalanced mind (Gunaratana, 1985), leading to dissatisfaction and unhappiness (Tsong-kha-pa, 2000).

Modern technology and erratic lifestyles contribute to heightened mental stress, impacting happiness (Joshi & Pandey, n.d.). According to the Bhagavad-Gita (Chinmayananda, 1992), psychological stresses originate from uncontrolled thoughts. Psychosomatic diseases, explained in Yoga Vashishtha (Chapter II, verses 709–723), stem from mind imbalances (Adhi) affecting vital life force and the physical body (Vyadhi), disrupting metabolic activities and leading to unhappiness (Kavuri et al., 2015). These diseases result from disturbances above the mind level, affecting prana channels and bodily functions (Yoga Vibhushan &Bhavanani, n.d.).

Health, harmony, and bliss are innate, but external factors often derail us. While we are meant to interact with the external world using our 'mind instrument' (Manas, buddhi, Chitta, ahamkara), we get entangled, losing touch with our true nature, characterised by health, harmony, bliss, and peace, as habits and tendencies overshadow it.

Motivated by ignorance, we persistently pursue these attributes from external sources, only to find temporary solace in peace, harmony, or fleeting health. Yet, these prove transient, leaving us in an endless chase or resigned to accept life's cycle of misery, discontent, and disharmony.

Health is inherent, dynamically adapting and nurturing us. Swastha [swa + stha], our natural state, defines health as Swaasthya. Conversely, straying from this state, termed Aswatha [a + swastha], evolves into habit, progressing into Vasana, solidifying into samskara, perpetuating a chronic departure from our true nature (Raghuram, 2023).

The World Health Organisation (WHO) defines health as "complete physical, mental, and social well-being, not merely absence of disease." Ayurvedic texts, including those by Acharya Susrutha, echo this, emphasising a disease-free state of well-being.

Susrutha highlights health as balanced Tridoshas (humours), proper Agni (Fire) function, equilibrium in Sapta Dhatus (body tissues), and regular excretion (urine, stool, sweat). Mental, sensory, and spiritual well-being and happiness are integral in Ayurveda (Guruprasad, 2017).

Promoting health and happiness involves prioritising physical, mental, and emotional well-being. Maintain a balanced diet andexercise regularly, including yoga, for physical health. Embrace mindfulness, like meditation, to reduce stress. Nurture relationships, pursue hobbies, and set realistic goals for emotional well-being. Ensure sufficient sleepand manage stress effectively. Spend time in nature, practice gratitude, and kindness. Keep learning, maintain a growth mindset. Purpose, balance, and community foster sustained health and happiness.

Yoga for Health to Cultivate Happiness

Yoga transcends mere exercise; it is a holistic lifestyle aiming for self-realisation and liberation from suffering. It embraces diverse studies for overall body, mind, and spirit growth, spanning physical, mental, cultural, aesthetic, classical, linguistic, religious, spiritual, medical, and scientific realms. Embedded in our culture, yoga is integral to our way of life.

In today's world, there's a rising call for holistic development, and yoga is poised to meet this demand. It holds promise as the ideal path for fostering a balanced nation, blending the dynamic Western mindset with our spiritual essence, promoting happiness and grounding people in harmonious living (Trivedi, 2022). 'Yoga' stems from the Sanskrit root 'Yuj' (Basavaraddi, 2015), connoting unity, control, and integration.

Chitta Prasadanam Practice for Healthy Living and Happiness

Upon introspection of one's inner self, envy (Irshya) towards joyful individuals, disdain for the suffering (Dukhi), critical attitude (Vidwesha) towards virtuous souls (Punyatma), and fascination (Papatma) towards sinful actions become evident. In Yoga Darshana, Patanjali proposes overcoming this mindset through Chitta Prasadanam, where "Chitta" denotes mind and "Prasadanam" signifies happiness.

Chitta Prasadanam practices unlock the mind's facets related to joy, sadness, honour, and dishonour. To nurture positivity, one can cultivate Maitri (Friendliness) towards joyous individuals, Karuna (Compassion) for the sorrowful, Mudita (Goodwill) towards the virtuous, and Upeksha (Detachment) towards the perceived wicked. These sentiments purify the mind. Maitri, Karuna, Mudita, and Upeksha are regarded as the four components of genuine love and the four immeasurable virtues. Developing these virtues enhances happiness (Nimbal et al., 2021).

Uncontrolled reactions to the human environment are often primary sources of mental disturbance. People's impulsive responses to situations based on moods lead to constant disturbances fuelled by intense emotions. The Bhagavad-Gita stresses being non-hateful (Advesta) and fostering friendship and compassion.

As articulated in the Bhagavad-Gita, "AdvestaSarvabhutanamMaitrah Karuna Eva ca," one devoid of hatred towards all beings sees them as extensions of oneself. Such individuals act with friendliness and compassion towards those in sorrow, reflecting empathy anda proactive response to extend kindness and alleviate others' stress and suffering. Empathy facilitates nurturing healthy social bonds, emphasising interconnectedness (Sinha, n.d.).

Ashtanga Yoga (8 Limbs) of Yoga for Healthy Living and Happiness

Yoga presents a structured pathway delineated by Maharshi Patanjali, comprising ethical disciplines (Yama), purifying practices (niyama), postures (asanas), controlled breathing (pranayama), inward reflection (pratyahara), concentration (dharana), meditation (dhyana), and self-awareness (samadhi). The ultimate aim is self-realisation or inner tranquillity. Though integral, Asanas, pranayama, and meditation have garnered popularity as therapeutic aids. Patanjali's Ashtanga yoga offers a comprehensive approach for holistic healthcare, emphasising global well-being, peace, and environmental sustainability.

Research validates yoga's therapeutic efficacy, aiding in health promotion and disease prevention. Its principles and practices are increasingly employed in preventative, complementary, and alternative medicine for diverse physical and mental ailments. Yoga's societal inclusivity, cost-effectiveness, and eco-consciousness foster its sustainability, yielding manifold health benefits across various domains.

Yoga Sutras elucidate ethical groundwork, mind-body regulation, and spiritual evolution. Ashtanga Yoga practice purges impurities and fosters profound wisdom. Key components include:

1. Yama: Universal morality
2. Niyama: Personal observances
3. Asanas: Body postures
4. Pranayama: Breathing exercises and control of prana
5. Pratyahara: Control of the senses
6. Dharana: Concentration
7. Dhyana: Devotion, meditation on the divine
8. Samadhi: Union with the divine

The foundational principles of yoga, encompassing both social conduct and personal observances, are pivotal in purifying moral behaviour and addressing societal well-being. These principles enhance willpower, intellect, and emotional stability, fostering social harmony and self-acceptance for societal wellness.

Integral to yoga are Asanas and Pranayama, which are known for improving physical aspects like strength, flexibility, blood circulation, and gastrointestinal health. These practices boost bodily systems, including digestion, lung capacity, and peristalsis.

In mental health, inner conflicts and stress greatly affect well-being. Yogic principles, pranayama, and meditation promote mental harmony, peace, and serenity, serving as a preventive measure against mental health issues.

Conclusion

The core tenets of yoga, addressing individual conduct and societal welfare, are vital in refining moral behaviour and promoting community well-being. By adhering to these principles, individuals can strengthen willpower, intellect, and emotional resilience, fostering social cohesion and self-acceptance to benefit society.

Central to yoga practiceare Asanas and Pranayama, renowned for enhancing physical attributes such as strength, flexibility, circulation, and digestive health. These techniques optimise bodily functions, including digestion, respiratory capacity, and intestinal movement.

In mental well-being, inner conflicts and stress profoundly impact overall health. Yogic principles, complemented by pranayama and meditation, facilitate mental equilibrium, tranquillity, and inner peace, offering a preventive approach to mental health challenges.

References

Basavaraddi, I. V. (2015). Yoga: Its origin, history and development. *Public Diplomacy, 1*, 1–5.

Chase, S. (2016). *Yoga and the pursuit of happiness: A guide to finding joy in unexpected places.* New Harbinger Publications.

Chinmayananda, S. (1992). *The holy Gita* (9th ed.). Central Chinmaya Mission Trust.

Gunaratana, H. (1985). *The path of serenity and insight: An explanation of the Buddhist jhanas.* South Asia Books.

Guruprasad, K. (2017). Concept of health in Ayurveda. *World Journal of Pharmaceutical Research, 7*(2), 1061–1074.

Joshi, S., & Pandey, K. (n.d.). Yoga and meditation makes life happy. *Life Skills, 81.*

Kavuri, V., Raghuram, N., Malamud, A., & Selvan, S. R. (2015). Irritable bowel syndrome: Yoga as remedial therapy. *Evidence-Based Complementary and Alternative Medicine, 2015.*

Nimbal, A. A., Katti, A., & Kulkarni, A. (2021). Chitta parikarma of yoga darshana and its application in the Ayurveda.

Raghuram, N. V. (2023). Yoga therapy principle-Role of yoga therapy in healing. *Vidwat, 16*(1), 6–8.

Seligman, M. (2002). *Authentic happiness.* Free Press.

Sinha, M. (n.d.). Enhancing life skills through yoga to meet the challenges of life. *Mssv Journal Of Humanities & Social Sciences, 86.*

Trivedi, D. (2022). Importance of Ashtanga-yoga in education.

Tsong-kha-pa,. (2000). The great treatise on the stages of the path to enlightenment. Snow Lion.

Webster's dictionary. (1993). 10th ed. Merriam-Webster.

Chapter 19
Ethnomedical Practices Among the Rongmei Naga Tribes of Manipur, India

Nikita Das and Aribam Bijayasundari Devi

Abstract Ethnomedicine is defined as the totality of health, knowledge, values, beliefs, skills, and practices of members of society, including all the clinical and non-clinical activities that relate to their health (Foster and Anderson, 1978, *Medical Anthropology*). Ethnomedicinal plants are a vital aspect of the Rongmei Naga way of life, embodying a harmonious connection between humans and the natural world. The present study tries to explore the different ethnomedicinal plants used among the Rongmei Naga tribes of Manipur. For the present research, data were collected through fieldwork conducted among the Rongmei Naga Tribe living at Tarung village, Imphal- West District, Manipur, India. A total of 79 ethnomedicinal plants have been recorded, which are being used for treating various diseases among the Rongmei Naga tribes. The study revealed the extensive use of different plant life forms, including trees, shrubs, herbs, climbers, and understory plants. The Rongmei Naga community has very rich ethnomedicinal knowledge in the field of herbal medicine. The ethnomedicinal plants serve as a reliable and readily available resource when most secondary and tertiary health care centers are very far away from the village.

Keywords Medical anthropology · Tarung village · Ethnomedicine · Manipur

---

N. Das
Research Scholar, Department of Anthropology, Dr. Harisingh Gour Vishwavidyalaya, Sagar, Madhya Pradesh, India
e-mail: nikita.das.anth@gmail.com

A. Bijayasundari Devi (✉)
Assistant Professor, Department of Anthropology, Dr. Harisingh Gour Vishwavidyalaya, Sagar, Madhya Pradesh, India
e-mail: bsaribam14@gmail.com

R. C. Sobti et al. (eds.), *Traditional Knowledge Systems in India: Sustainability, Ecology, and Resource Governance*, People, Cultures and Societies: Exploring and Documenting Diversities,
https://doi.org/10.1007/978-981-92-1346-7_19

Introduction

Ethnomedicine is defined as the totality of health, knowledge, values, beliefs, skills, and practices of members of society, including all the clinical and non-clinical activities that relate to their health needs, according to George Foster and Barbara Anderson in their 1978 essay Medical Anthropology (Foster & Anderson, 1978).

Ethnomedicinal plants are a vital aspect of the Rongmei Naga way of life, embodying a harmonious connection between humans and the natural world. Embedded in their cultural ethos, the knowledge of these medicinal plants has been carefully preserved and transmitted through oral traditions, folklore, and the teachings of tribal elders. The Rongmei Naga's profound understanding of their environment has fostered a profound respect for the healing properties of various flora, culminating in a diverse pharmacopeia that continues to amaze and intrigue modern researchers and botanists alike. For generations, these communities have relied upon the bounty of nature to treat a myriad of ailments and maintain their overall well-being. At the heart of their healing practices are ethno-medicinal plants, which have been passed down through the ages as precious treasures of the Rongmei Naga culture.

The unique geographical and climatic conditions of Manipur, characterized by its hilly terrain, fertile valleys, and tropical climate, have facilitated the growth of a wide range of ethnomedicinal plants. These plants have become a fundamental part of the Rongmei Naga's daily lives, shaping their customs, rituals, and remedies for various health conditions. Whether it be the treatment of common ailments, chronic illnesses, or spiritual healing, these plants have stood as stalwart companions in the tribe's journey toward holistic health. This research endeavors to delve deep into the diverse world of ethno-medicinal plants among the Rongmei Naga tribes, exploring their traditional uses, preparation methods, and cultural significance.

A study on the Jantia Tribes of North Cachar Hills District reported that the people inhabiting the area use 39 different plant species to cure various common illnesses and diseases (Sajem & Gosai, 2006). Another study by Shil & Chaudhury (2009) found that the Reang tribes of Tripura, in North East India, use 16 species of medicinal plants to cure different diseases. These plants were mostly found on roadsides and in moist areas. One of the studies conducted on the use of traditional knowledge and medicinal plants by the Patra tribe community in the North-Eastern region of Bangladesh found that 33 medicinal plants were used to cure various diseases (Rahman et al., 2011).

A study of the Ethnobotany of the Mao-Naga Tribe of Manipur, India, signifies that 63 taxa and 38 families of medicinal plants were used to cure different diseases (Lokho & Narasimhan, 2013). One study found about comparative study of Ethnomedicine among the Tribes of North East India using common medicinal plants that cure many diseases such as cough, fever, abortion, dysentery, urinary disorder, skin diseases, and stomach aches (Bhuyan, 2015). In an Ethnobotanical Study of traditional knowledge and uses of medicinal wild plants among the Marakwet Community in Kenya, it was found that they use various traditional plants to cure 41 diseases. It was also found that traditional knowledge was decreasing among their healer, Wanjohi et al. (2020).

By shedding light on the Rongmei Naga's ethnomedicinal heritage, this study seeks not only to document their valuable practices but also to foster greater awareness and appreciation for the intricate relationship between humans and nature. Furthermore, it aims to encourage dialogue between traditional knowledge and contemporary scientific research, ultimately paving the way for a harmonious coexistence between the ancient wisdom of the Rongmei Naga tribes and the advancements of the modern world. The present study tries to explore the different ethnomedicinal plants used among the Rongmei Naga tribes of Tarung village, Imphal West District, Manipur, India.

Area and People

Manipur is the land of ethnomedicinal plants. This is one of the seven states known as the "Seven Sisters" of India. Manipur was a sovereign state from 33 AD. From 1892 onwards, Manipur became a large state under the political control of the British government of India. According to the Census 2011, the total area of Manipur is 22,327 sq. Km. There were 9 districts in Manipur, which are divided into two parts, viz., Valley districts and Hill districts. In 2016, there were another 7 new districts added. Presently, the total number of districts in Manipur state is 16.

Imphal West is one of the major districts of the state. The total population of Imphal West is 514,683. The study was carried out in the Tarung village of Imphal West district, Manipur. The Rongmei Naga tribe is one of the sub-tribes of the Naga Tribe. They are distributed in Manipur, Assam, and Nagaland. They are also called the Kabui Tribe. The Rongmei word is derived from two words, "Ruang" and "mei". The meanings of these words are "south" and "people" respectively; these stand for the southerners who migrate to the Tamenglong district of Manipur. The ancestors of the Rongmei Naga belong to the hilly area of the Tamenglong district. They fall under the Tibeto-Burman family of the Mongolian race. The Rongmei community follows patrilineal and patriarchal system (Fig. 19.1).

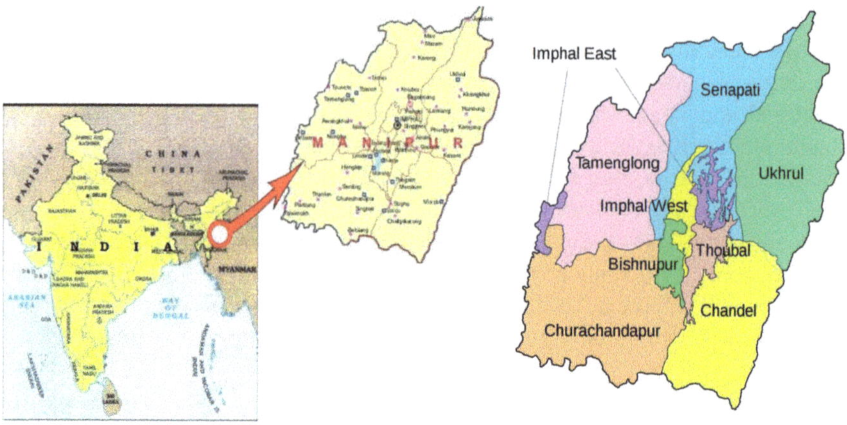

Fig. 19.1 Map of India showing Manipur state and the Tarung village (field site). *Source* File: https://www.Mapsofindia.Com/maps/manipur/

Material and Methods:

Study Area

For the present research, data were collected through fieldwork conducted among the Rongmei Naga Tribe living at Tarung village, Imphal-West District, Manipur, India. The selection of these locations was based on their accessibility and representation of the traditional practices of the Rongmei Naga community.

Data Collection

Field surveys were conducted to identify potential informants and medicinal plant resources in the study area. The surveys involved traversing the villages and surrounding areas to document the presence of medicinal plants and their use in the traditional healthcare system. Qualitative data were collected by using participant observation, in-depth interviews, key informant interviews, case studies, and group discussions.

Interviews

Semi-structured interviews were conducted with knowledgeable tribal elders, traditional healers, and community members. These interviews aimed to gather information about ethnomedicinal plant species, their traditional uses, preparation methods, and application for specific health conditions. In-depth interview was undertaken with all the elderly and knowledgeable persons of the village and a few key informant interviews were conducted with the traditional and spiritual healers.

Ethical Considerations

Prior informed consent was obtained from all participants before conducting interviews. The research complied with the ethical guidelines and regulations of research involving indigenous communities.

Plant Identification and Collection

Ethnomedicinal plants reported by the informants were identified by consulting local floras, herbarium specimens, and expert botanists.

Data Analysis

The data obtained from interviews were analyzed to categorize medicinal plants based on their uses and medicinal properties. The information on plant species, parts used, preparation methods, and application was compiled for further analysis.

Result

The Rongmei community relies extensively on traditional healers and herbalists who use locally available medicinal plants, ritualistic practices, and spiritual healing to treat physical and psychological ailments. Illness is often perceived as both a biological imbalance and a spiritual disturbance caused by supernatural forces or breach of taboos, thus requiring both herbal remedies and ceremonial interventions. The transmission of ethnomedical knowledge is largely oral, passed down through generations, though it faces threats from modernization and declining interest among the younger population. Despite these challenges, traditional healthcare continues to play a crucial role, especially in remote areas with limited access to modern medical facilities, emphasizing the need for documentation, preservation, and possible integration with mainstream healthcare systems.

Table 19.1 represents the 79 ethnomedicinal plants that are used for various diseases among the Rongmei Naga tribes in Tarung village of Imphal West district, Manipur. The medicinal plants are represented by the local name, family name, scientific name, life form, plant parts used, cured diseases, and mode of preparation and administration.

Table 19.1 Ethnomedicinal plants which are used for various diseases among the Rongmei Naga Tribes in Tarung Village of Imphal West District, Manipur

S.N.	Name of the plants (Local) Rongmei and Manipuri	Name of the plant (English)	Scientific name	Family name	Life forms	Parts	Disease	Mode of preparation and administration
1.	*Rarouga* (Ok-hidak)	Sweet flag	*Acorus calamus*	Acoraceae	Herb	Roots	Cough and chest congestion	The roots of the plant are used to make a necklace out of the piece of root and make the patient wear it for Cough and chest congestion.
2.	*Rangpong Bangmei* (Nongmangkhaangouba)	Malabar nut	*Adhatoda vasica*	Acanthaceae	Shrub	Leaves, flowers	Jaundice and muscular pains, asthma, diarrhoea, and dysentery	The leaves are boiled with water, and the decoction is given to relieve symptoms.
3.	*Ganam* (Chanam)	Garlic	*Allium sativum*	Amaryllidaceae	Bulb	Bulb	Cough and skin issues	Extract of the bulb mixed with mustard oil is applied to the chest, throat in case of cough and other chest complaints, juice is used for skin troubles.
4.	*Pajongthai* (Theibong)	Jack fruit	*Artocarpus integrifolia*	Moraceae	Tree	Roots	Diarrhoea.	Used in diarrhoea
5.	*Kamrangathai* (Heinoujom)	Carambola	*Averrhoea carambola*	Oxalidaceae	Tree	Fruit	Jaundice& Piles	Fruit extract is given for jaundice and unripe fruits in bleeding piles.
6.	*Chingthraoangouba*	White bauhinia	*Bauhinia acuminata*	Fabaceae	Shrub	Leaves, bark	Leprosy & asthma	Decoction of bark or leaves is useful in leprosy, asthma
7.	*Chingthraonganleinaba*	Pink bauhinia	*Bauhinia purpurea*	Fabaceae	Shrub	Bark	Insect bite, leprosy, and irregular menstruation.	Bark extract is used in insect bites, leprosy, and irregular menstruation.
8.	*Joupungma* (Torbot)	Ash gourd	*Beninca sahispida*	Cucurbitaceae	Fruit	Fruit	Stomach ulcers and jaundice	A boiled decoction of fruit is given for stomach ulcers and jaundice

(continued)

Table 19.1 (continued)

S.N.	Name of the plants (Local) Rongmei and Manipuri	Name of the plant (English)	Scientific name	Family name	Life forms	Parts	Disease	Mode of preparation and administration
9.	Tera	Silk cotton tree	*Bombax malabaricum*	Malvaceae	Tree	Flowers	Snake bites	Crushed flowers are applied to snake bites.
10.	*Joupungma* (Pang gong)	Flame of the forest	*Butea monosperma*	Leguminasae	Tree	Bark, leaves	Snake bites	Leaves are tonic, and bark is used in snake bites.
11.	*Taseoupai* (Mairong bi)	Red gram	*Cajanus cajan*	Fabaceae	Shrub	Seeds	Snake bites and loose motion.	Used in snake bite, and the beans are given raw to relieve stomach ailments and loose motion.
12.	Ganja	True hemp	*Cannabis sativa*	Cannabaceae	Herb	Leaves	Diarrhea, toothache ache and dysentery	Diarrhea and dysentery relieve a toothache.
13.	*Moupingthai* (Awathabi)	Papaya	*Carica papaya*	Caricaceae	Tree	Fruit	Constipation, abdominal disorders, dengue, and dysentery	Constipation, abdominal disorders, and dysentery. Juices of the leaves are given for dengue fever.
14.	*Koklei* (Perook)	Indian penny wort	*Centella asiatica*	Mackinlayaceae	Herb	Whole plant	Ulcer, enhance memory, and urinary troubles	Fresh plant juice with honey is given for ulcers and urinary troubles. Fresh leaves are chewed to enhance memory.
15.	*Tengphaita* (Tejpat)	Indian cassia	*Cinnamomum tamala*	Lauraceae	Tree	Bark, leaves	Diarrhoea and snake bite.	Boiled leaves are used for diarrhea and snake bites.
16.	*Mansugi* (Ushingsha)	Cinnamon	*Cinnamomum zeylanicum*	Lauraceae	Tree	Bark, leaves	Cough, cold, and vomiting	Boiled leaf extract given for cough, bark given for cold and vomiting
17.	*Gachangrim* (Yaingang)	Turmeric	*Cucurma domestica*	Zingiberaceae	Herb	Rhizome	Wounds and skin infections	Applied to wounds and skin infections

(continued)

Table 19.1 (continued)

S.N.	Name of the plants (Local) Rongmei and Manipuri	Name of the plant (English)	Scientific name	Family name	Life forms	Parts	Disease	Mode of preparation and administration
18.	*Noongaiperuk*		*Desmodium microphyllum*	Leguminasae	Herb	Whole plant	Urinary troubles due to stones	Plant decoction is used for urinary troubles due to stones.
19.	*Ru Bangmei* (Haaangouba)	Asiatic yam/ Purple yam	*Dioscorea alata*	Dioscoreaceae	Bulb	Bulbs	Leprosy and piles	Leprosy and piles
20.	*Num Ru* (Lam haa)	Potato yam	*Dioscorea bulbifera*	Dioscoreaceae	Vine	Bulbs	Ulcers	Dried bulbs are pounded and applied to ulcers.
21.	*Ru Hangmei* (Haaangangba)	Red Potato yam	*Dioscorea pentaphylla*	Dioscoreaceae	Vine	Bulbs	Swellings	Bulbs can be applied to swellings
22.	*TabangThing* (Khanggra)	English gurjun tree	*Dipterocarpus bulbulatus*	Dipterocarpaceae	Tree	Trunk	Ulcers	Oleoresin from the trunk is an antiseptic and is applied to ulcers
23.	*Lankithai* (Chorphon)	Olive	*Elaeocarpus floribundus*	Oleaceae	Shrub	Leaves, Bark	Mouth wash	Infusion of bark and leaves is used as a mouthwash for inflamed gums
24.	*Arikrang* (Awa phadigom)	False Coriander	*Eryngium foetidum*	Umbelliferae	Herb	Leaf, Roots	High blood pressure& stomach trouble	Leaf used in high blood pressure; Roots used in stomach trouble
25.	Jaam	Java plum	*Eugenia jambolana*	Myrtaceae	Tree	Bark	Diarrhea and dysentery	Decoction of bark is used in diarrhoea and dysentery
26.	Tengnoumanbi	Antique Euphorbia	*Euphorbia antiquarum*	Euphorbiaceae	Succulent	Stem	Gout	Stem decoction is used in gout
27.	*Ram Bang* (Khongnang)	Peepal tree	*Ficus retusa*	Moraceae	Shrub	Leaves	Headache	Powdered dry leaves are applied to the headache
28.	*Jadonangpu* (Tera paibi)	Hill Gynura	*Gynura cusimba*	Asteraceae	Herb	Leaf, Stem	Wounds	Stem and leaf juice applied to fresh wounds to stop blood

(continued)

Table 19.1 (continued)

S.N.	Name of the plants (Local) Rongmei and Manipuri	Name of the plant (English)	Scientific name	Family name	Life forms	Parts	Disease	Mode of preparation and administration
29.	*Kambuipuang Bangmei* (Takhelleiangouba)	Common ginger lily	*Hedychium coronarium koenig*	Zingiberaceae	Herb	Leaves	Throat complaints	Leaf decoction used as a gargle in throat complaints
30.	*Ganplu/Tamplu* (Tokningkhok)	Ginger Lily	*Hedychium marginatum*	Zingiberaceae	Herb	Rhizome	Stomach complaints	Decoction of the rhizome is given for stomach complaints
31.	*Kharotthai/Joukunathai* (Heijuga)	Walnut	*Houttuynia cordata*	Saururaceae	Herb	Leaves	Dysentery and eye troubles	Used for dysentery, used for memory enhancement, and eye troubles
32.	*Banamloi* (Thirei)	Lantana	*Lantana camara*	Verbenaceae	Shrub	Leaves	Injuries& fever	Leaf extract applied to injuries to stop bleeding, mixed with honey, given for fever
33.	*Abathai* (Heinou)	Mango	*Mangifera indica*	Anacardiaceae	Tree	Fruits	Constipation	Unripe fruit is given for constipation
34.	*Thingru* (U- mangra)	Tapioca	*Manihot esculenta*	Euphorbiaceae	Shrub	Bulb & leaves	Fever and stomach complaints also sores, and scabies	Bulb flour given in fever and stomach complaints, leaves used for sores and scabies
35.	Khe—U	Burmese lacquer tree	*Melanorrh oeausitata*	Anacardiaceae	Tree	Trunk	Leprosy	Oleoresin obtained from the plant is used in leprosy's
36.	*Nungshitpu* (Noogshihidak)	Field mint	*Mentha arvensis*	Lamiaceae	Herb	Shoot, leaves	Diarrhea & gum complaints	Fresh shoot juice mixed with salt or honey is given for diarrhea, and leaf powder is used for gum complaints
37.	*Chanungthai* (Heibi)		*Meyna laxiflora*	Rubiaceae	Shrub	Leaves	Blood purification	Fresh leaves are used as a blood purifier

(continued)

Table 19.1 (continued)

S.N.	Name of the plants (Local) Rongmei and Manipuri	Name of the plant (English)	Scientific name	Family name	Life forms	Parts	Disease	Mode of preparation and administration
38.	*Chuangpong* (Leihao)	Champak	*Michelia champaca*	Magnoliaceae	Shrub	Leaves, roots	On the head for removing lice	Extract of leaves is applied to the head for eradicating lice, and roots are used in stomach troubles.
39.	*Chuangpong Bangmei* (Leihaoangouba)	White champa	*Michelia montana*	Magnoliaceae	Shrub	Bark and leaves	Fever	The bark and leaves are boiled and applied to the body and forehead for relieving fever.
40.	*Ngaak* (Laphu)	Banana	*Musa paradisiaca*	Musaceae	Herb	Fruit	Dysentery, Stomach ailments, and diarrhea.	Unripe fruit is used in dysentery and diarrhea, and stems are cooked and eaten for stomach ailments.
41.	*Maikhou Nui* (Hidak mana)	Tobacco	*Nicotiana tabacum*	Solanaceae	Herb	Leaves	Insect bite and toothache	Leaves are applied to an insect bite and given for a toothache.
42.	Shamba	Midnight horror or Indian trumpet flower	*Oroxylum indicum*	Bignoniaceae	Tree	Leaves and Bark	Epilepsy & ulcer	Leaves in epilepsy, bark, both in fresh and dried used for mouth ulcer
43.	*Napgum* (Fou)	Rice	*Oryza sativa*	Poaccae	Grass	Seeds	Intestines	Boiled rice water is given to heal inflammatory states of the intestines. The raw rice water is boiled with herbs and used as a natural shampoo (*chinghi*).
44.	*Tupid* (Yensil)	Indian Sorrel	*Oxalis corniculata*	Oxalidaceae	Herb	Whole plant	Stomach complaints	Plant extract given in the stomach for complaints
45.	*Rangpong* (Nongmangkha)	Vasaka	*Phlogacan thusthyrsiflorus*	Acanthaceae	Herb	Flowers, leaves	Fever and cold	Used in fever and cold
46.	Uningthou	Bonsum, Angaria	*Phoebe hainesiana*	Lauraceae	Tree	Berries	Sores	Ash of berries applied to sores

(continued)

Table 19.1 (continued)

S.N.	Name of the plants (Local) Rongmei and Manipuri	Name of the plant (English)	Scientific name	Family name	Life forms	Parts	Disease	Mode of preparation and administration
47.	*Aloo thingbang* (Uchaan)	Khasi pine	*Pinus insularis*	Pinaceae	Tree	Wood	Headaches	Wood paste is applied to the forehead as a remedy for headaches.
48.	Yellang	Joint Weed.	*Polygonum barbatum*	Polygonaceae	Herb	Shoot	Ulcer	Shoot extract as a wash for ulcer, and shoots cooked and given for good intestinal health.
49.	*Tamu* (Heimaang)	Nutgall tree	*Rhus semialata*	Anacardiaceae	Tree	Fruits	Dysentery and indigestion	Dysentery and indigestion
50.	*Ngei* (Ee)	Thatch grass	*Saccharum spontaneum*	Poaceae	Grass	Roots	Low blood pressure	The juice of the grass is used for low blood pressure.
51.	*Atoi* (Thoiding)	Sesame	*Sesamum indicum*	Pedaliaceae	Herb	Seed	Burns	Seed paste is used as plaster to be applied to the burn
52.	*Alu* (Alu)	Potato	*Solanum Bulbosum*	Solanaceae	Herb	Leaf	Burn & cough	Leaf extract used for cough; potato paste applied to burns
53.	*Sanarei* (Sanarei)	African Marigold	*Tagetes erecta*	Asteraceae	Herb	Leaves	Fresh wounds for stopping bleeding	Crushed leaves applied to fresh wounds for stopping bleeding; believed to keep away snakes from around the house.
54.	*Tanaang* (Samapakpi)	Aramina	*Ureta lobata*	Malvoideae	Shrub	Leaves	Cough and fever	Leaf juice with honey is given for cough and fever
55.	Urikshibi	Chinese chaste tree	*Vitex negundo*	Lamiaceae	Shrub	Leaves	Swellings	Crushed leaves are applied to painful swellings; the decoction of the leaves boiled with rice water is applied to the hair to get rid of lice.
56.	Hamengsampaki	Cockle bur	*Xanthium strumarium*	Asteraceae	Herb	Leaf	Fever and cough	Leaf extract with honey is given for fever and cough.

(continued)

Table 19.1 (continued)

S.N.	Name of the plants (Local) Rongmei and Manipuri	Name of the plant (English)	Scientific name	Family name	Life forms	Parts	Disease	Mode of preparation and administration
57.	*Tangaang* (Mukthrubitingkhangpanbi)	Pricky winged leaf	*Zanthoxylum acanthopodium*	Rutaceae	Tree	Seeds and Leaves	Indigestion, cough, and bronchitis	Paste of Seeds and fresh Leaves used in indigestion, cough, and bronchitis
58.	*Gu* (Sing)	Ginger	*Zingiber officinale*	Zingiberaceae	Herb	Rhizome	Asthma and cough	Rhizome extract mixed with honey is given for asthma and cough.
59.	Yensil	Indian sorrel	*Zingiber zerumbet*	Oxalidaceae	Herb	Whole plant	Stomach complaints	Plant extract given in the stomach for complaints.
60.	Agar	Eagle Wood	*Aquilaria agallocha Roxb.*	Thymeliaceae	Tree	Whole plant	Diarrhea, vomiting, snake bite	The whole plant is used for severe diarrhea. vomiting and a snake bite
61.	Leteku	Burmese Grape	*Baccaurea ramiflora Lour*	Euphorbiaceae	Herb	Whole plant	Constipation	Constipation
62.	Kuthab	Hill glory bower	*Clerodendrumin fortunatum L.*	Verbenaceae	Herb	Leaves	Boils	Pound the fresh leaves and put them on the boil, and keep them overnight. Pidika (boils)
63.	Khongbamtakhelei	Costus/ Spiral Ginger	*Costus specious*	Zingiberaceae	Herb	Whole plant	Urinary complaints and snake Bite	Urinary complaints and snake Bite
64.	Lai-kham		*Goniothalamus squipedalis Hook.F.& Thoms*	Annonaceae	Herb	Leaves	Evil spirits	The fresh leaves of the plant will be boiled, and the water after cooling down is used for bathing the baby. The dried leaves are burned and make the room smoke so that the evil spirits will go away.
65.	*Yai Thamna Manbi*	Kaempferia	*Kaempferia Rotunda L.*	Zingiberaceae	Herb	Roots	Wounds	Roasted roots are applied hot for tumors, and dried leaf powder is used for healing wounds.

(continued)

Table 19.1 (continued)

S.N.	Name of the plants (Local) Rongmei and Manipuri	Name of the plant (English)	Scientific name	Family name	Life forms	Parts	Disease	Mode of preparation and administration
66.	*Nungsham*		*Lemanea australis Atkins*	Rhodophyceae	Herb	Whole plants	Abdominal pain	The whole plant is boiled with water, and then it is used to cure abdominal pain
67.	*Yensil*	Creeping wood sorrel	*Oxalis corniculata*	Oxalidaceae	Herb	Whole plant	Hair lotion and piles	The whole plant is used for hair lotion, and fresh leaves are used for piles.
68.	*Khoiju*	Little spur flower	*Plectranthus ternifolius D. Don*	Lamiaceae	Herb	Leaves	Hair care lotion, an antidote to smallpox and skin diseases	Boiled leaves or fresh leaves juice are used as a hair care lotion, an antidote to smallpox, and for skin diseases
69.	*Khimpui (Nongmangkha)*	Malabar Nut	*Adhatoda vasika*	Acanthaceae	Herb	Leaves	Cough, cold, and fever	Boil the fresh leaves and drink one glass of the boiled juice 2 or 3 times a day to cure cough, cold, and fever. Also used as a vegetable.
70.	*Kaboklei*	Flowering plant	*Angiosperm*	Rubiaceae	Herb	Whole plant	Constipation	Drink the flower juice mixed with sugarcane Juice for stomach/ Constipation
71.	*Tulsi nui*	Holy basil	*Ocimum tenuiflorum*	Lamiaceae	Herb	Leaves	Cold and cough	Fresh leaves are used with tea for Cold and cough
72.	*Ganmaakhui nui (Khuthap-ukabi)*	East Indian Glory Bower	*Clerodendrum colebrookianum*	Lamiaceae	Herb	Whole plant	High BP	Reduce high BP
73.	*Neem nui*	Divine tree neem	*Azadirachta indica*	Meliaceae	Herb	Leaves	Fever, cleaning a blood, Skin infection, and eczema itch	Fresh leaves are boiled with water and used for Fever, cleaning blood, skin infection, and eczema itch.

(continued)

Table 19.1 (continued)

S.N.	Name of the plants (Local) Rongmei and Manipuri	Name of the plant (English)	Scientific name	Family name	Life forms	Parts	Disease	Mode of preparation and administration
74.	*Tairen Nui*	Tree of heaven	*Ailanthus Exceels*	Meliaceae	Herb	Leaves	Threshing out the evil spirits.	A bunch of leaves is used for threshing out the evil spirits.
75.	*Ta thing*				Herb	Leaves	Threshing out the evil spirits.	A bunch of leaves is used for threshing out the evil spirits.
76.	*Banamroinui*				Herb	Whole plant	Gases	Releasing gases from our bodies.
77.	*Peruk*	Brahmi	*Bacopa monnieri*	Apiaceae	Herb	Leaves	To smooth hair and make it strong, and used for improving memory power	Leaves are boiled with rice water and used as shampoo for hair, smooth and strong, and used for improving memory power.
78.	*Bhubati*	Creat	*Paniculata Wall. ex Nees*	Acanthaceae	Herb	Roots	Fever, diabetes, piles, jaundice and dysentery.	Leaf decoction paste seed powder is used for fever, diabetes, piles, jaundice and dysentery.
79.	*Loklei*	Ginger Lily	*Hedychium coronarium*	Zingiberaceae	Herb	Roots	Coughing	Rhizome crushed and mixed with honey and used for improving cough and cold.

Figure 19.2 shows the percentage of Life form (Growth Habit): Herb 41 (51.9%), Tree 17(21.51%), Shrub 13 (16.45%), Bulb, Grass, and Vine are 2 (2.53%), Succulent and Climber are 1(1.27%).

Figure 19.3 shows the percentage of plants parts used Leaves 21(26.6%), Whole Plants 12(15.18%), Leaves and Bark 7(8.9%), Fruit and roots are 6 (7.6%), Seed, Rhizome and Tuber are 3 (3.8%), Bark, Leaves and flowers, Trunk, Leaf and Roots are 2 (2.53%), Stem, Leaf and Stem Tuber and Leaves, Seed and Leaves, Shoot, Wood, Shoots and Leaves, Berries, Bulb, Flowers are 1 (1.26%).

Figure 19.4 shows the percentage of Family Zingiberaceae 6 (8.1%), Acanthaceae, Oxalidaceae, and Lamiaceae are 4 (5.4%), Anacardiaceae, Fabaceae, Dioscoreaceae, Lauraceae and Asteraceae are 3 (4.05%), Moraceae, Leguminasae, Euphorbiaceae, Verbenaceae, Rubiaceae, Magnoliaceae, Solanaceae, Poaccae and Meliaceae are 2 (2.7%), Myrtaceae, Acoraceae, Cannabaceae, Saururaceae, Dipterocarpaceae, Oleaceae, Umbelliferae, Amaryllidaceaeb, Cucurbitaceae, Musaceae, Malvaceae, Bignoniaceae, Caricaceae, Pinaceae, Polygonaceae, Pedaliaceae, Malvoideae, Rutaceae, Thymeliaceae, Annonaceae, Rhodophyceae, Mackinlayaceae and Apiaceae are 1 (1.35%).

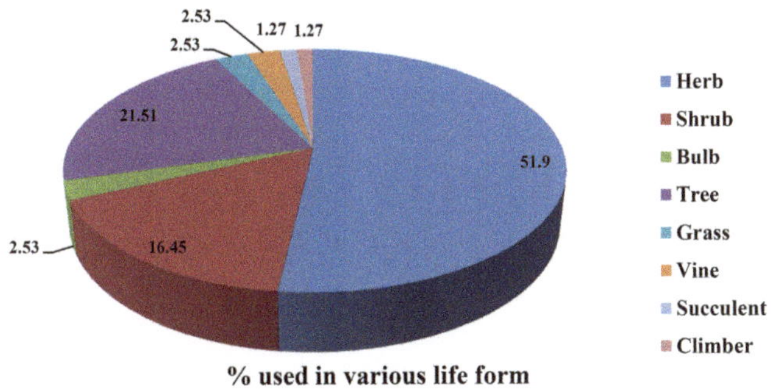

Fig. 19.2 Percentage of life forms of various medicinal plant species. *Source* The author

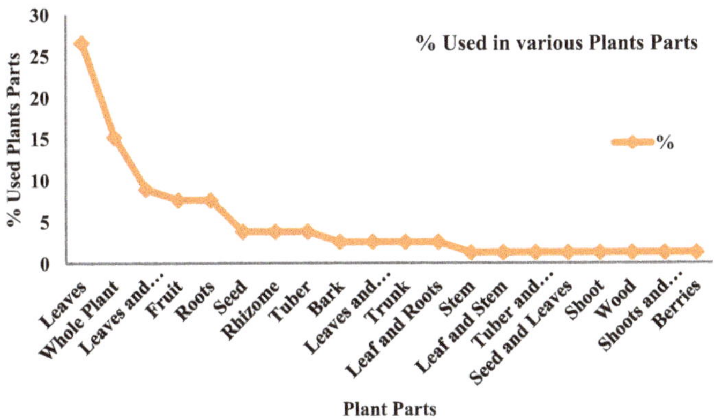

Fig.19.3 Percentage of different plant parts used as ethnomedicine. *Source* The author

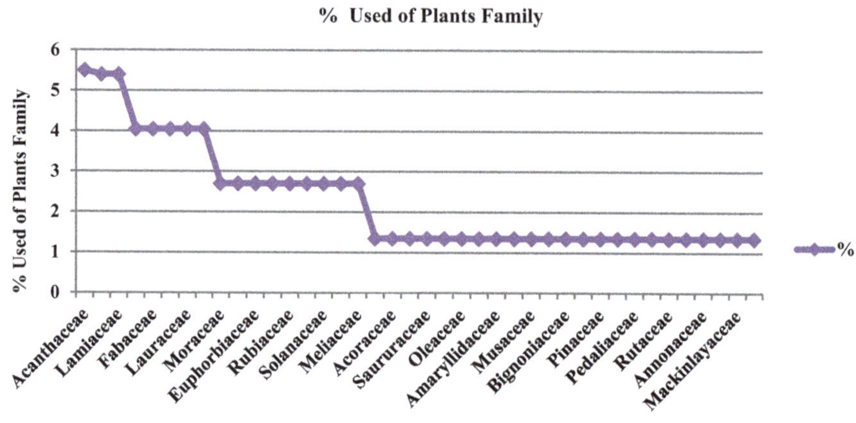

Fig. 19.4 Family of the plants used as ethnomedicines. *Source* The author

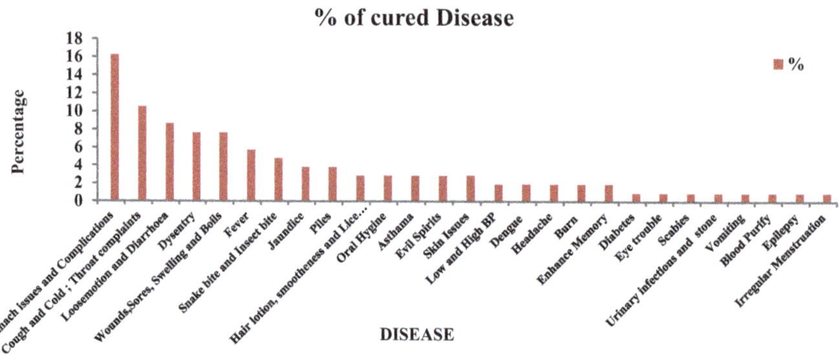

Fig.19.5 Percentage of cured disease. *Source* The author

Figure 19.5 shows the percentage of cured diseases name Stomach issues and Complications 17 (16.2%), Cough and Cold; Throat complaints 11(10.5%), Loose motion and Diarrhea 9 (8.6%), Dysentery and Wounds, Sores, Swelling and Boils are 8(7.61%), Fever 6(5.71%), Snake bite and Insect bite 5 (4.76%), Jaundice and Pile are 4(3.8%), Hair lotion, smoothness and Lice treatment, Oral Hygiene, Asthma, Evil Spirits and Skin Issues are 3 (2.9%), Low and High BP, Dengue, Headache, Burn and Enhance Memory are 2 (1.9%), Diabetes, Eye trouble, Scabies, Urinary infections and stone case, Vomiting, Blood Purify, Epilepsy and Irregular Menstruation are 1(0.95%).

1. *(Yensil) Oxalis corniculata*

2. *(Nongmangkha) Phlogacanthus thyrsiflorus*

3. *(Toningkhok) Houttuynia cordata*

4. *(Sougri) Oxalis corniculata*

5. *(Peruk) Centella asiatica*

6. *(U-mangra) Manihot esculenta*

Images of some ethnomedicinal plants used by the Rongmei Naga tribe of Tarung Village, District Imphal West, Manipur. *Source* The Author

Discussion

The Rongmei Naga tribes of Manipur, India, possess a profound knowledge of ethnomedicinal plants and their traditional uses. This study aimed to document and preserve this rich traditional knowledge, shedding light on the ethnomedicinal practices of the Rongmei Naga community. The findings from this research provide valuable insights into the diverse range of medicinal plants used by the tribe for various health conditions, as well as the preparation methods and applications.

Rich Ethnomedicinal Knowledge

The Rongmei Naga tribes have a close association with their natural environment, which has enabled them to develop a deep understanding of medicinal plants. The extensive use of ethnomedicinal plants in their traditional healthcare system reflects the long-standing relationship between the community and the surrounding biodiversity. Similar findings have been observed among other Naga tribes, emphasizing the importance of preserving and promoting indigenous knowledge (Kakati et al., 2018). Such traditional knowledge is often passed down through generations, making it imperative to document and safeguard it for future generations.

Medicinal Plant Diversity

The present study has identified a wide variety of medicinal plants used by the Rongmei Naga tribes. These plants belong to different families and are utilized to treat a diverse range of ailments. Studies conducted among other indigenous communities in India, such as the Garo and Khasi tribes, have also shown a remarkable diversity of ethnomedicinal plants, further emphasizing the significance of traditional medicine (Sanglyne et al., 2021; Kayang et al., 2005).

Use of Different Plant Forms

The study of ethnomedicinal plants among the Rongmei Naga tribes of Manipur, India, revealed a diverse range of plant life forms used for preparing traditional medicines. This discussion explores the different life forms of plants utilized and the specific plant parts employed in the preparation of medicinal remedies, backed by relevant citations.

Trees constitute a significant proportion of the ethnomedicinal plant resources used by the Rongmei Naga tribes. These tall plants provide a wide array of medicinal benefits, and their parts, such as barks, leaves, fruits, and roots, are employed in traditional medicine. For instance, the bark of *Terminalia chebula* is used to treat gastrointestinal disorders (Hassan et al., 2022). Additionally, the leaves of *Azadirachta indica* (neem) are known for their antiseptic and anti-inflammatory properties (Ghonmode et al., 2013).

Shrubs are another essential life form utilized in Rongmei Naga ethnomedicine. The leaves and stems of shrubs like *Andrographis paniculata* (kalmegh) are commonly used for their immunomodulatory and antipyretic effects (Wanjohi et al., 2020). Additionally, the whole plant of Solanum nigrum is employed as a folk remedy for various ailments, including respiratory disorders (Mandal et al., 2023).

Herbs are abundant in the region and play a crucial role in traditional medicine. The aerial parts (leaves, stems, and flowers) of herbs such as *Ocimum sanctum* (holy basil) are used for their antibacterial and anti-inflammatory properties (Shukla et al., 2020). Similarly, the rhizomes of *Curcuma longa* (turmeric) are well-known for their antioxidant and anti-inflammatory effects (Chattopadhyay et al., 2020).

Understory plants or shade-loving plants, which grow beneath the canopy of larger trees, also contribute to the ethnomedicinal practices of the Rongmei Naga tribes. The whole plant of *Centella asiatica* (*Peruk*) is traditionally used to enhance memory and cognitive function (Qian & Ke, 2014).

The use of Fungi as medicinal plants was not very prevalent among the Rongmei Naga tribes of Manipur. However, in other studies, certain mushrooms, like *Ganoderma lucidum* (Reishi), are valued for their immunomodulatory and adaptogenic effects (Wachtel-Galor et al., 2011). In some cases, fungi are used externally for wound healing and skin conditions.

The ethnomedicinal plants used by the Rongmei Naga tribes of Manipur, India, exhibit a diverse range of life forms, including trees, shrubs, herbs, climbers, understory plants, and fungi. Different plant parts, such as leaves, stems, roots, and barks, are employed to prepare traditional medicines for various health conditions. The utilization of these ethnomedicinal plants reflects the community's profound knowledge of their natural surroundings and the valuable contributions of these plants to their traditional healthcare system.

Utilization of Different Plant Parts

Leaves of numerous ethnomedicinal plants are commonly used in the preparation of traditional remedies. For example, *Adhatoda vasica* (*Nongmangkha*) leaves are used for their antipyretic treatment of cold, cough, chronic bronchitis, and asthma (Mahajan et al., 2022). The leaves of *Ocimum sanctum* (holy basil) are valued for their antibacterial and anti-inflammatory effects (Shukla et al., 2020). The roots of several ethnomedicinal plants are highly sought after for their medicinal properties. In the case of *Curcuma longa* (turmeric), the rhizomes (underground stems) are used for their antioxidant and anti-inflammatory effects (Chattopadhyay et al., 2020). Stems of certain ethnomedicinal plants play a significant role in traditional medicine.

The stem of *Musa paradisiaca* (banana), apart from the fruits, is used for its anti-inflammatory and analgesic effects, supporting its traditional use in the treatment of inflammatory conditions and pain management (Oke et al., 2014). The barks of certain trees are valuable components of Rongmei Naga ethnomedicine. For example, the bark of *Terminalia chebula* (*Manahi*) is used for its antimicrobial and antidiarrheal properties (Pebam et al., 2022). Similarly, the bark of *Quercus serrata or the Oak tree* (*U thanging*) is traditionally employed to treat respiratory disorders (Mishra et al., 2011).

The fruits of certain ethnomedicinal plants are utilized for their therapeutic benefits. The fruits of *Emblica officinalis (amla)* are commonly used for their antioxidant and immunomodulatory properties (Ahmed, 2018). Additionally, the fruits of *Ziziphus mauritiana* (Indian jujube) are employed for their hepatoprotective effects (Rathi et al., 2011). In some cases, the entire plant is used for medicinal purposes. For example, the whole plant of *Solanum nigrum* is used as a folk remedy for various ailments, including respiratory disorders (Chidambaram et al., 2022).

The utilization of different plant parts in Rongmei Naga ethnomedicine showcases the comprehensive understanding of the community about the medicinal properties of various plant components.

Traditional Medicinal Applications

The documented ethnomedicinal plants are used for various therapeutic purposes, including treating fever, pain, inflammation, gastrointestinal disorders, infectious diseases, respiratory issues, and skin ailments. The use of ethnomedicinal plants for similar health conditions has been reported among other tribal communities in India and Southeast Asia (Luitel et al., 2014; Kumar et al., 2021). These traditional practices have stood the test of time and continue to play a crucial role in the healthcare of the Rongmei Naga community.

Potential for Modern Medicine

The ethnomedicinal knowledge of the Rongmei Naga tribes holds immense potential for modern medicine and drug discovery. Many plant-derived medicines have already been integrated into mainstream healthcare systems. Additionally, studies have shown that traditional medicinal practices have led to the identification of bioactive compounds with promising pharmacological properties (Balbaa et al., 2021). Thus, the ethnomedicinal plants used by the Rongmei Naga tribes could serve as a valuable resource for future drug development and natural product research.

Challenges and Conservation

Despite the significance of ethnomedicinal knowledge, there are challenges in its preservation and conservation. Factors such as changing lifestyles, acculturation, and environmental degradation can impact the availability and transmission of this knowledge within indigenous communities (Aziz et al., 2018). Initiatives should be taken to protect traditional knowledge, promote sustainable use of medicinal plants, and respect the rights of indigenous communities over their resources (Sharma et al., 2021).

Conclusion

The Rongmei Naga community of Tarung village in Manipur has very rich ethnomedicinal knowledge in the field of herbal medicine. The findings highlight the importance of these medicines and the roles they play in healing many kinds of ailments in the absence or inaccessibility of primary health care. The ethnomedicinal plants serve as a reliable and readily available resource when most secondary and tertiary health care centers are very far away from the village. The present study also underlines the importance of preserving this knowledge and promoting the sustainable use of medicinal plants. However, the impact of modernization and acculturation has highly influenced the present community, as it is also situated amid Imphal City, the hub for cultural interactions. The younger generations of Rongmei Naga are not very interested in traditional healthcare practices. The present research contributes to the growing body of literature on ethnomedicine and underscores the potential for collaboration between traditional practices and modern medicine for the benefit of global healthcare. In the future, the pharmacognosy of these valuable ethnomedicines is required to be evaluated.

Acknowledgment The authors are truly thankful to all the informants, Traditional and Spiritual healers, and the Rongmei Naga community from Tarung Village, Imphal West District, Manipur, India, for their participation and valuable contributions to this research.

Declaration of conflict: The Authors declared that there is no conflict of interest.

References

Ahmed, H. M. (2018). Ethnomedicinal, phytochemical and pharmacological investigations of Perilla frutescens (L.) Britt. *Molecules*, *24*(1), 102.

Aziz, M. A., Khan, A. H., Adnan, M., & Ullah, H. (2018). Traditional uses of medicinal plants used by indigenous communities for veterinary practices at Bajaur agency, Pakistan. *Journal of Ethnobiology and Ethnomedicine*, *14*(1).

Balbaa, M., El-Zeftawy, M., & Abdulmalek, S. A. (2021). Therapeutic screening of herbal remedies for the management of diabetes. *Molecules*, *26*(22).

Bhuyan, M. (2015). Comparative study of ethnomedicine among the tribes of North East India. *International Research Journal of Social Sciences*, *4*(2), 27–32.

Chattopadhyay, I., Biswas, K., Bandyopadhyay, U., & Banerjee, R. K. (2020). Turmeric and curcumin: Biological actions and medicinal applications. *Current Science*, *87*(1), 44–53.

Chidambaram, K., Alqahtani, T., Alghazwani, Y., Aldahish, A., Annadurai, S., Venkatesan, K., Dhandapani, K., Thilagam, E., Venkatesan, K., Paulsamy, P., Vasudevan, R., & Kandasamy, G. (2022, September 21). Medicinal plants of Solanum species: The promising sources of phyto-insecticidal compounds. *Journal of Tropical Medicine*, *2022*, 4952221. https://doi.org/10.1155/2022/4952221 PMID: 36187457; PMCID: PMC9519333.

Foster, G. M., & Anderson, B. G. (1978). *Medical Anthropology*. Wiley.

Ghonmode, W. N., Balsaraf, O. D., Tambe, V. H., Saujanya, K. P., Patil, A. K., & Kakde, D. D. (2013). Comparison of the antibacterial efficiency of neem leaf extracts, grape seed extracts and 3% sodium hypochlorite against E. feacalis—an in vitro study. *Journal of International Oral Health*, *5*(6).

Hassan Bulbul, M. R., Uddin Chowdhury, M. N., Naima, T. A., Sami, S. A., Imtiaj, M. S., Huda, N., & Uddin, M. G. (2022, August 14). A comprehensive review on the diverse pharmacological perspectives of Terminalia chebula Retz. *Heliyon*, *8*(8), e10220. https://doi.org/10.1016/j.heliyon.2022.e10220 PMID: 36051270; PMCID: PMC9424961.

Kakoti, B. B., Dutta, B. K., & Khan, M. L. (2018). Traditional knowledge on herbal drinks among the Garo and Naga tribes of Northeast India. *Journal of Ethnobiology and Ethnomedicine*, *14*, 46.

Kayang, H., Kharbuli, B., Myrboh, B., & Syiem, D. (2005). *Medicinal plants of Khasi Hills of Meghalaya, India*. Bioprospecting & Ethnopharmacology.

Kumar, M., Rawat, S., Nagar, B., Kumar, A., Pala, N. A., Bhat, J. A., Bussmann, R. W., Cabral-Pinto, M., & Kunwar, R. (2021, February 5). Implementation of the use of ethnomedicinal plants for curing diseases in the Indian Himalayas and its role in sustainability of livelihoods and socioeconomic development. *International Journal of Environmental Research and Public Health*, *18*(4), 1509. https://doi.org/10.3390/ijerph18041509 PMID: 33562584; PMCID: PMC7915974.

Lokho, K., & Narasimhan, D. (2013). Ethnobotany of Mao-Naga Tribe of Manipur, India. *East Himalayan Society for Spermatophyte Taxonomy*, *7*(2), 314–324.

Luitel, D. R., Rokaya, M. B., Timsina, B., & Munzbergova, Z. (2014). Medicinal plants used by the Tamang community in the Makawanpur district of central Nepal. *Journal of Ethnobiology and Ethnomedicine*, *10*, 5.

Mahajan, S., Takarkhede, S., Tarmale, K., Thikekar, P., Terse, P., & Tiwari, K. (2022). Role of adulsa in respiratory disorders. *International Journal of Pharmaceutical Research and Applications*, *7*(1), 515–525. https://doi.org/10.35629/7781-0701515525

Mandal, S., Vishvakarma, P., Verma, M., Alam, M. S., Agrawal, A., & Mishra, A. (2023). Solanum Nigrum Linn: An analysis of the medicinal properties of the plant. *Journal of Pharmaceutical Negative Results*, *14*(2).

Mishra, N., Rout, S. D., & Panda, T. S. (2011). Ethno-zoological studies and medicinal values of similipal biosphere reserve, Orissa. *India. African Journal of Pharmacy and Pharmacology*, *5* (1).

Oke, O. A., Awopetu, I., Durotoye, L. A., Oyagbemi, A. A., Shode, F. O., & Yakubu, M. T. (2014). Anti-inflammatory and analgesic activities of the aqueous extract of Musa paradisiaca L. (Musaceae) in rodents. *Journal of Basic and Clinical Physiology and Pharmacology*, *25*(4), 423–430.

Pebam, M., Sushma, M. V., Sankaranarayanan, S. A., Thanekar, A. M., Koyande, N., & Rengan, A. K. (2022, August 29). Antiviral perspectives of economically important Indian medicinal plants and spices. *Proceedings of the Indian National Science Academy*, *88*(3), 392–416. https://doi.org/10.1007/s43538-022-00099-w PMCID: PMC9422945.

Prakash, N., Ansari, M. A., Punitha, P., & Sharma, P. K. (2014). Indigenous traditional knowledge and usage of folk bio-medicines among rongmei Tribe of tamenglong district of manipur, India. *African Journal of Traditional, Complementary, and Alternative Medicines (AJTCAM)*, *11*(3), 239–247.

Qian, Z. M., & Ke, Y. (2014). A is it an effective disease-modifying drug for alzheimer's disease? *Frontiers in Aging Neuroscience*, *6*.

Rahman, M. H., Fardusi, M. J., & Reza, M. S. (2011). Traditional knowledge and use of medicinal plants by the patra tribe community in the North-Eastern region of Bangladesh. *Proceedings of the Pakistan Academy of Sciences*, *48*(3), 159–167.

Rathi, B. S., Bodhankar, S. L., Mohan,, & Thakurdesai, P. A. (2011). Ameliorative effect of fruits of Ziziphus mauritiana on glucose tolerance in a mouse model of type-2 diabetes. *Food and Chemical Toxicology*, *49*(10), 2784–2789.

Sajem, A. L., & Gosai, K. (2006). Traditional use of medicinal plants by the Jaintia tribes in North Cachar Hills district of Assam, northeast India. *Journal of Ethnobiology and Ethnomedicine*, *2* (33).

Sanglyne, M. W., Dirborne, C. M., & Nongkynrih, D. N. (2021). Ethnomedicinal plants used by the tribal inhabitants of two of the wettest places on earth: An overview. *International Journal of Botany*, *6*(2).

Sharma, A., Patel, S. K., & Singh, G. S. (2021). Traditional knowledge of medicinal plants among three tribal communities of Vindhyan highlands, India: An approach for their conservation and sustainability. *Environmental Sustainability*, *4*(4).

Shil, S., & Dutta Choudhury, M. (2009). Ethnomedicinal importance of pteridophytes used by reang tribe of Tripura. *North East India. Ethnobotanical Leaflets*, *13*, 634–643.

Shukla, M., Gupta, K., & Rasheed, Z. (2020). Ocimum sanctum Linn. (Tulsi): An ethnomedicinal, phytochemical and pharmacological review. *Journal of Herbal Medicine*, *5*(4), 158–167.

Wachtel-Galor, S., Yuen, J., Buswell, J. A., & Benzie, I. F. F. (2011). Ganoderma lucidum (Lingzhi or Reishi): A medicinal Mushroom. In I. F. F. Benzie & S. Wachtel-Galor (Eds.), *Herbal medicine: Biomolecular and clinical aspects* (2nd ed.). CRC Press/Taylor & Francis.

Wanjohi, B. K., Sudoi, V., Njenga, E. W., & Kipkore, W. K. (2020). An ethnobotanical study of traditional knowledge and uses of medicinal wild plants among the marakwet community in Kenya. *Evidence-Based Complementary and Alternative Medicine*, 1–8.

Chapter 20
Indigenous Medical Wisdom: A Case Study of Rung Community of Kumaun Himalayas

Rishav Rawal, Kritika Bora, and Sanjay Ghildiyal

Abstract The present conducted work is an effort for documentation and analyzing the indigenous wisdom of the Rung Shauka community residing the higher valleys of Pithoragarh district in Kumaun Himalayas. This research has been conducted in the remote villages situated in high-altitude region. The primary methodological tools used for gaining insights into the indigenous wisdom of Rung community are ethnographic in nature, including interviews, participant observations and in-depth field surveys. The findings of this study reveals that the healthcare practices of these people are deeply inter-woven with their cultural and spiritual beliefs, with diverse range of local medicinal plants, spiritual healing rituals, and sustainable practices. This research finally tries to highlight the importance of preserving such traditional knowledge systems (TKS) by providing useful insights for integration of indigenous and modern medical practices for improving healthcare outcomes in remote and marginalized areas.

Keywords Indigenous · Traditional knowledge · Medicinal plants · Rung community

Introduction

According to census of India 2011, more than 8% of the population in India is tribal. The tribal communities are known to have close relationship and tremendous appraisal capacities for its surrounding. Their distinctive way of survival with the nature distinguishes them from the rest of the world. Indigenous communities have

R. Rawal · S. Ghildiyal
Department of History, Kumaun University, Nainital, India

K. Bora (✉)
Department of Geography, Kumaun University, Nainital, India
e-mail: kritikabora@kunainital.ac.in

R. C. Sobti et al. (eds.), *Traditional Knowledge Systems in India: Sustainability, Ecology, and Resource Governance*, People, Cultures and Societies: Exploring and Documenting Diversities,
https://doi.org/10.1007/978-981-92-1346-7_20

extensive knowledge of local flora and their medicinal properties. Of late indigenous medical wisdom has increasingly been recognized for its value in promoting health and treating disease, often offering alternatives or complements to western medicine.

The Indian Himalayan region is home to rich and diverse ethnic communities and medicinal plants. The appraisal of surrounding vegetation for medical value has traditionally been an integral part of tribal lifestyle of these regions (Malik et al., 2015). Traditional medical practices are a result of knowledge accumulated through generations and which are passed on to newer generations. The indigenous groups continuously work to augment their preserved knowledge (Gadgil et al., 1993). Efforts to preserve this knowledge are vital, as many indigenous cultures face threats from modernization, loss of biodiversity, and cultural assimilation. Collaboration between traditional healers and modern medical practitioners can enhance healthcare outcomes and foster respect for diverse healing traditions.

Deeply rooted in the natural environment and cultural practices of these communities. Indigenous medical wisdom of tribes in India represents a profound body of knowledge accumulated over centuries. This wisdom encompasses a diverse range of healing practices, including the use of medicinal plants, spiritual healing, dietary regimes, and physical therapies. Each tribal community has developed unique methods tailored to their specific ecological contexts, resulting in a rich tapestry of traditional medicine that offers valuable insights into holistic health and well-being. Despite the advent of modern medicine, these traditional practices continue to play a crucial role in the healthcare of many tribal populations. The importance of ethno-medical knowledge and practices has been recognised by the Ministry of AYUSH which now funds the exploration of indigenously flourishing traditional healing systems (Joshi, 2019).

However, the rapid pace of modernization and cultural assimilation poses significant challenges to the preservation of this knowledge. Lately, there seems a transformation in the traditional ecological knowledge and practices caused due to environmental degradation and substantial changes in economic, social and environmental systems. Indigenous communities worldwide are increasingly adjusting and adapting their traditional practices to align with the pace of development, leading to the erosion of indigenous knowledge over time (Rai & Lalramnghinglova, 2010). This alteration in the practices and behaviour needs quick check and documentations.

Efforts to document and integrate indigenous medical practices into broader healthcare systems are essential for maintaining this cultural heritage and exploring its potential contributions to contemporary medicine. This paper aims to delve into the indigenous medical wisdom of Indian tribes, highlighting its historical significance, core components, and the pressing need for its preservation and ethical integration into modern healthcare frameworks. Indigenous medical practices in India predate documented history, with traditions like Ayurveda tracing back thousands of years.

The Indian Central Himalaya is known for its dominance of tribal communities. These communities have served to be the knowledge base for ethnic medical practices. Amongst many ethnic groups, is a sub clan of Bhotia tribe called Rung community. Bhotia tribe is of Mongoloid origin inhabiting the high-altitude region of Indian central Himalayas. Their body of knowledge has helped them to live sustainably in close association with their surroundings. This knowledge not only serves the locals but also greatly helps in conserving bioresources in and around the community (Sammal et al., 2010). The present study deals with ethno-medicinal know how Rung community of Uttarakhand.

Study Area

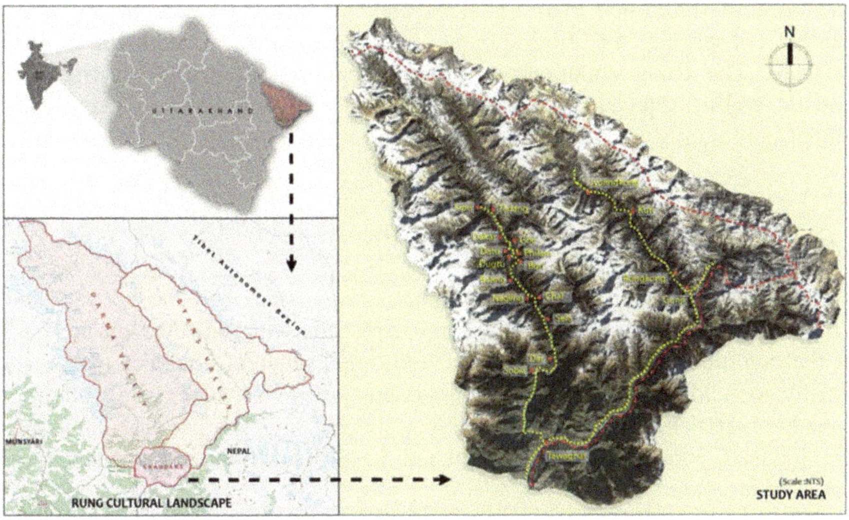

This study was conducted in the Rung community settlement located on the eastern outskirts of Pithoragarh district, Uttarakhand, India. This region includes the Darma, Beas and Chaudans valleys located in the Kali River system. The region is known for its enormous cultural and religious significance, and the Rung community plays an important role in preserving and promoting the region's ethno medical heritage.

The Dharma, Byas and Chaudans valleys are characterized by high altitude environment, beautiful rivers and unique flora and fauna. These natural features form the basis of the traditional healing practices of the Rung community, which are passed down from generation to generation and are deeply rooted in local culture.

The role of the community as guardians of local ethno medical knowledge is very important. A broader understanding of the medicinal properties and health applications of native plants contributes significantly to the well-being of local communities. Rather than being just anecdotal, this knowledge is based on centuries of empirical observation, experimentation, and improvement. The community's experience includes medicinal herbs, healing practices, and spiritual healing techniques that are integral to their way of life. The diversity, rarity and uniqueness of the traditional knowledge-based medicinal practices of the Rung community are noteworthy. The plants and herbs used in medicinal practices are often indigenous to the area and thrive in the specific climatic and geographical conditions of the Pithoragarh region. This endemicity ensures that knowledge and practice are closely linked to the local environment and contribute to the conservation of local biodiversity. The Rung community's sustainable methods of identifying, collecting and using these plants play an important role in maintaining the ecological balance of the region.

Rung community settlements in these valleys, are an important repository of traditional ethno medical knowledge. Deeply rooted in local biodiversity and cultural traditions, their practices play an important role in promoting health, conserving the environment and preserving the region's cultural heritage. The Rung community's commitment to sustainable practices and holistic wellness provides valuable information and lessons to the broader global community.

Methodology

This study villages from Rung cultural landscape were selected on the basis of the community's representation of traditional medical practices and accessibility to field work. Initially, a comprehensive review of existing literature on ethno medicine, indigenous health practices, and the cultural context of the focused community was conducted to ground the study and identify knowledge gaps.

The survey utilised Semi-structured interviews conducted with various groups of community including traditional healers, experienced community members and other knowledgeable people, for collecting detailed information about traditional health practices, medicinal plants and spiritual healing rituals.

In this process in-depth discussions, and interviews were recorded. The survey also utilised ethnographic approach, including participation health promotion practices, and ritual observations. Detailed field notes were kept to document observations and interactions, with particular attention to the context and application of community health practices.

The results were compiled in order to comprehensively highlight the importance of preserving traditional knowledge systems. The community's cultural sensitivity and intellectual property rights were respected, including access to results and recommendations for improving health care. This research aims to improve health outcomes in remote areas and contribute to the preservation of valuable traditional knowledge systems. the information gathered through survey's has been summarised in Table 20.1.

Table 20.1 List of medicinal plants found in the study region

S. no.	Name of plant	Local name	Uses/ Cure	Altitude m	Status
1	Angelica glauca Edgew	Chippi	Indigestion & Dyspepsia	3000	CR
2	Berberis aristata DC	Kilmoru/ Daruhald	Eye Trouble	15,000–3000	EN
3	Berginiastracheyi (Hk.f. & Th.) Engler	Silphora	Kidney/ Urinary trouble	3600–4200	VU
4	Dactylorrhizahatagirea D. Don	Salampanja/ Hathazari	Mensutrational issue/ Cuts and wound	2800–4000	CR
5	Nardostachys grandiflora DC	Masi/ Jatamas	Epilepsy /Rheumatism	3200–5000	CR
6	Polygonum nepalensisMesin.	Bhotia Chai	Pregnancy Throat pain /Cough	1200–4100	CR
7	Picrrorhizakurooa Royle ex Benth	Kutki	Stomachache and Fever	3300–4800	EN
8	Rubia cordifolia Linn	Manjitha	Skin Care/Acne	1200–2000	EN
9	Rheum australe D. Don	Dolu	Fracture	3000–4200	VU
10	Orchis habenarioides King & Pant	Salam Mishri	Pregnancy	3000–4500	EN
11	Fagopyrum esculentum Moench (Polygonaceae)	Palti/Ogal	Stomach ache	2400–3800	EN
12	Saussureacostus (Decne.) Sch.-Bip. (Asteraceae)	Kuth	Cold and cough, asthma, bronchitis, dysentery	2500–3600	EN
13	Origanum vulgare L. (Lamiaceae)	Ban Tulsi	Bronchtites, Cough & Cold	2000–3800	NT
14	Geranium wallichianumD.Don ex Sweet. (Geraniaceae	Ratan Jyoti	Stomach ache, cold and cough, cuts and wounds	2400–4000	NT
15	Hippophae salicifolia D. Don (Elaeagnaceae)	Chook	Stomach ache, cold and cough	3400–3500	VU
16	Pyrus pashiaBuch.-Ham. ex D.Don [Rosaceae	Mehal	Fever	2700–3000	VU
17	SaussurealanicepsHand.-Mazz. [Asteraceae]	Ghughii	Gynaecological, mental & menstrual disorders	3200	EN
18	Potentilla fulgens Wall. ex Seems [Rosaceae	Bajardanti	Stomatitis, aphthae	1800–3500	NT
19	Parnassia nubicola Wall. ex Royle [Celastraceae]	Phutkya	Food Poisning	3000	NT
20	Podophyllum hexandrum Royle [Berberidaceae]	Bankakri	Headache, high fever, abdominal pain	2000–4000	CR

CR—Critically rare, EN: Endangered, VU: Vulnerable, NT: Non Thretent.

Results and Discussion

Proper natural resource management is long past practices followed globally by indigenous societies. These practices consisting age old knowledge usage of phyto-resource contributes to the conservation of these resources. In this context utilization of plant resources is a traditional practice having rich history and worldwide acceptability (Bhatt et al., 2009). In present rural populations in isolated mountain regions of IHR continue to employ medicinal plants to treat diseases, despite advancements in modern healthcare (Dhar et al., 2000).

This work advocates on the importance of utilization patterns and therapeutic applications of medicinal plants for communities living in Himalayas, by focusing on rich system of traditional healing practices prevailing amongst Rung Shauka living in higher altitude of Kumaun Himalayas.

Higher altitude regions are recognized to give superior products; thus, they are often the favored choice of local consumers for rapid therapy and pharmaceutical businesses as precious ingredients (Mikage & Mouri, 2000), these species are identified crucial part local medicinal system. These ethnic plants are used to address systematic issues like fevers, digestive disorders, and respiratory issues. Different parts of plants are used to cure different issues due to their topical applications shedding light on multiple potential of ethno botanical tapestry of the region. The region consists versatile ethnic knowledge base for utilization of these threpetic plants. This shows diverse nature of ethnic medicinal practices in addressing health concerns within local communities.

In addition of medicinal usage, these botanical treasures have great cultural and spiritual significance regarding indigenous healing practices. Traditional healers Integration of spiritual rituals into the collection of medicinal plants and formation of medicinal properties is utilized by traditional healers since long past. Chanting of mantras and offering prayers during preparation of medicine reflect a holistic intertwining spiritual well-being with physical health. This interconnectedness intermingling of cultural beliefs and healing practices clarifies that the broad nature of traditional medicine, deeply rooted in community identity and values. It clarifies that healing practices are not just about physical ailments but also about maintaining cultural heritage and spiritual connections within communities. The practices of these communities have proved that biological diversity and traditional knowledge are intensively interlinked (Negi, 2010, 2012).

The study also explores the fact that despite of cultural importance and efficiency, there are numerous challenges in continuation of traditional medicinal knowledge due to changing Socio-economic dynamics, globalization, and urbanization due to which coming remain unaware of traditional remedies. Making it difficult for transmission of traditional healing towards younger generations.

In addition, there is lack of scientific validation and regulatory frameworks for traditional medicines becoming barriers to for wider acceptance and integration into formal healthcare systems. This challenge undermines efforts to preserve and promote traditional healing practices as viable healthcare options.

Environmental factors also contribute to the challenges faced by traditional medicine. The dwindling availability of medicinal plants due to environmental degradation and over-harvesting threatens the sustainability of traditional healing practices. Addressing these challenges requires sustainable harvesting practices and community-led conservation efforts to ensure the long-term viability of medicinal plant resources.

Integrating traditional medicinal knowledge with modern healthcare practices presents significant opportunities to enhance healthcare outcomes and promote health equity. Scientific research validating the efficacy of medicinal plants can bridge the gap between traditional wisdom and evidence-based medicine. Collaborative efforts between traditional healers, scientists, and healthcare professionals can lead to the development of standardized protocols for the cultivation, harvesting, and use of medicinal plants.

Furthermore, integrating traditional medicine into healthcare systems can expand treatment options, particularly for conditions where conventional medicine may be limited or inaccessible. By acknowledging and incorporating traditional knowledge, policymakers can support cultural diversity in healthcare delivery and promote community resilience.

Future research efforts should prioritize documenting and validating the efficacy of medicinal plants through rigorous scientific studies. This includes identifying bioactive compounds, elucidating their mechanisms of action, and exploring potential synergies with modern pharmaceuticals. Such endeavours are crucial for enhancing the credibility and acceptance of traditional medicine within mainstream healthcare systems.

Policy initiatives should focus on creating supportive regulatory frameworks that respect and integrate traditional healing practices into national healthcare strategies. This involves promoting sustainable harvesting practices, protecting traditional knowledge rights, and fostering collaboration between traditional healers and healthcare providers.

In traditional societies, sustainable natural resource management is motivated by the beliefs and attitudes of human communities, and local cultures are strengthened by their close linkages to the natural environment that sustains them (Rist et al., 2003)

The study's findings underscore the profound significance of medicinal plants in traditional healing practices, highlighting their therapeutic versatility and cultural importance. Addressing the challenges and leveraging the opportunities presented by traditional medicine can contribute to more inclusive and effective healthcare systems globally.

Conclusion and Way Forward

Study explores the rich history and diversity of the Rung, Shauka communities in the Kumaon Himalayas, focusing on indigenous healing techniques that have been passed down from generation to generation. Using a variety of methods, the study sheds light on the community as healthcare practices and how they relate to cultural beliefs. This study explores the residents deep understanding of the natural environment and its resources by examining in detail their use of local plant resources in the Himalayan countryside. The deep knowledge and practices of the Rung, Shauka communities provide a window into how medicine and modern medicine can interact to create healthier, more productive lives. This integration is not about preserving ancient practices, but understanding and utilizing the rich knowledge held by ethnobotanical communities. Ethno botanical practices are an important part of the heritage of Rung, Shauka communities and have the potential to contribute to modern health. Writing down these practices ensures they don't fade away over time, especially as routines lead to daily life. Additionally, this information can serve as a valuable resource for researchers and practitioners seeking health-related health and cultural practices. First, there must be a collective effort to collect and preserve ethno botanical information. This requires collaboration between communities, scientists, government and non-governmental organizations. By creating knowledge about medicinal plants and their uses, we can pass this knowledge on to future generations. This includes understanding and validating the effectiveness of traditional practices through scientific research. Collaborative research that combines the experiences of Rung, Shauka communities with modern science can lead to the development of new and effective treatments. This collaboration can also help create policies that support the use of traditional medicine in modern healthcare. Educating young people in the community about the importance of their traditional knowledge can lead to an interest and responsibility in preserving their heritage. Additionally, increased awareness among physicians about the beneficial effects of traditional medicine may promote a more integrated approach to patient care. Community culture should be encouraged to encourage the conservation and collection of medicinal plants. This ensures that ecological balance is maintained while society continues to benefit from available resources. Integrating multicultural knowledge with modern healthcare systems can pave the way for improved health and well-being in the Kumaon Himalayan region. By recognizing, preserving and integrating these practices, we can not only improve health outcomes but also respect and preserve the heritage of the Rung, Shauka communities for future generations.

References

Bhatt, D., Joshi, G. C., & Tiwari, L. M. (2009). Culture, habitat and ethno-medicinal practices by Bhotia tribe people of Dharchula region of Pithoragarh District in Kumaun Himalaya, Uttarakhand. *Ethnoboptanical Leaftlets*.

Dhar, U., Rawal, R. S., & Upreti, J. (2000). Setting priorities for conservation of medicinal plants—A case study in the Indian Himalaya. *Biological Conservation*, 57–65.

Gadgil, M., Berkes, F., & Folke, C. (1993). Indigenous knowledge for biodiversity conservation. In *Biodiversity: Ecology, economics, policy* (pp. 151–156).

Joshi, P. (2019). Ethnomedical practices and Indian scheduled tribes. *Romanian Journal of Sociological Studies*, 111–124.

Malik, Z. A., Bhat, J. A., Ballabha, R., Bussmann, R. W., & Bhatt, A. B. (2015). Ethnomedicinal plants traditionally used in health care practices by inhabitants of Western Himalaya. *Journal of Ethnopharmacology*, 133–144.

Mikage, M., & Mouri, C. (2000). Altitudinal variations of berberine contentin the bank of Berberis plants from West Nepal. *News Letter Himalayan Bot*, 16–19.

Negi, C. (2010). Traditional culture and biodiversity conservation: Examples from Uttarakhand, Central Himalaya. *Mountain Research and Development*, 259–265.

Negi, C. (2012). Culture and biodiversity conservation: Case study from Uttarakhand, Central Himalaya. *Indian Journal of Traditional Knowledge*, 273–278.

Rai, P. K., & Lalramnghinglova, H. (2010). Ethnomedicinal plant resources of Mizoram, India: Implication of traditional knowledge in health care system. *Ethnobotanical Leaflets*, 274–305.

Rist, S., Burgoa, F. D., & Wiesmann, U. (2003). The role of social learning processes in the emergence and development of Arymara land use systems. *Mountain Research and Development*, 263–270.

Sammal, P. K., & Dhyani, P. P, & Dollo, M. (2010). Indigenous medicinal practices of Bhotia tribal community in Indian Central Himalaya. *Indian Journal of Traditional Knowledge*, 256–260.

GPSR Compliance
The European Union's (EU) General Product Safety Regulation (GPSR) is a set of rules that requires consumer products to be safe and our obligations to ensure this.

If you have any concerns about our products, you can contact us on

ProductSafety@springernature.com

In case Publisher is established outside the EU, the EU authorized representative is:

Springer Nature Customer Service Center GmbH
Europaplatz 3
69115 Heidelberg, Germany

www.ingramcontent.com/pod-product-compliance
Ingram Content Group UK Ltd.
Pitfield, Milton Keynes, MK11 3LW, UK
UKHW020631260926
475829UK00003BA/23
* 9 7 8 9 8 1 9 2 1 3 4 5 0 *